Advanced Technology of Waste Treatment

Advanced Technology of Waste Treatment

Editors

Daniel Vollprecht
Renato Sarc

MDPI • Basel • Beijing • Wuhan • Barcelona • Belgrade • Manchester • Tokyo • Cluj • Tianjin

Editors
Daniel Vollprecht
Montanuniversität Leoben
Austria

Renato Sarc
Montanuniversität Leoben
Austria

Editorial Office
MDPI
St. Alban-Anlage 66
4052 Basel, Switzerland

This is a reprint of articles from the Special Issue published online in the open access journal *Processes* (ISSN 2227-9717) (available at: https://www.mdpi.com/journal/processes/special_issues/waste_treat).

For citation purposes, cite each article independently as indicated on the article page online and as indicated below:

LastName, A.A.; LastName, B.B.; LastName, C.C. Article Title. *Journal Name* **Year**, *Volume Number*, Page Range.

ISBN 978-3-0365-3755-9 (Hbk)
ISBN 978-3-0365-3756-6 (PDF)

Contents

About the Editors

Daniel Vollprecht

Dr. Daniel Vollprecht studied Mineralogy at TU Bergakademie Freiberg, Germany, and graduated in 2008. He started his career at Graz University of Technology, Austria, and received his doctoral degree in Earth Sciences in 2013. Since then, he has been the leader of the Working Group "Mineral Wastes, Landfills, and Contaminated Sites", which recently changed its name to "Material-oriented Waste Technology". His research focuses on the interaction between waste and the environment, especially with the leaching of heavy metals. He has conducted several national and international research projects, e.g., on landfill mining (national project LAMIS (2013–2015) and the EU project NEW-MINE (2016)), the mineralogy and leachability of steel slags (projects MiLeSlag (2016–2019) and MiLeSlag 2.0 (2019–2022)), in situ remediation (projects ChromSan (2013–2015), LISA (2015–2019) and InnoBLA (2020–2022)) and metal recovery from waste waters (projects RECOMET (2014–2015) and RECOMET 2.0 (2015–2018)). He is a member of the ISWA and ÖWAV working groups on landfills and a board member of the German and Austrian Mineralogical Societies. In 2020 he received the "venia docendi" in Waste Technology for his Habilitation Thesis entitled "Exploration, Mobilization and Fixation of Constituents of Mineral Wastes, Landfills, Contaminated Sites and Waste Waters". Finally, he has authored more than 40 peer-reviewed publications at the intersection between waste management and geosciences and has been a scientific reviewer for various international journals from both fields.

Renato Sarc

Dr. Renato Sarc studied Industrial Environmental Protection, Waste Management and Recycling at Montanuniversität Leoben, Austria, and graduated in 2010. He started his career at the same university, Montanuniversität Leoben, and received his doctoral degree in Waste Management in 2015. Since then, he has been the leader of the Working Group "Waste Fuel", which recently changed its name to "Process-oriented Waste Technology". Since 2015 he has been the Deputy of the Chair of Waste Processing Technology and Waste Management and since 2017 he has worked as a Post Doc Senior Scientist and Project Leader of the COMET—Competence Centers for Excellent Technologies "ReWaste4.0" (2017–2021) and "ReWaste F" (2021–2025) at Montanuniversität Leoben. Since 2020 he has been a member of the national Committee 157 of the Austrian Standards Institute and the Austrian Delegate in ISO/TC 300 (Working Group 5), and since 2022 he has been the Chairman of the Advisory Board of the V.EFB—the association responsible for awarding certificates to waste management companies in Austria. He has gained additional qualifications in waste management such as "waste manager" and "waste management officer". He has conducted several national and international research projects in cooperation with different national and international companies and institutions. Finally, he has authored a wide range of peer-reviewed publications in the scientific area of waste management, treatment and technology; has been a scientific reviewer for various international journals; and has received several prestigious awards for his work.

Preface to "Advanced Technology of Waste Treatment"

The protection of human health and the environment as well as the sustainable use of natural resources requires the chemical, biological, and physical treatment of waste. This refers to the conditioning (e.g., drying, washing, comminution, rotting, stabilization, neutralization, agglomeration, homogenization), conversion (e.g., incineration, pyrolysis, gasification, dissolution, evaporation), and separation (classification, direct and indirect (i.e., sensor-based) sorting) of all kinds of wastes following waste hierarchy principles (i.e., prevention (not addressed by this issue), preparation for re-use, recycling, other recovery and landfilling). Longstanding challenges include the increase in the yield and purity of recyclable fractions and the removal or destruction of pollutants from the circular economy.

Daniel Vollprecht and Renato Sarc
Editors

processes

Editorial

Special Issue on "Advanced Technology of Waste Treatment"

Daniel Vollprecht * and Renato Sarc

Chair of Waste Processing Technology and Waste Management, Department of Environmental and Energy Process Engineering, Montanuniversitaet Leoben, Franz-Josef-Str. 18, 8700 Leoben, Austria; renato.sarc@unileoben.ac.at
* Correspondence: daniel.vollprecht@unileoben.ac.at

The protection of human health and the environment (representing the main reason for waste management), as well as the sustainable use of natural resources, requires chemical, biological, physical and thermal treatment of wastes. This refers to the conditioning (e.g., drying, washing, comminution, rotting, stabilization, neutralization, agglomeration, homogenization), conversion (e.g., incineration, pyrolysis, gasification, dissolution, evaporation), and separation (classification, direct and indirect (i.e., sensor-based) sorting) of all types of wastes to follow the principles of the waste hierarchy (i.e., prevention (not addressed by this issue), preparation for re-use, recycling, other recovery, and disposal). Longstanding challenges include the increase of yield and purity of recyclable fractions and the sustainable removal or destruction of contaminants from the circular economy.

This Special Issue on "Advanced Technology of Waste Treatment" of *Processes* collects high-quality research studies addressing challenges on the broad area of chemical, biological, physical and thermal treatment of wastes.

The mechanical treatment of municipal solid wastes (MSW, including separately collected fractions (i.e., paper, glass, plastics) and mixed municipal (i.e., residual) wastes, as well as wastes from landfill mining projects) is a key step in the circular economy as it produces "concentrates" of specific secondary raw materials from heterogeneous wastes. Digitalization and intelligent interconnection of mechanical waste processing plants become increasingly important to optimize the process, and especially to improve yield and purity of the produced concentrates which are subsequently utilized as recyclates, when substituting for primary raw materials, and as energy carriers. For this purpose, approaches such as sensor-based material flow characterization (SBMC) [1], sensor-based sorting (SBS) [2], and intelligent robotics [3] are applied more and more in mechanical waste treatment plants. Sarc et al. [4] developed the vision of a "Smart Waste Factory" in which these approaches are combined using digital communication and interconnection. For the realization of this vision, a fundamental understanding of waste properties and their evolution along the waste treatment chain is required. Therefore, Khodier and Sarc [5] developed a distribution-independent model of particle size distributions and applied it successfully to the shredding of mixed commercial waste.

Besides the production of concentrates, the removal of contaminants from the circular economy is the second task of mechanical waste treatment. Currently, the recyclability of MSW is limited by presence and leachability of contaminants, especially when source separation did not occur and/or material interactions and alterations have taken place, e.g., during use, treatment and—in the case of landfill mining [6]—disposal. Schwabl et al. [7] developed a wet-mechanical process to purge polyolefin concentrates from different waste streams, simultaneously removing surface contaminations.

Lithium-ion batteries (LIBs) represent a valuable secondary raw material when collected separately, but are a contaminant when disposed of in the residual MSW. Nigl et al. [8] conducted a risk assessment of LIB-caused fires in waste treatment processes, highlighting their role as ignition sources.

Industrial and mining wastes differ from MSW, as it is not so much the level of the particle but the level of the mineralogical phase which determines their recyclability—both

Citation: Vollprecht, D.; Sarc, R. Special Issue on "Advanced Technology of Waste Treatment". *Processes* **2022**, *10*, 217. https://doi.org/10.3390/pr10020217

Received: 13 December 2021
Accepted: 11 January 2022
Published: 24 January 2022

with respect to separability and contaminant immobilization capacity [9]. Consequently, comminution along phase boundaries is a prerequisite to obtain high-value concentrates of individual mineral phases for subsequent recycling. Seifert et al. [10] demonstrated the feasibility of an innovative comminution technology, electrodynamic fragmentation, to disintegrate spent refractory ceramics along phase boundaries which were subsequently separated using SBS technology.

Since mechanical waste treatment does not change the phase composition of the material, it is a necessary but not a sufficient step in the circular economy. Therefore, the concentrates produced in mechanical waste treatment have to undergo (thermo-/hydro-)chemical waste treatment for conversion into new products or for safe disposal.

Chemical recycling of plastic waste is an emerging technology which allows closing the loop for plastic waste fractions which cannot be recycled at the material level via the established "mechanical" route which, however, also includes the thermal process of melting and regranulation. Lechleitner et al. [11] used lumped kinetic modelling for the development of a pyrolysis process for the chemical recycling of polyolefins, i.e., polypropylene (PP) and polyethylene (PE). In a complementary study, Rieger et al. [12] focused on plastics from waste electrical and electronic equipment (WEEE) and the comprehensive chemical characterization of the pyrolysis products. Finally, Hee et al. [13] used marine litter waste as feedstock for chemical recycling (pyrolysis, gasification) and energy recovery (incineration) with special emphasis on the potential of the pyrolysis condensate for subsequent upcycling.

Chemical recycling of other types of waste is already more established when considering that, e.g., metal recycling involves chemical reactions and is therefore a thermochemical, not a thermal process. In this field, Windisch-Kern et al. [14] demonstrated how slagging of lithium can be reduced in pyrometallurgical battery recycling when using the InduCarb reactor concept.

Waste treatment and wastewater treatment leave behind secondary wastes, such as MSW incineration ashes and sewage sludge, respectively. The recovery of resources from these secondary wastes is a key challenge in the circular economy, as these materials are often also a sink for contaminants from primary wastes. Sewage sludge incineration is a process which allows energy recovery, but also ensures destruction of organic contaminants. However, the moisture content of sewage sludge hinders the thermal valorization. Therefore, Ekanthalu et al. [15] applied hydrothermal carbonization (HTC) to produce energy-rich hydrochar products and to enable phosphorous recovery. Based on an inventory of MSW incineration fly ash in Switzerland by Zucha et al. [16], Weibel et al. [17] studied metal recovery from these materials by acid leaching, whereas Wolffers et al. [18] investigated co-leaching of MSW incineration fly ash and waste wood fly ash. These leaching processes yield aqueous solutions which are used by Hettenkofer et al. [19] for copper recovery using polymer-assisted ultrafiltration.

In summary, this Special Issue presents an overview on recent international developments in the treatment of primary municipal and industrial, as well as secondary, wastes, covering both mechanical and (hydro-/thermo-)chemical processes. We thank all the contributors, as well as the editorial staff of *Processes*, for their efforts.

Author Contributions: Writing—original draft preparation, D.V.; writing—review and editing, R.S. All authors have read and agreed to the published version of the manuscript.

Funding: This research received no external funding.

Acknowledgments: The authors thank Roland Pomberger for his support.

Conflicts of Interest: The authors declare no conflict of interest. The funders had no role in the design of the study; in the collection, analyses, or interpretation of data; in the writing of the manuscript, or in the decision to publish the results.

References

1. Kroell, N.; Chen, X.; Maghmoumi, A.; Koenig, M.; Feil, A.; Greiff, K. Sensor-based particle mass prediction of lightweight packaging waste using machine learning algorithms. *Waste Manag.* **2021**, *136*, 253–261. [CrossRef] [PubMed]
2. Küppers, B.; Hernández Parrodi, J.C.; García Lopez, C.; Pomberger, R.; Vollprecht, D. Potential of sensor-based sorting in enhanced landfill mining. *Detritus* **2019**, *8*, 24–30. [CrossRef]
3. Sarc, R.; Curtis, A.; Kandlbauer, L.; Khodier, K.; Lorber, K.E.; Pomberger, R. Digitalisation and intelligent robotics in value chain of circular economy oriented waste management–A review. *Waste Manag.* **2019**, *95*, 476–492. [CrossRef]
4. Sarc, R. The "ReWaste 4.0" Project—A Review. *Processes* **2021**, *9*, 764. [CrossRef]
5. Khodier, K.; Sarc, R. Distribution-Independent Empirical Modeling of Particle Size Distributions–Coarse-Shredding of Mixed Commercial Waste. *Processes* **2021**, *9*, 414. [CrossRef]
6. Vollprecht, D.; Machiels, L.; Jones, P.T. The EU Training Network for Resource Recovery through Enhanced Landfill Mining–A Review. *Processes* **2021**, *9*, 394. [CrossRef]
7. Schwabl, D.; Bauer, M.; Lehner, M. Advanced Plastic Recycling by Wet-Mechanical Processing of Mixed Waste Fractions. *Processes* **2021**, *9*, 493. [CrossRef]
8. Nigl, T.; Baldauf, M.; Hohenberger, M.; Pomberger, R. Lithium-Ion Batteries as Ignition Sources in Waste Treatment Processes–A Semi-Quantitate Risk Analysis and Assessment of Battery-Caused Waste Fires. *Processes* **2021**, *9*, 49. [CrossRef]
9. Neuhold, S.; Algermissen, D.; Drissen, P.; Adamczyk, B.; Presoly, P.; Sedlazeck, K.P.; Schenk, J.; Raith, J.G.; Pomberger, R.; Vollprecht, D. Tailoring the FeO/SiO_2 Ratio in Electric Arc Furnace Slags to Minimize the Leaching of Vanadium and Chromium. *Appl. Sci.* **2020**, *10*, 2549. [CrossRef]
10. Seifert, S.; Dittrich, D.; Bach, J. Recovery of Raw Materials from Ceramic Waste Materials for the Refractory Industry. *Processes* **2021**, *9*, 228. [CrossRef]
11. Lechleitner, A.E.; Schubert, T.; Hofer, W.; Lehner, M. Lumped Kinetic Modeling of Polypropylene and Polyethylene Co-Pyrolysis in Tubular Reactors. *Processes* **2021**, *9*, 34. [CrossRef]
12. Rieger, T.; Oey, J.C.; Palchyk, V.; Hofmann, A.; Franke, M.; Hornung, A. Chemical Recycling of WEEE Plastics—Production of High Purity Monocyclic Aromatic Chemicals. *Processes* **2021**, *9*, 530. [CrossRef]
13. Hee, J.; Schögel, K.; Lechthaler, S.; Plaster, J.; Bitter, K.; Blank, L.M.; Quicker, P. Comparative Analysis of the Behaviour of Marine Litter in Thermochemical Waste Treatment Processes. *Processes* **2021**, *9*, 13. [CrossRef]
14. Windisch-Kern, S.; Holzer, A.; Ponak, C.; Raupenstrauch, H. Pyrometallurgical Lithium-Ion-Battery Recycling: Approach to Limiting Lithium Slagging with the InduRed Reactor Concept. *Processes* **2021**, *9*, 84. [CrossRef]
15. Ekanthalu, V.S.; Narra, S.; Sprafke, J.; Nelles, M. Influence of Acids and Alkali as Additives on Hydrothermally Treating Sewage Sludge: Effect on Phosphorus Recovery, Yield, and Energy Value of Hydrochar. *Processes* **2021**, *9*, 618. [CrossRef]
16. Zucha, W.; Weibel, G.; Wolffers, M.; Eggenberger, U. Inventory of MSWI Fly Ash in Switzerland: Heavy Metal Recovery Potential and Their Properties for Acid Leaching. *Processes* **2020**, *8*, 1668. [CrossRef]
17. Weibel, G.; Zappatini, A.; Wolffers, M.; Ringmann, S. Optimization of Metal Recovery from MSWI Fly Ash by Acid Leaching: Findings from Laboratory- and Industrial-Scale Experiments. *Processes* **2021**, *9*, 352. [CrossRef]
18. Wolffers, M.; Weibel, G.; Eggenberger, U. Waste Wood Fly Ash Treatment in Switzerland: Effects of Co-Processing with Fly Ash from Municipal Solid Waste on Cr(VI) Reduction and Heavy Metal Recovery. *Processes* **2021**, *9*, 146. [CrossRef]
19. Hettenkofer, C.; Fromm, S.; Schuster, M. Municipal Solid Waste as Secondary Resource: Selectively Separating Cu(II) from Highly Saline Fly Ash Extracts by Polymer-Assisted Ultrafiltration. *Processes* **2020**, *8*, 1662. [CrossRef]

processes

MDPI

Review

The "ReWaste4.0" Project—A Review

Renato Sarc

Chair of Waste Processing Technology and Waste Management, Montanuniversitaet Leoben, Franz-Josef-Straße 18, 8700 Leoben, Austria; renato.sarc@unileoben.ac.at; Tel.: +43-3842-402-5105

Abstract: ReWaste4.0 is an innovative and cooperative K-Project in the period 2017–2021. Through ReWaste4.0 the transformation of the non-hazardous mixed municipal and commercial waste treatment industry towards a circular economy has started by investigating and applying the new approaches of the Industry 4.0. Vision of the ReWaste4.0 is, among others, the development of treatment plants for non-hazardous waste into a "Smart Waste Factory" in which a digital communication and interconnection between material quality and machine as well as plant performance is reached. After four years of research and development, various results have been gained and the present review article summarizes, links and discuss the outputs (especially from peer-reviewed papers) of seven sub-projects, in total, within the K-project and discusses the main findings and their relevance and importance for further development of the waste treatment sector. Results are allocated into three areas, namely: contaminants in mixed waste and technical possibilities for their reduction as well as removal; secondary raw and energy materials in mixed waste and digitalization in waste characterization and treatment processes for mixed waste. The research conducted in ReWaste4.0 will be continued in ReWaste F for further development towards a particle-, sensor- and data-based circular economy in the period 2021–2025.

Keywords: mixed waste; municipal waste; commercial waste; waste treatment; recycling; recovery; contaminants; plastics; digitalisation; smart waste factory

Citation: Sarc, R. The "ReWaste4.0" Project—A Review. *Processes* **2021**, *9*, 764. https://doi.org/10.3390/pr9050764

Academic Editor: Avelino Núñez-Delgado

Received: 31 March 2021
Accepted: 25 April 2021
Published: 27 April 2021

1. Introduction

ReWaste4.0 is a long-term oriented, innovative as well as cooperative K-Project (note: the Competence Centres for Excellent Technologies (COMET) programme is a national funding line of the Austrian Research Promotion Agency (FFG) that aims to carry out high-quality research in science–industry collaboration [1]) at the highest scientific, technical and economic level with two partners from science and eight from industry each. The overall innovative objective for the first time is to investigate and partially implement the new approaches of the "Industry 4.0" (i.e., digital networking, communication between waste quality and plant performance, dynamic process control and optimization and others). This innovative development of waste treatment of non-hazardous mixed waste will transform the branch towards a circular economy and enable high-quality recycling and recovery processes. Through ReWaste4.0 the experimental data-based development of waste treatment plants into the so-called "smart waste factory" will be supported. At a material level, the focus is set on the treatment of non-hazardous mixed waste (i.e., mixed municipal and mixed commercial waste as well as selected non-hazardous mixed fractions declared as the "output" of mechanical treatment processes like Solid Recovered Fuels (SRF) that are utilized in cement industry). ReWaste4.0 consists of one comprehensive strategic project and two scientific–technical areas with six subordinated specific projects and connected by the before mentioned approaches of the Industry 4.0, see Figure 1 [2].

At the time of the project application development (i.e., in 2016), the European Commission had presented an ambitious new Circular Economy Package (on 2 December 2015) including a proposal for recycling targets for municipal (65% by 2030) and packaging waste (75% by 2030) [3].

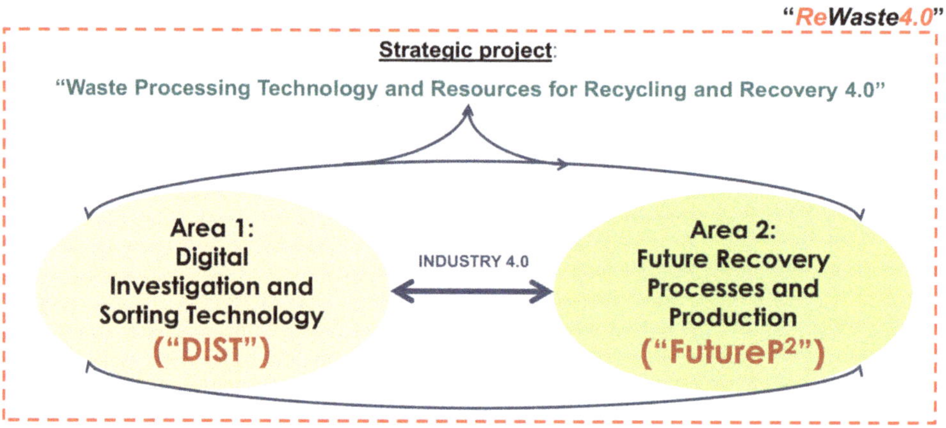

Figure 1. Holistic approach of ReWaste4.0—interdependencies of the strategic project and both areas connected by the approaches of Industry 4.0 [2].

In the meantime, the most important action in European waste management was the adoption of the Circular Economy Package that came into force in 2018 [4]. The revised legislative framework on waste entered into force in July 2018 and, among others, it sets clear targets for selected waste, namely:

- 65% recycling of municipal waste by 2035;
- 70% recycling of packaging waste by 2030.

The dynamic change in the performance of municipal waste management for the EU 28 and the development needed to meet the recycling rates set by the circular economy package 2018 are shown in Figure 2.

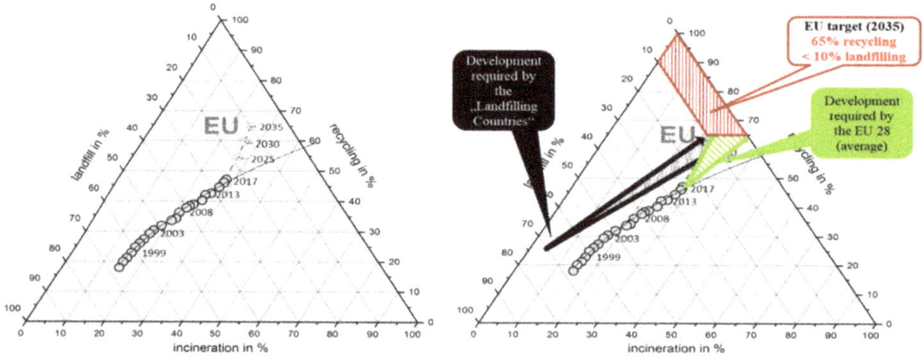

Figure 2. Left: development of municipal waste management in the EU 28 from 1995 to 2017 with the forecasts (linear extrapolation, and potential development towards the minimum requirements of the new circular economy package) until 2035. Right: required development to meet the recycling targets [5], modified from [6].

Next, in 2019, the European Green Deal [7] was published. It is a roadmap for making the EU's economy sustainable. Climate change and environmental degradation are an existential threat to Europe and the world. To overcome these challenges, Europe needs a new growth strategy that transforms the Union into a modern, resource-efficient and competitive economy where:

- there are no net emissions of greenhouse gases by 2050;
- economic growth is decoupled from resource use;

- no person and no place are left behind.

1.1. Municipal and Commercial Waste Management

Worldwide, 2.01 billion tons of municipal solid waste (MSW) from residential, commercial, and institutional origins were generated in 2016. This number is expected to increase substantially to 3.4 billion tons by 2050 [8].

In EU28, approximately 252 million tons of municipal waste were generated in 2018, i.e., 491 kg/capita and 47% of it was recycled [9]. Weißenbach et al. [10] report that an additional recycling potential of about 46.3 million tons of municipal waste by 2040 is available.

In Austria, on a federal level, there is no specific legislation on mixed commercial waste (MCW). The term "commercial waste" is not defined, neither in federal nor in provincial legislation. In the Austrian waste catalogue, ÖNORM S 2100 [11] based on Waste Catalogue Ordinance) commercial waste is assigned to waste group 91 "Solid municipal waste, including similar commercial waste". A clear definition of MCW is given, for example, in Germany that has stipulated the Commercial Waste Regulation [12]. There, MCW is a non-municipal residual waste that is not collected separately, such as office waste, industrial waste, for example. Typical MCW can be characterized by low moisture content, high calorific value, low organic content, and high content of recyclables [13,14]. Owing to its energetically usable calorific value higher than that of mixed municipal waste, it is used to produce SRF [15–18]. The composition of MCW varies widely and depends on the industry in which it is generated [14]. Weißenbach [13] gives aggregated results of MCW sorting analyses in 2018 and 2019, see Table 1.

Table 1. Results on composition of mixed commercial waste from investigations made by Weißenbach [13] performed in 2018 and 2019.

Material Fraction	Weight Percent (%)	
	Site 1 (2 Experiments)	**Site 2 (2 Experiments)**
Plastics	19.1	16.2
Paper/cardboard	16.8	12.0
Metals	4.3	4.0
Wood	7.1	9.8
Inert	4.3	
Textiles	3.6	58.0
Others	44.7	
Total	99.9	100.0

The definition of mixed municipal waste (MMW), that is relevant for this paper is given in § 2, number 4, point 2 of the Austrian Waste Management Act 2002 [19]. There unprocessed mixed municipal waste is *"waste from private households and other waste which, because of its nature or composition, is similar to waste from private households; the classification shall take into account the European Waste List as defined in Article 7 of Directive 2008/98/EC on waste. Mixed municipal waste within the meaning of the European Waste List shall continue to be regarded as mixed municipal waste even if it has undergone a treatment process which has not significantly altered its properties."* In contrast to MCW, MMW has a higher content of moisture, lower calorific value, higher organic content and quite lower recyclable content [20]. The composition of MMW depends on various factors including the available waste collection system, the socioeconomic structure of the population and/or the situation of urban or rural households.

The Federal Waste Management Plan (FWMP) of Austria provides extensive information on treatment plants as well as data on the composition of MMW and the contained share of potential recyclables for 2018, see Table 2 [21].

Table 2. Composition of mixed municipal waste and the contained share of recyclables in Austria for the year 2018 [21].

Material Fraction	Share (in %)	Sub-Total (in %)	Recyclable Quantity (in t)
Plastic—packaging	7.10	17.58	256,455
Other light packaging	1.10		
Plastic—no packaging	9.38		
Biowaste (incl. not avoidable food waste)	1.31	17.81	
Avoidable or partly avoidable food waste	16.50		
Paper and cardboard—packaging	2.20	13.96	203,647
Paper and cardboard—no packaging	11.76		
Sanitary products	9.64	9.64	
Textiles	7.79	9.79	
Shoes	2.00		
Inert	5.86	5.86	
Glass—packaging	3.80	4.86	70,897
Glass—no packaging	1.06		
Metal—packaging	2.50	4.70	59,373
Metal—no packaging	2.20		
Other waste	4.01	5.71	
Wood—no packaging	1.70		
WEEE	0.77	1.54	
Batteries, incl. accumulators	n.d.		
Hazardous household waste	0.77		
Others (not identifiable)	8.55	8.55	
Total	100.00	100.00	590,372

The shares of the potentially recyclable materials: plastic, paper/cardboard, glass and metal add up to 41.1%, equalling a total of 590,372 tonnes in 2018. Experience shows, however, that a lot of these materials cannot actually be recycled because of their low quality (e.g., contamination with dust, biodegradables, or moisture) [13]. According to the FWMP [21] mixed waste (MMW and MCW) and mixed outputs from treatment plants for source separated fractions like paper/cardboard, or plastics are treated in mechanical and mechanical–biological treatment plants and, there, SRF and other outputs are produced. The second option for treating of mixed waste is energy recovery in waste-to-energy plants.

As shown in Tables 1 and 2, a large amount of plastic is still available in mixed municipal and commercial waste as well as SRF and therefore it represents an important potential for further recycling processes.

1.2. Plastics and Their Importance in the EU

The European strategy for plastics [22] intends to transform the linear (make–use–dispose) economy of plastics into a more circular, resource-efficient system. Various policies have been put in place to achieve this goal:, e.g., banning certain single-use plastic products, limiting lightweight plastic bags or creating quality standards for secondary plastics. Among others, mandatory recycling rates for plastic packaging (i.e., 50% by 2025 and 55% by 2030) were introduced. In order to achieve these ambitious recycling targets, the recycling of separately collected plastic packaging is not enough. Plastics must also be recovered from other waste streams. Due to the high volume of waste and the high unused plastics content, MCW and MMW would be suitable for this [20].

1.3. Digitalisation in the Waste Management Sector

In order to have a common understanding of used definitions, acc. to Tschandl et al. [23]:

"Digitalisation" generally describes the integration of digital technologies into everyday life. This integration is called "Industry 4.0" because it embodies the fourth industrial revolution. The English term is "Internet of Things" (IoT) and is divided into two parts: "Industrial Internet of Things" and "Consumer Internet of Things".

Then, Tschandl et al. [23] show that no uniform definition for the term "Industry 4.0" has yet been established. However, the different definitions can be used to derive the following general definition:

> *"Industry 4.0 describes the widespread introduction of information and communication technology (ICT) as well as its connection to an Internet of Things, Services and Data with the goal of real-time control of production and value chain networks".*

Next, Tschandl et al. [23] summarize the term "smart factory" as follows:

> *"Individual companies or corporate groups that use ICT for product development, production, logistics and interface coordination with customers in order to respond more flexibly to incoming requests. A smart factory masters complexity, is less disruptive and enables a more efficient production. The communication between people, machines and resources is self-evident and comparable to a social network."*

Curtis and Sarc [24] as well as Sarc et al. [25] have introduced the term "Smart Waste Factory Network" (SWFN) that is part of the ReWaste4.0 project and defined as follows:

> *"The SWFN4.0 describes a system consisting of several waste treatment plants, which perform different tasks in the waste management system and are interconnected via data streams and logistics systems (e.g., sorting plants, production plants for Solid Recovered Fuels, etc.). The individual processes and machines within the plants as well as the individual plants are digitally connected with each other. This connection of the individual machines and systems and the real-time analysis of the waste streams enable dynamic process control and various actuator systems actively intervene in the processes. In addition, people can cooperate interactively with the technology around them."*

To define the state of the art of digitalization on the waste management sector, Sarc et al. [25] gathered relevant contributions in literature via databases with the following relevant search terms (keywords): "digitalisation", "robotics", "smart waste", "smart factory", "industry 4.0", "internet of things", "waste management" and "circular economy". These search terms were used individually and in all possible combinations for the search. For the evaluation of the results of this scientific contribution, relevant publications could be found over the period from 2001 to 2019, although the majority are from the last three years (2017 to 2019), see Figure 3. In addition, detailed research has been conducted on existing technologies in the environmental field, with a focus on waste management [25].

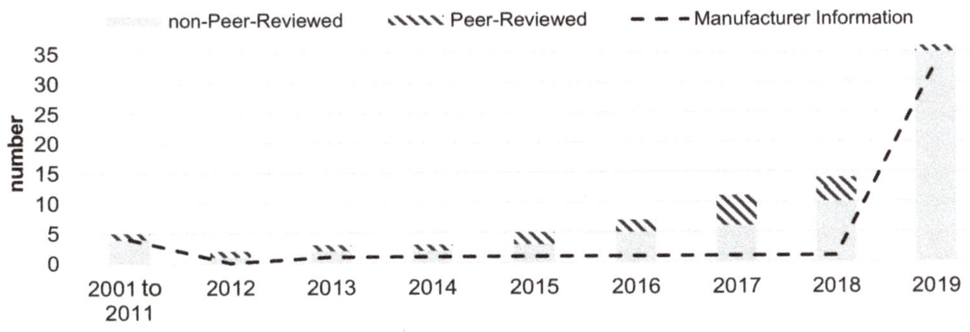

Figure 3. Number of reviewed literature sources by type, assigned to peer-reviewed or non-peer-reviewed papers over the years 2001–2019 [25].

A total of 115 literature sources could be utilized. It should be noted that only sources with content relevant for waste management were considered in the graphics, resulting in a total of 85 relevant literature sources. In addition, legal regulations were used for drafting the contribution, which are not considered in the illustrations shown. The results show that the topics digitalization and intelligent robotics in waste management, have yet not intensively been discussed in peer-reviewed papers and most of the information comes from technology- and platform-manufacturers (websites, brochures and others) [25].

Sarc et al. [25] report that "Big Data" is a fundamental element of digitalization and already a valuable raw material for many industries. In combination with "Artificial Intelligence" (AI), it is possible to structure, analyze, evaluate and use large amounts of data as a basis for software programs that can generate new (or extended) knowledge together with the aspects of "Machine Learning" as a part of AI. From this, future forecasts can be derived as well as used in optimization measures. Often "Deep Learning" is used, which is based on the human brain and uses artificial neural networks to mimic the learning processes of humans. This makes it possible to use data volumes meaningfully across the entire value chain (see Figure 4).

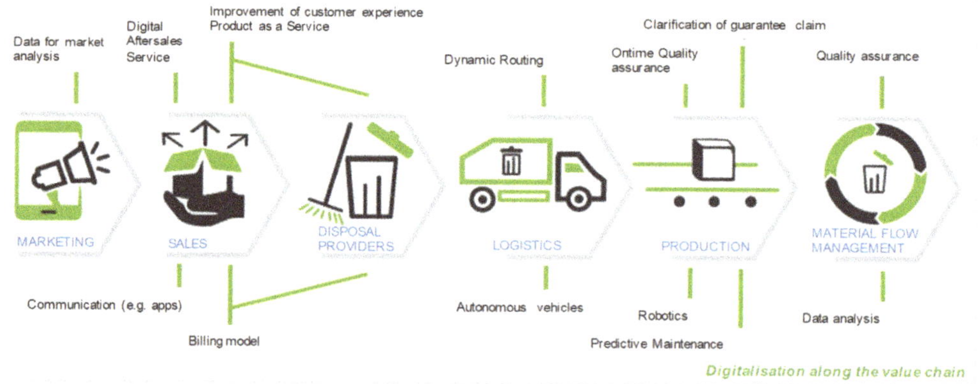

Figure 4. Digitalization along the value chain [26].

Abdallah et al. [27] have extensively investigated artificial intelligence (AI) applications in solid waste management and have found similar conclusions as Sarc et al. [25]. Abdallah et al. [27] report that MATLAB was the most common simulation software tool utilized in the waste sector. Next, Abdallah et al. [27] conclude that Artificial Intelligence-based solid waste management systems are still mostly in the research and development phase. Additionally, the most important limitations or challenges in waste management are, among others, insufficient amounts of data and its quality as well as the slow shift of waste management business entities in adapting towards the utilization of artificial intelligence versus traditional methods.

Finally, to simulate or carry out digital processes in waste management plants, large amounts of high-quality real, online/ontime data are needed as a basis. These data were gained within ReWaste4.0 based on practical, experimental as well as large scale tests in waste treatment plants.

As described, ReWaste4.0 is mainly focused on the treatment of mixed waste by applying Industry 4.0 approaches and the main focus of the present review-article is set on three categories of investigated topics:

1. Contaminants in mixed waste and technical possibilities for their reduction as well as removal;
2. Secondary raw and energy materials in mixed waste;
3. Digitalization in waste characterization and treatment processes for mixed waste.

Finally, the objective of the present review article is to summarize, link and discuss the main topics and results of the K-project and to discuss the main findings and their relevance as well as importance for further development of waste treatment sector. The review is executed mainly based on the peer-reviewed papers that have been published within the project.

2. Materials and Methods

A literature review mainly based on the peer-reviewed publications and selected conference contributions published within the project ReWaste4.0 (time period: 2017–2021) was carried out to focus on and discuss selected important results and findings of four years of cooperative research and development. Additionally, certain conference proceedings and peer reviewed papers not originating from ReWaste4.0, but with relevant information and impact on the topics of ReWaste4.0, were considered too.

3. Results and Discussion

As shown before, mixed waste such as Mixed Municipal Waste (MMW) and Mixed Commercial Waste (MCW) are the focus of research and development within the ReWaste4.0 project. Knowledge on the properties of such waste that serves as an input for mechanical waste treatment plants, however, is important for defining the number and processing depth of the required treatment steps as well as assessing the feasibly achievable quality of the output fractions. In order to characterize waste (plant's input and output) and to determine important parameters, e.g., heavy metal contents, material composition, or the mass share of valuable materials, sampling is still unavoidable for the state of the art. The main criterium that needs to be fulfilled by a sampling procedure is representativeness. With increasing material heterogeneity and particle sizes (note: they are large for MMW and MCW), more elaborated sampling procedures are required. Sampling procedures shall therefore follow the principles of the theory of sampling (TOS), which is an integral component of many internationally recognized waste sampling standards (e.g., EN15442:2011) [28,29].

Khodier et al. [30] and Viczek et al. [31] have determined the relative sampling variability (RSV) for material classes and chemical elements in different particle size classes of coarsely shredded MCW with a replication experiment. A total of 10 representative samples were taken and screened to yield 9 particle size classes. RSVs for material classes in different particle size classes ranged up to 231%, while RSVs for chemical elements ranged up to 203.5%. The RSVs for different chemical elements are depicted in Figure 5 and show that higher RSVs are tendentially observed for chemical elements with a higher constitutional heterogeneity, i.e., elements that occur in largely different concentrations in different particles. In general, far better RSVs were achieved when calculated for the original, mixed waste stream instead of the nine particle size classes.

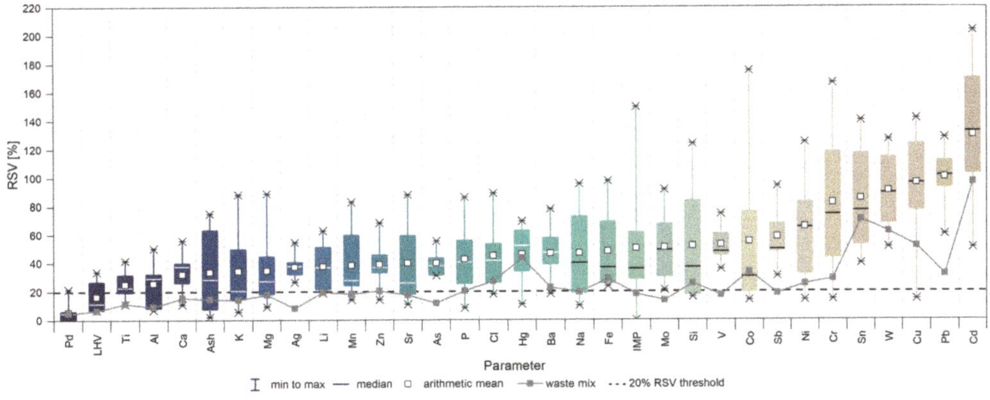

Figure 5. Relative sampling variabilities (RSV) for different elements, lower heating value (LHV), ash content, and hard impurities (IMP) in different particle size classes. The solid grey line represents RSVs calculated for the whole waste mix, i.e., for the united particle size fractions [31].

The procedure for sampling and analysis, which was elaborated and evaluated in these publications almost completely eliminates incorrect sampling errors and suggests that grouping and segregation errors causing distributional heterogeneity significantly influence the results. Consequently, increasing the number of increments taken to reach the target sample mass is expected to have a larger beneficial effect than increasing the mass of the composite sample, which advantageously does not affect the efforts for analyzing the composite sample [30,32]).

Furthermore, Viczek et al. [31] combined the results obtained for the material composition and chemical analyses to assess whether a mathematical relationship between these parameters can be derived. A preliminary modeling approach shows that it may be possible to predict most element concentrations based on the material composition of MCW, which may be promising for quality assurance in MCW treatment plants, e.g., SRF and other waste fuel production plants.

Despite the mentioned challenges when dealing with coarse and heterogenous waste like MCW, MMW and SRF, high-quality and representative results have been obtained and are presented in the following chapters.

3.1. *Contaminants in Mixed Waste and Technical Possibilities for Their Reduction as Well as Removal*

Contaminants play an important role for all waste treatment options, independent of the classification of treatment process as recovery (R) or disposal (D) operations according to the EU [33]. For the Austrian SRF industry, especially As, Cd, Co, Cr, Hg, Ni, Pb, Sb and Cl are relevant due to legal limit values [34] or technical requirements [35]. Viczek et al. [36] have identified the material fractions or products ending up in MCW or MMW that carry significant amounts of the above-mentioned nine elements, i.e., contaminant carriers. Their review showed that contaminant carriers in MCW and MSW are highly versatile. With respect to the Austrian situation, relevant contaminant carriers include PVC (Cl, Cd, Sb, Pb), flame-retardant plastics or textiles (Sb), rubber (Sb, Cl), glass (As, Co, Pb, Ni), leather (Sb, Cr), specific wood (As, Pb, Hg, Ni), electronic equipment and batteries (As, Cd, Cl, Cr, Co, Pb, Hg, Ni), shoes (Cd, Cl, Cr, Pb), or metals (Cr, Co, Pb, Ni). As many of these elements (Cd, Sb, Cr, Co, Pb) are used as pigments for ceramics or plastics, these products can be contaminant carriers as well. Furthermore, the literature review of [36] often identified the fine fraction with different particle sizes as important contaminant carriers for various elements, e.g., Pb, Hg, or Ni. This indicates that the relevant contaminants are not only enriched in specific materials, but also in specific particle size classes, which is why the distribution of various chemical elements among different particle size fractions of MCW was investigated [37].

Based on their particle size-dependent distribution, Viczek et al. [37] divided different elements into three groups:

- A: Negative linear correlation—higher concentrations in smaller particle size classes,
- B: No linear correlation—low concentration in smallest and largest particle size classes,
- C: Positive linear correlation—higher concentrations in medium to large particle size classes.

While As, Cr, Co, Pb, Hg and Ni belong to group A, Cd belongs to group B, and Cl, Sb, as well as the LHV were assigned to group C. This implies that these elements influence the output of a waste processing plant into different extents depending on the design of the process. The patterns for Hg and the LHV are depicted in Figures 6 and 7, respectively. Similar findings were also reported by Curtis et al. [38] who investigated four particle size classes of different samples of MCW and compared the concentrations in these particle size classes to the limit values of the Austrian WIO [34].

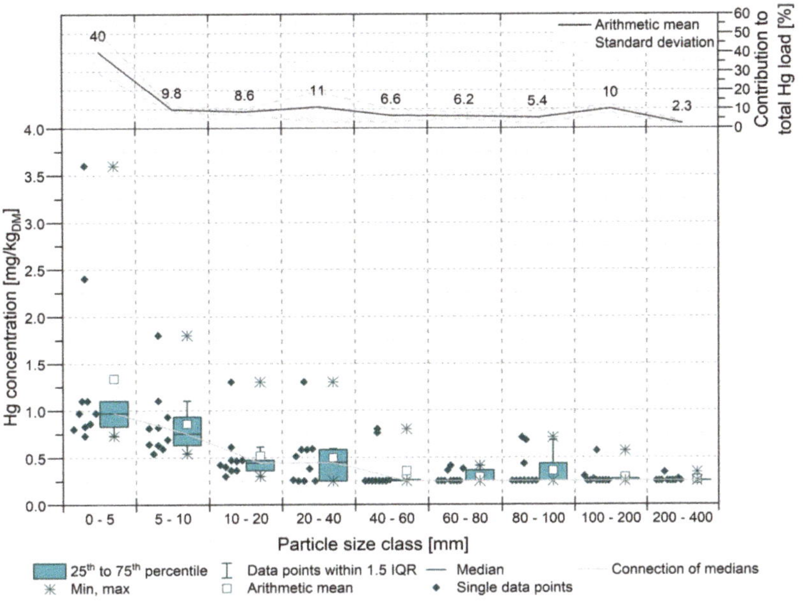

Figure 6. Concentrations of Hg in different particle size classes of mixed commercial waste and contribution of each particle size fraction to the total Hg load [37].

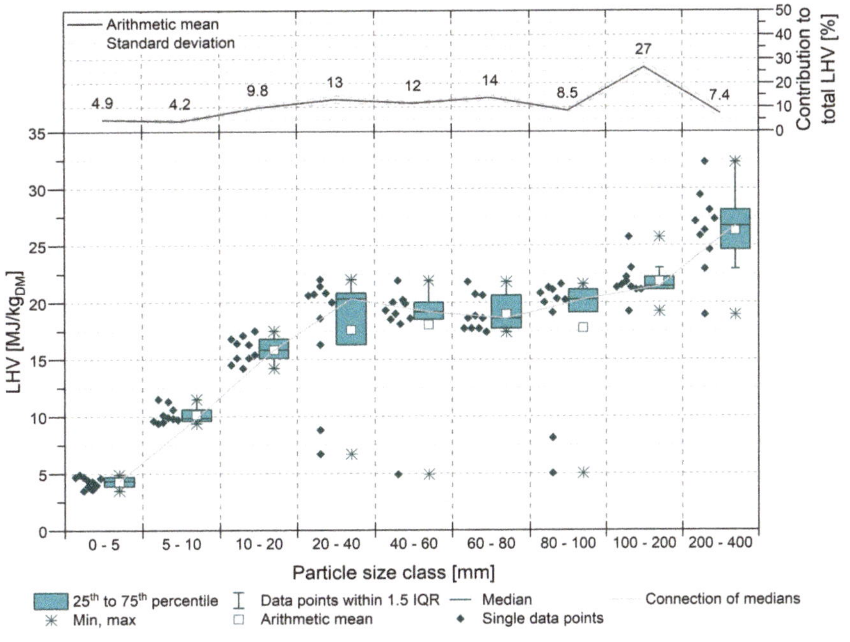

Figure 7. Lower heating value of different particle size classes of mixed commercial waste and contribution of each particle size fraction to the total LHV [37].

The results for the LHV, Cl, and Sb reflect the fact that plastics (which are rich in LHV and often contain Cl and Sb) are rather present in large particle size fractions, while smaller particle size fractions rather contain larger amounts of biogenic or inorganic materials. Generally, Cl, Sb, and the LHV showed negative correlations with most other elements, while being positively correlated with each other. Furthermore, a strong positive correlation was observed for Cr and Ni, which are typical constituents of metal alloys.

The reported results on contaminants in different particle sizes from Viczek et al. [37] indicated that screening can be a suitable technique to (technically simply and financially reasonably) remove relatively large amounts of contaminants from the waste stream. Calculations showed that a theoretical removal of the fine fraction <5 mm or <10 mm results in significant decreases in the concentrations of Hg, Co, Ni, As, Cr, and Pb, but the Cd, Sb, and Cl concentration in the remaining waste is increased (Figure 8). However, this effect is compensated by the increasing LHV if the concentrations are considered in mg/MJ. As Cd, Sb, and Cl frequently occur in plastics, a combination of screening and targeted NIR sorting seemed to be a promising approach to decrease the concentrations of all relevant contaminants, which was tested by Viczek et al. [39].

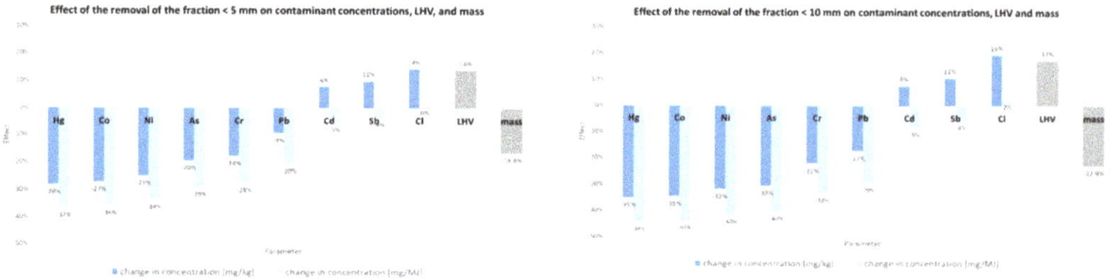

(a) (b)

Figure 8. Effect of the removal of the fine fraction <5 mm (**a**) and <10 mm (**b**) on analyte concentrations in mg/kg$_{DM}$ and mg/MJ [40].

As noted before, the experimental investigations of Viczek et al. [39] demonstrated that a combination of screening with NIR sorting can decrease the Sb, Cl, and Cd content in the remaining waste stream. Especially the removal of the PVC fraction in combination with removing the fine fraction, e.g., <20 mm, can give good results with respect to the analytes. The removal of PET, in comparison, had rather low effects on the contaminant concentrations in the remaining waste stream. Removing PET and PVC, but not removing the fine fraction, in contrast, can result in an increase in the concentrations of some contaminants. Another fraction that contained large amounts of Cl, Sb, and Co were black and grey materials, but they cannot be identified or detected with conventional NIR sorters working in the range of 900–1700 nm. However, other technologies that may be able to remove these materials exist, and these materials require more attention and research in future [39].

However, while the quality of the screen overflow is increased concerning several parameters, the screen underflow represents a waste fraction with poor quality and a low LHV, exceeding selected limit values, e.g., those for SRF defined by the Austrian WIO [34]. Nevertheless, at the same time, it contains mineral matter that can substitute primary raw materials and be recycled in the cement industry (see Section 3.2), which leads to a so-called "conflict of interests" between environmental protection and resource conservation/utilization.

In conclusion, contaminants frequently occur in a broad range of products and therefore waste management needs to deal with them on a daily basis. For this reason, existing, arising and future recycling concepts, e.g., chemical recycling, need to enable high recycling rates while also providing quality oriented and assured recycling.

3.2. Secondary Raw and Energy Materials in Mixed Waste

Here, the focus is given to plastics and SRF as well as their co-processing, mainly in the clinker production process of the cement industry.

3.2.1. Plastics

The content of plastics in MMW and MCW acc. to Weißenbach et al. [14] and Möllnitz et al. [20] is about 15% for MMW and between 15% and 23% for MCW. Both types of waste are treated in splitting plants or mechanical–biological plants for, among others, the production of SRF or in thermal plants (with and without energy utilization) (note: in selected EU countries including Austria, direct landfilling of untreated MMW and MCW is legally forbidden). For a further, more efficient, recovery of the plastics contained in these wastes, it is necessary to know and understand in which geometric dimension (two-dimensional/three-dimensional) and in which particle size range they are present after pre-shredding—which is usually the first processing step in mechanical processing, targeting the comminution and liberation of waste particles according to Khodier et al. [41].

The investigations on total composition of plastic content of MMW and MCW after pre-shredding and screening of fine fraction have shown that a screen cut at 20 mm removes about one third (MCW: 33%; MMW: 37%) of the total material after pre-shredding. This fine fraction has a high inert, contaminant and organic content [20,36] and is therefore unsuitable for plastics recovery.

The results given in Figure 9 show that MCW has a twice as high plastics-3D content as MMW. The plastics-2D content (e.g., foils), on the other hand, is similar in both mixed waste types. A particle size dependence of the dimensionality of the plastic particles can also be seen for both wastes: the share of 3D particles is higher in small particle size classes, while in large size classes more 2D-plastics are found. This information is relevant for treating the waste stream in a targeted and efficient way and to be able to separate the desired fraction at the proper process step. For further processing, the particle shape is also of high importance for the separation and sorting success and the associated purity of the targeted fraction (e.g., flight behaviour of films in NIR-sorting) [20].

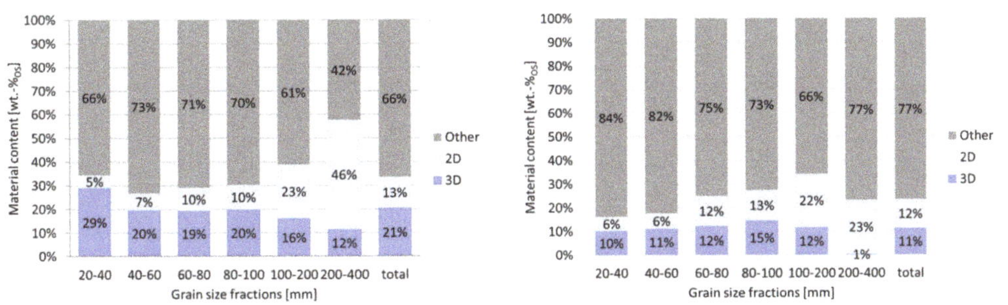

(a) (b)

Figure 9. Content of plastics-2D and plastics-3D in mixed commercial waste (**a**) and mixed municipal waste (**b**) in different particle sizes [20].

Figure 10 shows the results of the sensor-based sorting by NIR of each individual particle size class into its plastic types and remaining components using MCW as an example [20]. The results show that certain types of plastics are more common in distinct particle size classes than in others and that this is also dimensionality dependent. This information is important for further increasing the purity of extracted plastics by efficiently routing those material flows that have high concentrations of plastics or specific plastic types.

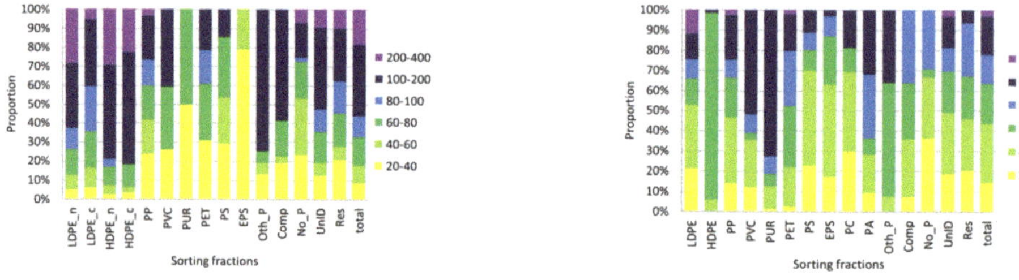

(a) (b)

Figure 10. Particle size distribution in investigated material types for plastics-2D (**a**) and plastics-3D (**b**) normalized to 100% for mixed commercial waste [20].

The presented results clearly show that there is a high, as yet unused potential of plastics in the two mixed waste types investigated. For the efficient recovery of certain types of plastics, early screening of the fine fraction after pre-shredding makes sense in order to remove contaminants and impurities (see also Möllnitz et al. [42]). A targeted material flow division according to dimensionality (e.g., using a ballistic separator as it is proposed by Möllnitz et al. [42]) as well as further screen classifications into specific particle size classes are necessary to achieve the targeted concentration of plastics and to make sensible use of plant capacities. In order to be able to implement this processing step in a real waste processing plant, it is therefore necessary to know the respective waste in detail.

The plastic concentrates produced in the waste sorting plants usually still have a high amount of impurities and contaminants. These can be of organic (e.g., other plastics, adhering oils or fats) or inorganic (e.g., metallic coatings) origin [43]. In order to meet the quality requirements of the recycler for the flakes and those of the plastics processor for the granulate produced, subsequent wet processing at the plastics recycler site is usually necessary [44]. A benchmark analysis [44] was carried out with data from experts and stakeholders in Austria and Germany to determine the correlation between different quality features and how they affect the pricing policy for recyclates. Several quality assurance measures are carried out along the value chain from plastic waste to final plastic products. The most important quality assurance parameters and how they (would) affect prices of recyclates were investigated. Friedrich et al. [44] report that pricing correlates with different quality parameters such as degree of mixing, degree of degradation and presence of impurities and contaminants and that the origin of waste affects the assessment of the sorted plastic waste quality. Next, the physical, rheological and mechanical properties of the recyclates are of great interest and the following characteristics were analyzed in the course of a random sample inspection:

1. physical properties like density determination;
2. rheological properties like melt-mass flow rate;
3. mechanical properties like tensile properties, especially modulus of elasticity and notch impact strength.

Frequently, further parameters of the recyclates are determined. These include [44]:

- melting temperature;
- colour distribution and colour composition;
- size and form of the granulated material (e.g., lenses, cylinder);
- moisture content;
- filtration fineness;
- ash content;
- heavy metal content.

To investigate plastics from MMW and MCW regarding the aforementioned requirements, the processability of different polymer fractions from mixed waste was examined and the determination of material properties for recycling was carried out by Möllnitz et al. [45]. For the investigations, commercially available SRF and two mixed polyolefin (PO) fractions (polyethylene—PE and polypropylene—PP) obtained from MMW and MCW treatment processes were used. In addition to a different processing depth (washed and unwashed) of the inputs, the focus of the investigations was the processability of the recovered plastic types (PE, PP, polyethylene terephthalate (PET), polystyrene (PS) see Figure 11) and plastic mixtures (PO and SRF) in a compression molding process (with and without previous homogenization in an extruder) and their characterization regarding the following parameters:

- Thermal properties: determination of the crystallization temperature (T_C) with the respective crystallization enthalpy (ΔH_C), melting temperatures (T_{m1} and T_{m2}) with the respective melting enthalpy (ΔH_{m1} and ΔH_{m2}), and the glass transition temperature (T_g) with differencial scanning calorimetry (DSC);
- Mechanical properties: impact strength and notched impact strength, tensile test (Young's modulus (E), tensile strengths (σ_M), elongations at break (ε_B)), bulk density of flakes after shredding and granulate after extrusion, determination of ash content;
- Rheological properties: melt flow rate (MFR) [45].

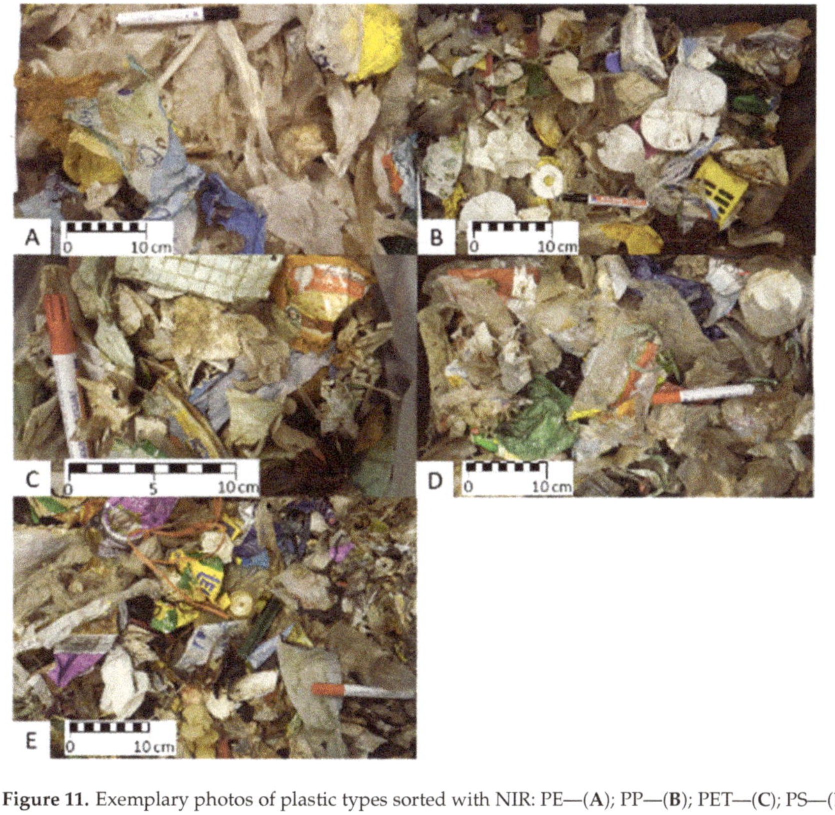

Figure 11. Exemplary photos of plastic types sorted with NIR: PE—(**A**); PP—(**B**); PET—(**C**); PS—(**D**); and others—(**E**) [45].

The main results of the investigations in Möllnitz et al. [45] are: except for PET, all plastic types and mixtures were processable (extrudable and/or compressible). Both the flakes and granulates showed good feeding and conveying behavior. The thermal properties were generally good and indicate little material damage but organic impurities present. The heating and cooling curves were particularly reproducible for the heterogeneous materials. The results of the mechanical properties show a clear material embrittlement due to existing impurities (especially for the PS materials). Furthermore, it was found that washing does not always lead to a significant improvement of the mechanical properties. It could also be shown that sorting into certain plastic types, such as PE and PP, is not necessary for certain applications, as the PO and mixed plastic fractions investigated have a good mechanical property profile.

Regarding processability of the mixed waste (MCW and MMW) with focus on plastics recovery, the following technologies were investigated and the results were obtained:

1. Dry mechanical waste treatment with pre-screening prior further processing by ballistic separation and sensor-based sorting for generation of a 3D plastics pre-concentrates for recycling [42].

Large-scale experiments with mobile machines were conducted to investigate the influence of an upstream drum screen on the downstream process and sorting quality of 3D-plastics when processing MCW and MMW. For each input waste, two tests were carried out with and without a drum screen. A mass balance was determined for each test run and the screening efficiencies (drum screen and ballistic separator) were calculated. All generated outputs were sampled and subjected to an extensive screening and sorting analyses. The main results of the investigations in [42] are:

The mass balances (see Figure 12) show that for both wastes, a slightly increased 3D yield (2–5%) with a simultaneously reduced 2D yield (approx. −20%) due to upstream drum screening on the ballistic separator result were produced. Pre-screening (doubling the active screen area) improved the overall screening efficiency up to 0.99 and increased the total fines yield (<80 mm) by 10–15%. In addition, it was found that the combination of selective comminution and pre-screening resulted in the screen fines having a higher contaminant, inert and organic content per lower calorific value than the ballistic separator fines [31]. The ballistic separator fines consisted of more near mesh size particles of plastics and other valuable materials with higher calorific value. As Möllnitz et al. [20] show, this particle size fraction still has a considerable plastics potential and could be used for plastics recovery for recycling or SRF production.

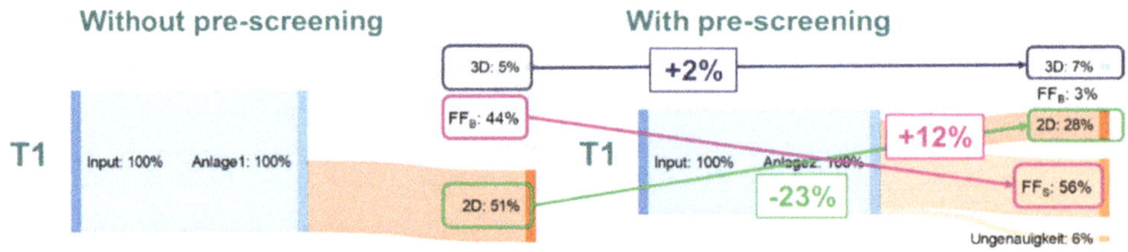

Figure 12. Mass balance for test run 1 with mixed commercial waste [42].

The manual sorting analyses of the 2D fractions show a reduction in inert (up to −50%) and fines (<80 mm) (up to −75%) by pre-screening. Especially in the test runs with MCW, tail formations were observed in the drum screen, which led to material losses, false discharge into the 2D-fraction and plant downtimes. The manual sorting analyses of 3D fractions show a very low amount of 2D and fines. This confirms the optimum operation type for producing a 3D fraction with high purity. Applied pre-screening improves the NIR-sorting efficiency (yield) of 3D fraction by 6%, see Figure 13.

2. Wet mechanical processing of plastic rich 2D-fractions with a focus on polyolefins (POs) from mixed waste for chemical recycling [46]:

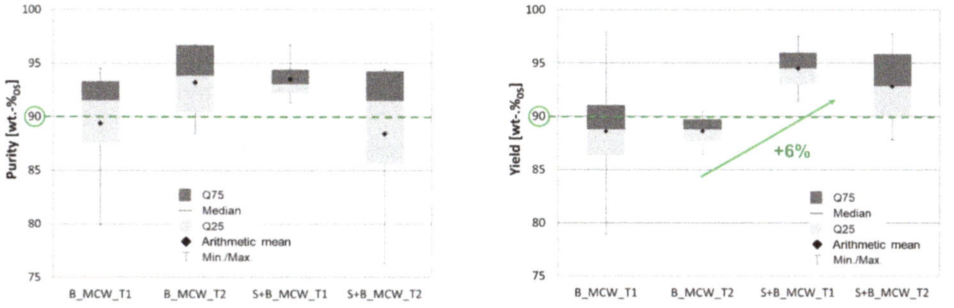

(**a**) (**b**)

Figure 13. Effects of pre-screening on purity (**a**) and yield (**b**) of the 3D-plastics out of the 3D-fraction from the ballistic separator of the investigated mixed commercial waste [42].

The potential and applicability of a dry-mechanical (materials from investigations presented in Möllnitz et al. [46]) and subsequently wet-mechanical processing with the aim of generating a PO concentrate for chemical recycling purposes was assessed. The focus of the investigation was the wet density separation by using a centrifugal force separator (CFS) [47] as the core element of the wet-mechanical process. The fed material is separated according to its density in the separation medium water in a feed-related air/water vortex and discharged as heavy and light material. The light fraction out of the first run was fed into the centrifugal force separator (CFS) once again, simulating a cascade connection of two CFS plants to achieve a further concentration of the POs. The required process water circulated throughout the entire experiment and was only replaced by freshwater when the material was changed. The process water was collected and decoupled from the treatment line and fed to a static drum filter to separate the liquid and solid phases. The mass balances for both input materials and both test runs were calculated. The input material, all output materials (light and heavy fractions, sediment) and the process waters were sampled and chemically and physically characterized to estimate potential treatment or recycling paths.

For an evaluation of the suitability of the PO-fraction (light fraction two-LF II) produced as input material for the thermochemical conversion process (ReOil process), the measured values were compared with the quality limit values in Table 3. The specified particle size and bulk density range were achieved through targeted pre-treatment for both waste fractions. The limit value for the moisture content was not met as only one linear vibrating screen was available. The limit values for the calorific value and the chlorine content were met for both waste fractions. The required content of PE, PP, PS is also met for both waste fractions within the assumption made. Since the total contaminants contained in both fractions exceed 3%, compliance with the required limit values for polymer impurities, inorganic and organic contaminants is not guaranteed.

In summary, all the results from this chapter show that the recovery, treatment, and processing of plastics from non-hazardous, mixed, solid waste for mechanical or chemical recycling is possible. By transferring these plastics from thermal treatment to recycling purposes, an important contribution is made for achieving the recycling targets, resource conservation, and reducing greenhouse gases and waste.

3.2.2. Solid Recovered Fuels (SRF)

SRF has been defined by the CEN/TC343 in the standard EN 15359:2011 [49] as fuel that fulfils following four criteria:

1. It is solid fuel;
2. It is prepared from non-hazardous waste only;
3. It is to be utilized for energy recovery in incineration or co-incineration plants;

4. It must meet certain quality criteria (i.e., lower heating value (expressed as mean), chlorine content (expressed as mean) and mercury content (expressed as median and 80th percentile value) and to be allocated in one of total five classes depending on the measured values for each mentioned parameter.

Table 3. Comparison of the measured values with the quality requirements for the ReOil process (MMW: mixed municipal waste, MCW: mixed commercial waste, CFS: centrifugal force separator) [46].

Parameter	ReOil-Process [48]	LF II MCW	LF II MMW	Comments
Particle size (mm)	<30–40	<20	<20	required particle size for CFS processing.
Bulk density (kg $m^{-3}{}_{DM}$)	50–100 (mainly 2D-objects)	approx. 80	approx. 60	was determined in [46].
Moisture content (%)	<20	49.5	67.1	additional dewatering equipment is required.
Calorific value (MJ $kg^{-1})_{DM}$	>30	38.3	35.1	
Chlorine content $(\%)_{DM}$	<2	0.2	0.47	
PP, PE, and PS content (%)	>90	95.8	94.9	assumption that the LFII consists only of PE, PP and PS.
Polymer impurities $(\%)_{DM}$	PET: ≤3; PVC: ≤2			
Inorganic contaminants $(\%)_{DM}$	≤3	4.2	5.1	total contamination content.
Organic contaminants $(\%)_{DM}$	≤5			

In Austria, SRF is defined as " ... *waste that is used entirely or to a relevant extent for the purpose of energy generation and which satisfies the quality criteria laid down* ... " [34,50]. Quality criteria that are limits for antimony, arsenic, lead, cadmium, chromium, cobalt, nickel and mercury and are expressed as amount per energy content (i.e., pollutant content per net calorific value related to dry matter—mg/MJ_{DM}) [34].

Sarc et al. [15–18] and Lorber et al. [35] report that the production, quality and quality assurance as well as utilization of SRF in the cement industry has become state of the art and even thermal substitution rates of up to 100% [51] can be technically reached. The data for 28 years (1998–2018) on the average thermal substitution rate (TSR) of SRF in the cement industry in different countries, the EU and worldwide show very positive and dynamic development [29,52], see Figure 14. As shown for 2018, the following TSR were reached: worldwide 18.5%, in the EU28 47.7% and in selected countries like Germany (68.6%) and Austria (81.1%).

Sarc et al. [17] report regarding two main types of SRF for the cement industry, namely, SRF PREMIUM Quality and SRF MEDIUM Quality and Viczek et al. [53] additionally give the definition of these two types as follows:

- *SRF for secondary firing (SRF "secondary"): SRF with a lower heating value between 12 and 18 MJ/kg*$_{OS}$ *(corresponding to class NCV 3 or 4 in EN 15359) suitable for the use in secondary firing (calciner, kiln inlet, or hot disc combustion chamber, etc.) in the kiln system of cement manufacturing plants. Particle sizes can range up to 80 mm when used in a calciner or at the kiln inlet and up to 300 mm for a hot disc combustion chamber.*
- *SRF for primary firing (SRF "primary"): SRF with a lower heating value between 18 and 25 MJ/kg*$_{OS}$ *(corresponding to class NCV 1, 2, or 3 in EN 15359), and particle sizes below 30 (35) mm suitable for the use as a main burner fuel in the rotary kiln of cement manufacturing plants.*

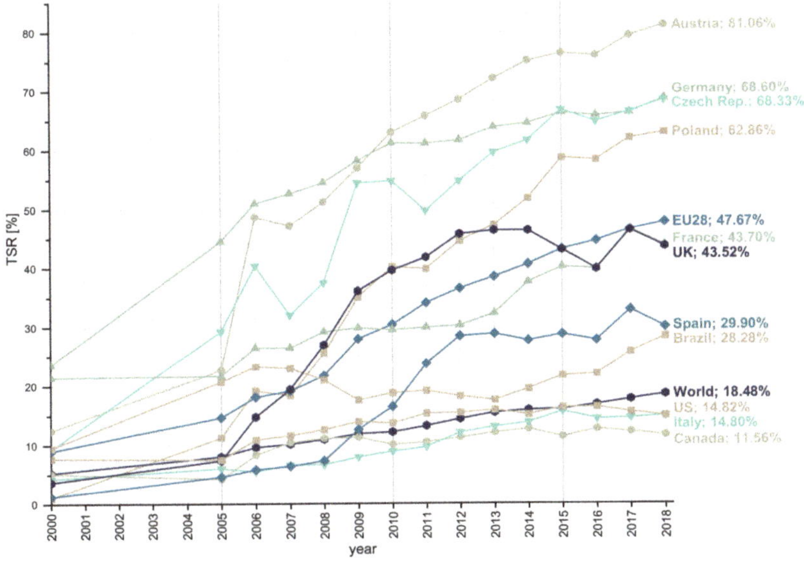

Figure 14. Development of the thermal substitution rate (in %) of SRF in the cement industry in different countries, the EU, and worldwide for the time period 1990–2018 [29,52].

Sarc et al. [17] extensively report the chemical–physical quality of SRF "secondary" and SRF "primary" currently available on the middle European market and conclude that SRF coming from multi-stage waste processing plants fulfil the Austrian legal and international SRF and co-incineration market requirements. On one side, SRF contributes to the energy production in the clinker process and, on the other side, as its ash can provide selected secondary raw materials for the production of clinker, SRF can be seen as "co-processed" in the cement industry. Co-processing is a term that comprises industrial processes that simultaneously enable energy recovery and recycling of the constituents [53].

3.2.3. Co-Processing of SRF–Ash Constituents of SRF as a Valuable Secondary Raw Material

The main raw materials required for the production of cement clinker are CaO, SiO_2, Al_2O_3, and Fe_2O_3. A typical composition of raw meal for production of Portland cement clinker is: 77.36 wt% of $CaCO_3$, 13.73 wt% of SiO_2, 2.93 wt% of Al_2O_3, 1.84 wt% of Fe_2O_3, 1.83 wt% of MgO, 1.08 wt% of SO_3, 0.85 wt% of K_2O, 0.14 wt% of Na_2O, 0.02 wt% of P_2O_5, 0.15 wt% of TiO_2, 0.06 wt% of Cl and 0.01 wt% of ZnO [54]. A part of the required raw materials can also be provided by SRF ash, which is incorporated into the clinker and thereby recycled when SRF is co-processed in the cement industry.

To determine the material-recyclable share (R-index) of SRF, Aldrian et al. [55] have developed and validated an analytical method for the determination of the ash composition and have proposed a formula to calculate this R-index. Average ash composition (arithmetic means) of SRF primary and SRF secondary is given in Figure 15. Different R-indices are achieved when different element oxides are considered [53]: For example, when considering only Al_2O_3, CaO, SiO_2, and Fe_2O_3, R-indices of 13.9% (SRF secondary) or 13.3% (SRF primary) are achieved. When additionally, MgO, TiO_2, K_2O, Na_2O and SO_3 are considered, the R-indices rise to 16.2% for SRF secondary or 15.9% for SRF primary. When the whole ash content is considered as recycled, the R-index amounts for 17.7% (SRF secondary) and 17.6% (SRF primary).

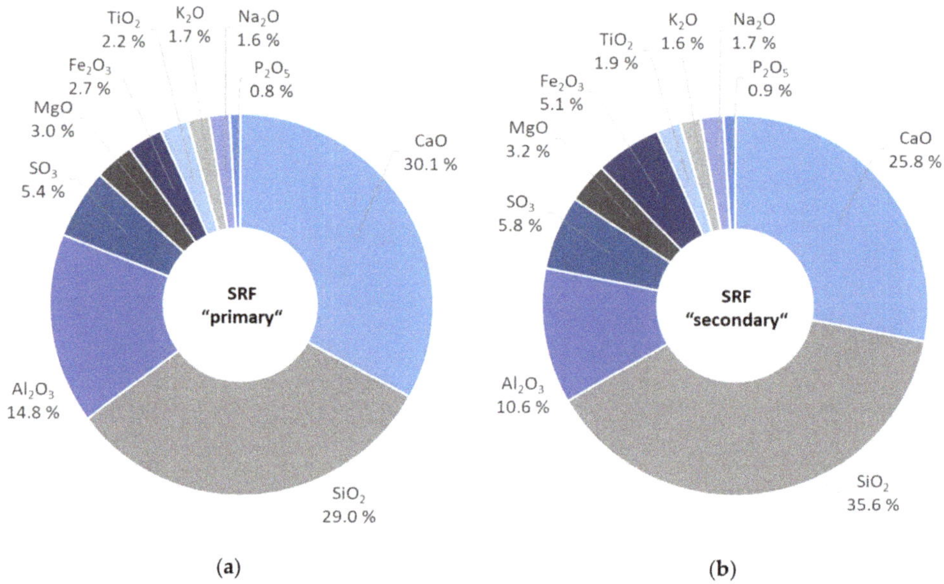

Figure 15. Average ash composition (arithmetic means) solid recovered fuel (SRF) primary (**a**) and secondary (**b**) in mass percent dry mass [53].

The average SiO_2: CaO: Fe_2O_3+Al_2O_3 ratio reported by Viczek et al. [53] is depicted in Figure 16. On average, the share of SiO_2 is slightly higher in SRF secondary than SRF primary. Generally, the ratio of these element oxides in SRF ash is similar to lignite coal, SRF primary is even closer to sewage sludge. However, the higher share of CaO shifts many SRF samples closer to the desired ratio in clinker.

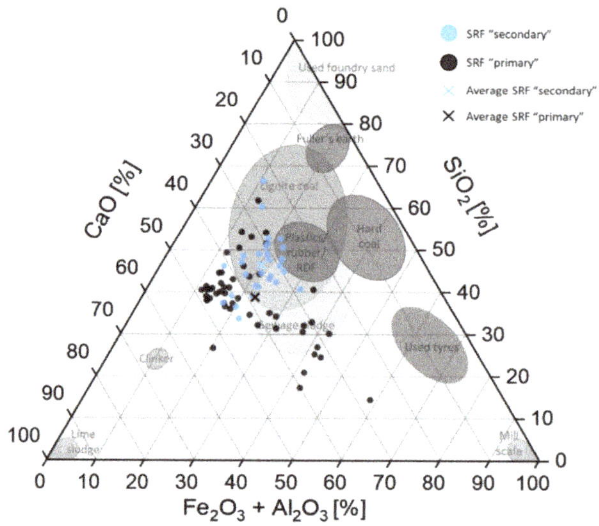

Figure 16. Ternary diagram illustrating the solid recovered fuel (SRF) ash composition of 80 investigated SRF samples and their average composition compared with other relevant raw materials and fuels [53].

A more detailed analysis of the ash of different SRF fractions and specific material fractions extracted from SRF, performed by Viczek et al. [56], shows that the highest material-specific R-indices are achieved for the fine fraction <10 mm, composite materials, or the sorting residue, see Figure 17. All these fractions are highly heterogeneous, and data from other established recycling processes are not yet available.

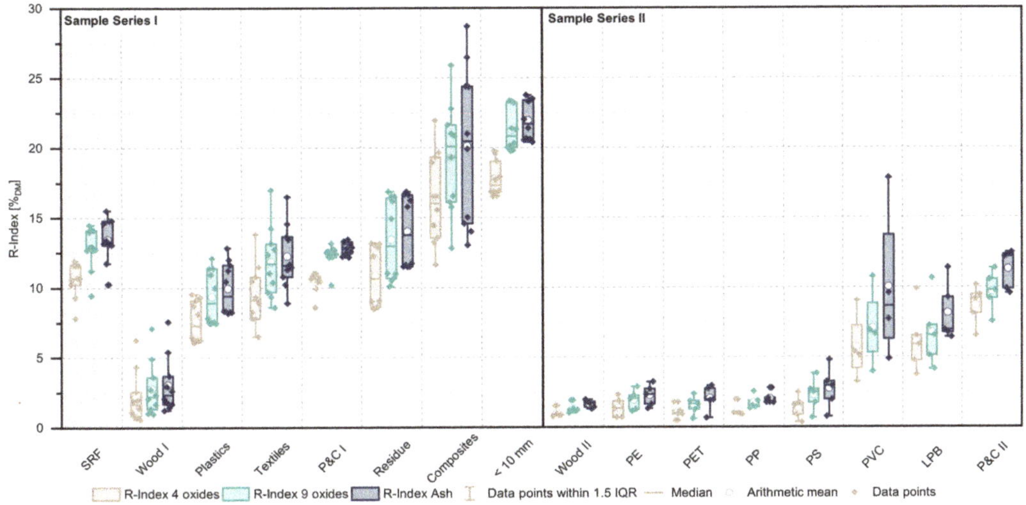

Figure 17. Calculated R-Indices for solid recovered fuel (SRF) sorting fractions (left) and specific materials extracted from SRF (right) for three scenarios considering different element oxides [56].

Consequently, when SRF is utilized in the clinker production of the cement industry, not only is the energy recovered, but also minerals from SRF ash are recycled by being incorporated into the cement clinker. This implies that the locally or regionally organized recycling and thermal recovery of SRF in the cement industry, commonly referred to as SRF co-processing, is a significant contribution of waste management for the globally producing cement industry.

3.3. Digitalisation in Waste Characterization and Treatment Processes for Mixed Waste

As mentioned before, Sarc et al. [25] report that digitalization in the waste management sector in comparison with other industrial sectors is still in its infancy. Nevertheless, "Digitalisation" and "Industry 4.0" approaches are of high interest in waste management and a survey carried out in 2017 shows that 63% of companies see digitalization as a chance for their further development [57].

Here, selected results at different levels of digitalization are given, namely:

- Digitalization as a modern tool for innovative data-based development of smart processes;
- Online–Ontime material particle characterization and quality assurance;
- Experimental monitoring of waste flows and machine performance in waste treatment plants.

3.3.1. Digitalization as a Modern Tool for Innovative Data-Based Development of Smart Processes

Real-time (smart) process control for waste processing plants—but also in general—requires three fundamental elements (see Figure 18), according to Khodier [32]: *"metrology that measures the state and performance of the process, algorithms that calculate optimal factor settings from the measurements, and actuators, like a shredders gap width, that function as change-*

able factors for influencing the process". Due to the variety and complexity of waste as a material, the issue of online-measuring material qualities is a highly relevant topic in current research.

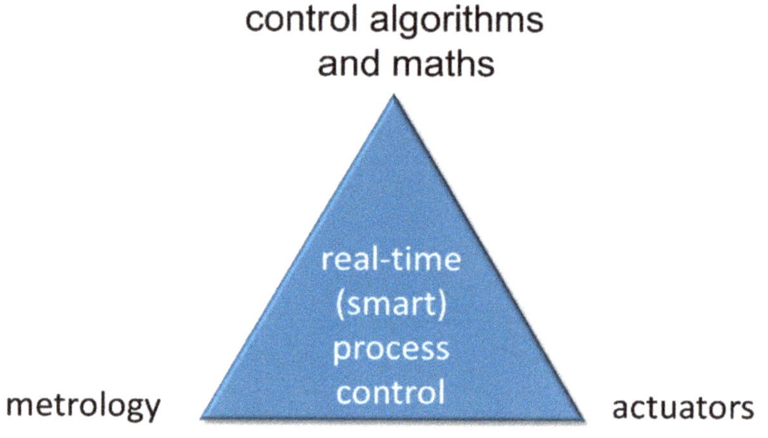

Figure 18. Trinity of real-time process control [32].

The mechanical treatment of mixed solid waste involves a variety of processing machines, e.g., shredders, screens and magnetic separators. Understanding the influence of their parameters (like the gap width of a coarse shredder) on the process is essential for the optimized operation of the treatment plants—in terms of process properties, like the throughput, but also concerning produced material qualities.

Nevertheless, waste processing machines are often tested and operated at fixed parameter settings, chosen by experience and intuition, without any physical or statistical proof of optimality, as Khodier [32] points out for coarse-shredding. A main reason for this unsatisfactory status quo is the complexity that is involved in reliable investigations on mixed solid waste processing.

Khodier et al. [41] discuss that physics-based numerical studies are hardly usable for investigations on the real-scale behavior of waste processing machines—the variability of waste and the diversity of materials and shapes hardly allows collecting the necessary information on the material to be processed. Therefore, while physical models are favorable in terms of the detailed process insights they provide, alternatives are needed.

Empirical modeling is a useful approach for, nevertheless, drawing reliable conclusions on mechanical treatment processes for mixed solid waste: a configurable set of parameters is chosen as factors to be investigated, concerning their effects on a set of chosen target values. Remaining parameters—like the composition of the waste, that cannot be controlled in detail—are treated as distortion in the data. In the presence of a sufficient amount of data, true effects become significant in a subsequent analysis of variance (ANOVA) and can hence be distinguished from data noise and quantified with corresponding confidence and prediction regions.

While empirical modeling itself is not novel, to the best of the author's knowledge, no industry-scale parameter studies on solid waste processing that comprehensively incorporate statistical analyses on significance were available when ReWaste4.0 started in 2017. Potential reasons include the tremendous efforts involved in generating a sufficiently large amount of real-scale data, the complexity of material analyses, that include considerations on waste sampling, and the non-triviality of modeling non-scalar target values, i.e., particle size distributions.

As a necessary precondition for generating the data to consider product qualities in empirical models, Khodier et al. [30] established a procedure for sampling coarsely shredded mixed commercial waste that comprehensively incorporates the theory of sampling, while referring to the Austrian standard ÖNORM S 2127 for the calculation of total sample masses. Different to existing standards and literature, they did not only define and evaluate the sampling procedure based on theory, but also empirically quantify sampling errors, by comparing the shares of particle size-material fractions, determined from 10 samples of the identical material in a so-called replication experiment. Their results obtained are shown in Table 4. As can be seen, sampling errors are relatively low for the shares of the overall particle size distribution and the overall material composition but become much higher as the level of detail is increases—which decreases the shares of the analytes.

Table 4. Relative sampling variabilities (RSV) for various particle size classes and for the material classes metal (ME), wood (WO), paper (PA), cardboard (CB), 2D plastics (2D), 3D plastics (3D), inert materials including glass (IN), textiles (TX), and a residual fraction (RE) [30].

Particle Class (mm)	ME (%)	WO (%)	PA (%)	CB (%)	2D (%)	3D (%)	IN (%)	TX (%)	RE (%)	Sum (%)
0–5	-	-	-	-	-	-	-	-	12.3	12.3
5–10	-	-	-	-	-	-	-	-	12.3	12.3
10–20	-	-	-	-	-	-	-	-	10.4	10.4
20–40	41.4	17.7	24.3	39.3	18.4	17.1	19.7	29.3	22.7	11.6
40–60	47.3	21.5	16.8	25.6	14.2	8.7	37.2	43.4	16.4	8.8
60–80	39.4	23.3	22.9	18.1	17.7	10.0	49.9	30.4	8.9	8.1
80–100	62.0	34.7	38.0	14.2	19.3	17.5	210.7	43.4	17.2	7.7
100–200	74.0	47.7	69.0	21.6	28.9	35.2	131.8	40.9	40.0	10.9
200–400	153.0	230.9	203.9	126.2	38.3	39.8	-	42.9	52.2	28.8
Sum	16.4	18.3	10.5	15.0	16.6	12.1	31.2	26.6	3.6	0.0
		RSV < 20%			20% ≤ RSV < 50%			RSV ≥ 50%		

Conclusive evaluations, that consider calculations of theoretical sampling errors, finally show that in the applied procedure, the so-called fundamental sampling error only explains a small share of the sampling variability. The authors conclude that grouping and segregation errors are likely to contribute a significant share—hence a higher number of increments at an unchanged total sample mass is expected to increase sampling quality.

Sampling errors, but also inter-experimental differences in the waste, lead to a considerable residual distortion of the data, when investigating influences of machine parameters. Considering the high efforts and costs of real-scale waste processing experiments, methods for increasing the efficiency of the data, in terms of extractable information per experimental run, are required. The publication of Khodier et al. [41] addresses this very interesting topic: they empirically model the influences of coarse shredders' radial gap width, shaft rotation speed, and three different cutting tool geometries, based on a 32-run Design of Experiments-based investigation, designed in consideration of a reduced cubic polynomial design model.

Design of Experiments targets increasing the efficiency of the data. The work of Khodier et al. [41] gives a general introduction into the method, while also discussing in detail the choice of the chosen experimental design (a so-called D-optimal design) and the design model. Their publication is hence an important reference as a "starting point" for incorporating empirical modeling in mechanical waste processing investigations.

Besides the provided proof of the potential of their chosen methods, the referenced work also provides quantitative conclusions on the influences of the described factors on the throughput (in terms of mass and volume), the throughput steadiness (in terms of the quotient of the 10th and 90th percentile mass flow and volume flow) and the specific energy demand. The results for each factor—under average settings of the corresponding other two factors—are shown in Figure 19, while the publication provides additional insights on an interdependence of the influences of the gap width and cutting tool geometry.

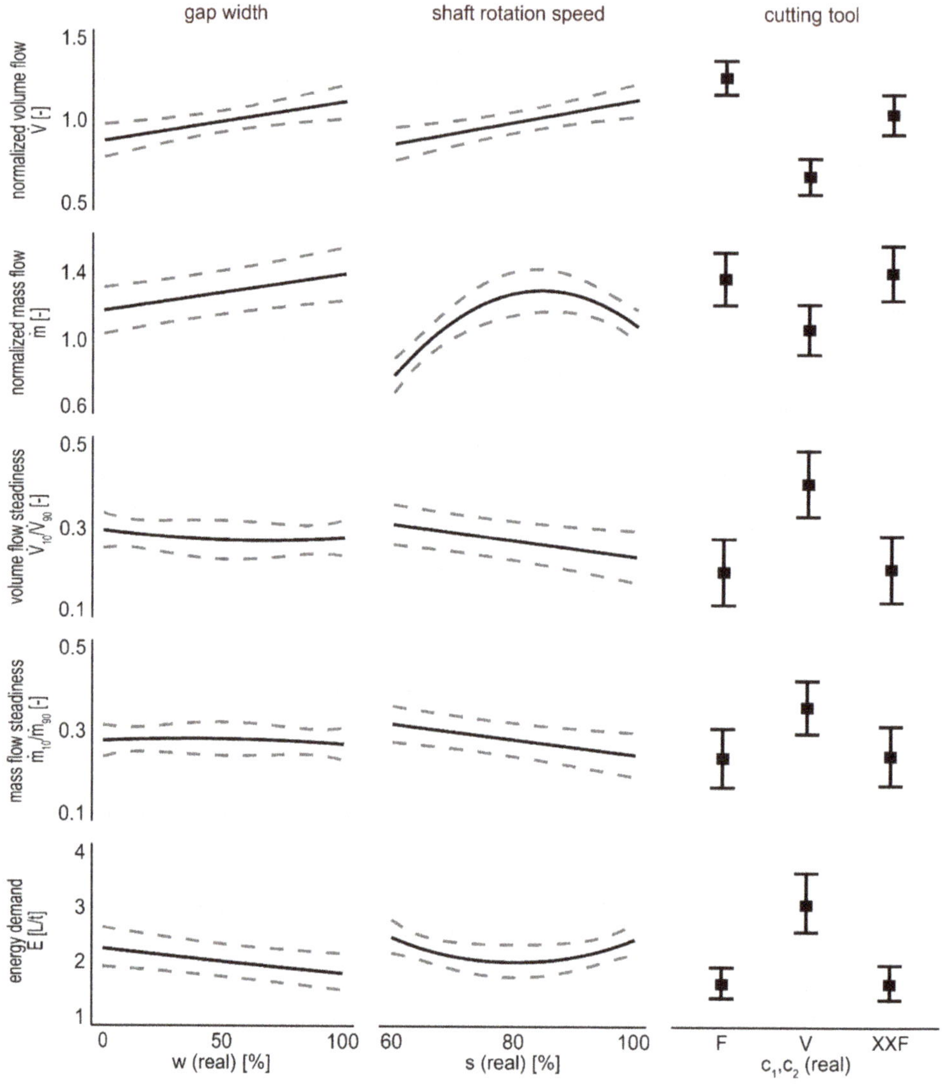

Figure 19. Effect plots with confidence bands for the influence of the gap width, shaft rotation speed, and cutting tool geometry on throughput, throughput steadiness, and specific energy demand, at average settings of the respective other factors [41].

Eventually, Khodier et al. [41] conclude that the cutting tool geometry is the factor with the highest influence and cannot be compensated by the easily changeable settings of the gap width and shaft rotation speed. Finally, they point out that an overall determination of economic optimality must evaluate the effects of the modeled target values on the (monetary) performance of the overall processing plant.

The modeled process properties must always be evaluated in combination with the properties of the processed material: the particle size distribution of the waste is an essential quality parameter of the processing product, and also influences the performance of the whole treatment process, as Khodier and Sarc [58] point out. They published an extension of the methods described by Khodier et al. [41] on non-scalar target values, i.e., particle size distributions. They first give a comprehensive overview on classical descriptive methods for particle size distributions, e.g., particle lists, summary values, or analytical probability density functions. However, they find out that none of these allows the distribution-independent modeling of particle size distributions—which is essential for covering all kinds of processes—while always preserving all relevant information.

As a solution, they transfer methods from the field of modeling and analysis of compositional data: they model the particle size distribution through a multivariate multiple linear regression of isometric log-ratio-transformed shares of a deliberate choice of particle size classes. Isometric log-ratios free the data from the constraints of a simplicial vector space (all shares are positive, and their sum is constant), through a bijective projection on the one-dimension lower real space, hence allowing the application of standard statistics. To account for the interdependence of the particle size classes' shares, they furthermore involve the multivariate character of the data in all evaluations, hence evaluating factor significances through a multivariate analysis of variance (MANOVA).

Concerning the evaluated process—coarse-shredding of mixed commercial waste—Khodier and Sarc [58] find that, from the three examined factors, only the cutting tool geometry significantly changes the shares of three particle size classes (>80 mm, 30–80 mm, 0–30 mm), which were chosen based on their relevance concerning SRF qualities [17]. One remarkable conclusion—in consideration of the results of Khodier et al. [41] is that larger gap widths lead to higher throughputs and lower energy demands (which is both desirable) while not (negatively) affecting the material quality.

3.3.2. Online-Ontime Material Particle Characterization and Quality Assurance

Kandlbauer et al. [59] address the field of metrology, targeting the online-measurement of shredded mixed commercial waste's particle size distribution, as defined by a drum screen. In their approach, particle sizes are determined through a partial least square regression (PLSR) model, that is based on geometric descriptors of the particles' shapes, which were derived from two dimensional RGB (red—green—blue) images of single particles. Such two-dimensional images can be obtained in real-scale processing plants, by accessing RGB sensors on the acceleration conveyor belt of sensor-based sorters. Potential geometric descriptors were collected in a literature research: they include, for example, smallest enclosing polygons (e.g., triangles, circles, rectangles), and Feret diameters. A Partial Least Squares Regression (PLSR) finally uses these in a combined model, that keeps the number of empirical model constants low.

Kandlbauer et al. [59] conclusively find, that—while there are still unanswered questions, e.g., concerning image quality—when individually considering different material fractions, the performance of the models is very promising, with average errors of 0.2% for the shares of the particle size classes of wood for example, in the case of a uniform distribution. However, they also motivate for follow-up research to increase the accuracy of the measurements of individual particles, to ensure a distribution-independent good performance of the method—e.g., by evaluating the performance of machine learning models.

As described before, real-online/ontime and high-quality data are important in development of digitalized solutions for the waste management sector and especially in case of a so-called smart waste factory of the future. Various sensors are applied to measure waste data. Therefore, through ReWaste4.0 a new level of waste material characterization has been introduced for mixed waste, namely the characterization of mixed waste at a waste particle level and the target is to apply this approach online (i.e., real-time) and ontime (i.e., directly before and after each processing step and in case that such intensive monitoring is not possible to be carried out in a waste treatment plant, then at least input and output qualities should be monitored). Therefore, on one side monitoring of waste treatment processes can be executed and on the other side a quality assurance concept for waste materials can be carried out. This approach would deliver data for "intelligent communication" between material quality and machine as well as plant performance as aimed by ReWaste4.0 [2].

Weißenbach [60] report that there is no general definition of the term "waste characterization" but this procedure is applied for a number of different purposes. Typical aspects of the waste characterization procedure are the definition of waste composition based at the material type level as well as chemical and physical investigations for definition of waste properties. Furthermore, more technological methods of waste characterization are realized by applying sensor technologies (cf. [25,61,62]) as well as RGB cameras (cf. [59]).

As described by Weißenbach and Sarc [60,63], the investigation has been carried out at the level of individual particles and a total of eight fractions have been characterized, namely: paper/cardboard, wood and liquid packaging board (LPB) as well as plastic fractions PE, PP, PS, PET and polyvinyl chloride (PVC). Selected parameters are the projected particle area and the particle mass. In total, 15,542 particles of fine SRF (<30 mm) have been investigated. Next, single particles from SRF with particle size 30–80 mm (1,078 particles investigated) as well as a pre-treated MCW fraction with particle size 80–500 mm (1,268 particles investigated) have been examined [63]. All data have been statistically assessed and, in summary, the following results have been gained, see Table 5. As shown, the projected area value for pre-treated MCW is 10 times bigger than the value for SRF secondary and this, in turn, is 3 to 4 times bigger than the value for SRF primary. For the mass, the values have factors of 13 to 14 (pre-treated MCW vs. SRF secondary) and of about 6 (SRF secondary vs. SRF primary) [13,60,63].

Table 5. Median of the projected area and particle mass, classified according to particle size classes and material fraction [13,60,63].

Material Fraction	SRF Primary		SRF Secondary		Pre-Treated MCW	
	Proj. Area (in cm^2)	Mass (in g)	Proj. Area (in cm^2)	Mass (in g)	Proj. Area (in cm^2)	Mass (in g)
Paper	4.1	0.14	13.8	1.17	168.9	15.4
Wood	1.4	0.14	6.4	1.62	72.9	57.7
LPB	3.8	0.17	19.5	2.66	131.4	24.0
PE	5.4	0.15	15.8	0.64	198.9	5.3
PP	4.2	0.23	11.8	1.37	124.6	9.3
PS	3.2	0.19	12.6	1.04	68.1	8.9
PET	4.4	0.30	9.0	1.46	135.4	21.1
PVC	3.4	0.23	11.0	2.17	64.4	24.9
Total median	3.6	0.19	11.5	1.21	114.2	16.65

3.3.3. Experimental Monitoring of Waste Flows and Machine Performance in Waste Treatment Plants

Sarc et al. [25] report that, in most waste treatment plants, a typical weighbridge is used for measuring of input and outputs but, in the technical process itself, there is no online and ontime mass or volume flow measurement introduced. Additionally, in many cases the shovels of wheel loaders are equipped with weighing devices for a rough determination of a throughput. However, these methods do not allow machine, process and/or plant real-time (ontime) monitoring. As stated in the paper, the retrofitting of machines, processes and/or plants with volumetric flow measuring systems seems to be a promising approach as its installation would not need process concept changes but relatively simple mounting over the conveyor belts. However, to be able to convert these measured volume flow data into mass (i.e., tones) data, the bulk density of the material at every installed position is to be determined. When processing mixed waste like MMW, MCW, SRF, it has to be noted, as described before, that the composition and fluctuations in its composition as well as the properties of such waste are constantly changing and therefore, a continuous measurement in order to calibrate the system in real time is required.

Curtis et al. [64] and Curtis and Sarc [65] have extensively investigated these fluctuations in a semi-large scale "Technikumslinie4.0" (technical production line 4.0) and large-scale experiments and based on results made a distinction in:

- Short-term fluctuations (expressed as throughput change in intervals <15 s);
- Mid-term fluctuations (expressed as throughput change in intervals of 15–600 s);
- Long-term fluctuations (expressed as throughput change in intervals >600 s).

As proven by experiments from Curtis et al. [64] and Curtis and Sarc [65], short-term fluctuations resulted mostly from material´s composition, constitution and its particle size characteristics and machine specific parameters (e.g., drum screen speed, reversing intervals of shredders). Mid-term fluctuations originated especially from a discontinuous feeding process with the wheel loader or other feeding machines. Long-term fluctuations would be in a range of >600 s or longer and result from changes in machine parameter—while testing, no such fluctuations were recorded as all shredder parameters were kept constant. Additionally, fluctuations were investigated regarding their influence on the performance of selected (subsequently positioned) machines like the sensor-based-sorting machine as well as a monitoring unit after a shredding step to combine the measured data on volume flow fluctuations with the power consumption of the shredder. Results show that fluctuations in throughput can be measured by volumetric flow measuring systems which ensure the availability of selected information for a better understanding of the treatment process and its conditions. Next, fluctuations have significant influence also on the performance of the subsequent machines and can directly influence the product quality and the therewith connected market value of the product. Finally, the combination of online and ontime material flow and composition monitoring systems (before and after a machine) with machine (note: conveyor belts are also to be considered as "machines") consumption (e.g., energy), machine settings and performance control, both systems equipped with proper data collection, analysis and management as well as digital interconnection tools, can contribute to the further development of the so-called Smart Waste Factory for the waste treatment sector.

4. Conclusions

Within the current FFG-funded COMET K-Project "ReWaste4.0", two scientific and eight company partners initiated in 2017 the ambitious and required paradigm shift for the transformation of the non-hazardous mixed municipal and commercial waste management towards a Circular Economy 4.0. ReWaste4.0 and the approved follow-up project "Recycling and Recovery of Waste for Future (ReWaste F—2021–2025), among others, are future oriented projects that are based on legal developments in the EU, especially Circular Economy Package from 2015 and 2018 and European Green Deal from 2019 as well as market developments regarding digitalization and Industry 4.0 approaches.

As shown in the present review-paper, Circular Economy 4.0 involves a better understanding of contaminants contained in products and waste (note: every product will finally become a waste material—it is just a question of time!), deeper know-how on secondary raw and energy materials as well as co-processing materials and application of digitalization and Industry 4.0 approaches in waste characterization and the waste treatment sector.

When dealing with mixed waste like MMW and MCW with high material heterogeneity and various as well as large particle sizes, representative sampling and analysis of such waste is of high priority as the results influence all further steps. Therefore, within ReWaste4.0 the principles of the theory of sampling and internationally recognized waste sampling standards were applied. Finally, based on the results it can be concluded that increasing the number of increments for forming composite samples seems to have a larger beneficial effect on sampling quality than increasing the mass of the composite sample.

The obtained results show that the content of valuable plastics in MMW and MCW is about 15% in MMW and 15–23% in MCW. From a processing point of view, for the efficient recovery of these plastics, screening of the fine fraction after pre-shredding makes sense to remove contaminants and impurities. Additionally, after a material flow division based on particle dimensionality, further screening classification into targeted particle size classes is required to achieve higher concentrations of plastics and efficiently use plant capacities. Next, results presented show that all plastic types and mixtures investigated are processable for production of flakes and granulates and finally can become a valuable secondary raw material after proper and quality focused treatment processes.

Regarding contaminants in the investigated waste, it can be concluded that they are not only enriched in specific materials, but also in selected particle size classes. Nevertheless, contaminants are present in a broad range of products and waste management needs to deal with them properly to produce quality assured secondary raw and energy materials from mixed waste for the recycling, recovery and co-processing sector.

Digitalization and Industry 4.0 approaches are at the very beginning of being introduced in the waste treatment sector. Especially, real-time (smart) monitoring, data analysis and management and process control for waste processing plants are to be developed in the future. The mixed waste treatment sector has specific challenges like the continuously changing waste quality as well as material and machine related fluctuations. These should be considered when developing smart treatment solutions. Finally, results gained within ReWaste4.0 support the development of further digitalized solutions for waste treatment sector and especially in case of a so-called smart waste factory.

ReWaste F—Recycling and Recovery of Waste for Future—is the logical high-quality R&D continuation for the period 2021 to 2025—building on extensive know-how and intensive results gained from ReWaste4.0, considering current and future waste streams (i.e., non-hazardous mixed waste), technology (machines, sensors, cameras) and digitalization (data analytics, simulation as well as intelligent material–machine–machine digital interconnection) developments.

ReWaste F has been developed with an enlarged consortium (i.e., four scientific and 14 company partners) and deepened data, sensor and digital interconnection expertise to create and implement a particle-, sensor- and data-based circular economy. In ReWaste F, partners will progress the branch towards a circular waste economy and technology according to the newly published European Green Deal 2019 for the sustainable and resource efficient development of the waste, secondary raw material and energy sectors. Furthermore, particle-, sensor- and data-based technologies (machines and processes become cyber–physical systems) and digital and intelligent interconnection (including the development of a manufacturer-independent "digital platform") are crucial to enable the online/ontime communication of Material–Machine–Machine within new "Smart Waste Factory solutions" and to generate added value throughout the whole value chain.

Funding: Partial funding for this work was provided by: The Center of Competence for Recycling and Recovery of Waste 4.0 (acronym ReWaste4.0) (contract number 860 884) under the scope of the COMET—Competence Centers for Excellent Technologies—financially supported by BMDW, BMK, and the federal states of Styria, managed by the FFG.

Acknowledgments: The author would like to thank the entire ReWaste4.0 team (especially his PhD-team: Alexander Curtis, Karim Khodier, Selina Möllnitz, Sandra Viczek and Thomas Weißenbach), as well as all project partners for the fruitful cooperation.

Conflicts of Interest: The author declares no conflict of interest.

References

1. Austrian Research Promotion Agency (FFG). COMET—Competence Centers for Excellent Technologies—General Information, Factsheets and Success Stories. 2021. Available online: https://www.ffg.at/en/comet-competence-centers-excellent-technologies (accessed on 18 April 2021).
2. Sarc, R. ReWaste4.0—Recycling and Recovery of Waste 4.0. Application K-Project, Project Description, Version 1.1. 9 February 2017.
3. EC (European Commission). Closing the Loop: Commission Adopts Ambitious New Circular Economy Package to Boost Competitiveness, Create Jobs and Generate Sustainable Growth. 2015. Available online: http://europa.eu/rapid/press-release_IP-15-6203_en.htm (accessed on 1 March 2021).
4. EC (European Commission). Circular Economy Package. 2018. Available online: https://ec.europa.eu/environment/circular-economy/first_circular_economy_action_plan.html (accessed on 1 March 2021).
5. Sarc, R. ReWaste F—Recycling and Recovery of Waste for Future. Application Comet-Projects, Project Description, Version 1.0. 22 June 2020.

6. Pomberger, R.; Sarc, R.; Lorber, K.E. Dynamic visualisation of municipal waste management performance in the EU using Ternary Diagram method. *Waste Manag.* **2017**, *61*, 558–571. [CrossRef]
7. EC (European Commission). A European Green Deal. 2019. Available online: https://ec.europa.eu/info/node/123797 (accessed on 1 March 2021).
8. Kaza, S.; Yao, L.C.; Bhada-Tata, P.; van Woerden, F. *What a Waste 2.0: A Global Snapshot of Solid Waste Management to 2050*; World Bank: Washington, DC, USA, 2018.
9. Eurostat. Municipal Waste by Waste Management Operations, Last Updated: 16-02-2021. 2021. Available online: http://appsso.eurostat.ec.europa.eu/nui/submitViewTableAction.do (accessed on 1 March 2021).
10. Weißenbach, T.; Graf, J.; Pomberger, R.; Sarc, R. Calculation of the additional recycling potential in the European Union by implementing the Circular Economy Package. *Environ. Waste Manag. Recycl.* **2020**, *3*, 1–9.
11. Austrian Standard Institute (ASI). ÖNORM S 2100:2005 10 01, 2005. Abfallverzeichnis [Waste List].
12. BMU. Verordnung über die Bewirtschaftung von Gewerblichen Siedlungsabfällen und von Bestimmten Bau- und Abbruchabfällen (Gewerbeabfallverordnung) [Commercial Waste Ordinance]. Amendment 2017: BGBl. I S. 896. Available online: https://www.gesetze-im-internet.de/gewabfv_2017/BJNR089600017.html (accessed on 1 March 2021).
13. Weißenbach, T. Development of an Online/Ontime Method for the Characterization of Waste Streams in Waste Pre-Treatment Plants. Ph.D. Thesis, Montanuniversitaet Leoben, Leoben, Austria, 2021.
14. Weißenbach, T.; Pomberger, R.; Sarc, R. Composition of mixed commercial waste with focus on recyclable fractions. In Proceedings of the 7th International Conference on Sustainable Solid Waste Management, Heraklion, Greece, 26–29 June 2019.
15. Sarc, R.; Lorber, K.E. Production, quality and quality assurance of refuse derived fuels (RDFs). *Waste Manag.* **2013**, *33*, 1825–1834. [CrossRef] [PubMed]
16. Sarc, R.; Lorber, K.E.; Pomberger, R.; Rogetzer, M.; Sipple, E.M. Design, quality and quality assurance of solid recovered fuels (SRF) for the substitution of fossil feedstock in the cement industry. *Waste Manag. Res.* **2014**, 32. [CrossRef]
17. Sarc, R.; Seidler, I.M.; Kandlbauer, L.; Lorber, K.E.; Pomberger, R. Design, quality and quality assurance of solid recovered fuels for the substitution of fossil feedstock in the cement industry—Update 2019. *Waste Manag. Res.* **2019**, *37*, 885–897. [CrossRef] [PubMed]
18. Sarc, R.; Kandlbauer, L.; Lorber, K.E.; Pomberger, R. Production and Characterisation of SRF Premium Quality from Municipal and Commercial Solid non-hazardous Wastes in Austria, Croatia, Slovenia and Slovakia. *Detritus* **2020**, *9*, 125–137. [CrossRef]
19. BKA (Bundeskanzleramt Österreich). Bundesgesetz über eine Nachhaltige Abfallwirtschaft (Abfallwirtschaftsgesetz 2002—AWG 2002) [Waste Management Act]. BGBl. I Nr. 102/2002, as Amended BGBl. I Nr. 24/2020. Available online: https://www.ris.bka.gv.at/GeltendeFassung.wxe?Abfrage=Bundesnormen&Gesetzesnummer=20002086 (accessed on 1 March 2021).
20. Möllnitz, S.; Khodier, K.; Pomberger, R.; Sarc, R. Grain size dependent distribution of different plastic types in coarse shredded mixed commercial and municipal waste. *Waste Manag.* **2020**, *103*, 388–398. [CrossRef]
21. BMK. *Die Bestandsaufnahme der Abfallwirtschaft in Österreich, Statusbericht 2020 (Referenzjahr 2018) [Inventory of Waste Management in Austria, Status Report 2020 (Reference Year 2018)]*; Federal Ministry for Climate Protection, Environment, Energy, Mobility, Innovation and Technology: Vienna, Austria, 2020.
22. European Commission. Communication from the Commission to the European Parliament, The Council, The European Economic and Social Committee and the Committee of the Regions a European Strategy for Plastics in a Circular Economy {SWD (2018) 16 Final}. 2018. Available online: https://eur-lex.europa.eu/legal-content/EN/TXT/?qid=1516265440535&uri=COM:2018:28:FIN (accessed on 1 March 2021).
23. Tschandl, M.; Peßl, E.; Sorko, S.; Lenart, K. Roadmap Industrie 4.0 für Unternehmen aus dem Umwelt- bzw. Abfallbereich (Roadmap industry 4.0 for companies in the environmental and waste management sector). 2019.
24. Curtis, A.; Sarc, R. Definition of the Term "Smart Waste Factory Network" within the Project ReWaste4.0, Montanuniversitaet Leoben on 8 October 2018.
25. Sarc, R.; Curtis, A.; Kandlbauer, L.; Khodier, K.; Lorber, K.E.; Pomberger, R. Digitalisation and Intelligent Robotics in Value Chain of Circular Economy Oriented Waste Management—A Review. *Waste Manag.* **2019**, *95*, 476–492. [CrossRef] [PubMed]
26. Green Tech Cluster. Digitale Abfallwirtschaft—Mehrwert Entlang der Gesamten Wertschöpfungskette (Digital Waste Management—Added Value along the Entire Value Chain). 2018. Available online: www.greentech.at (accessed on 1 March 2021).
27. Abdallah, M.; Abu Talib, M.; Feroz, S.; Nasir, Q.; Abdalla, H.; Mahfood, B. Artificial intelligence applications in solid waste management: A systematic research review. *Waste Manag.* **2020**, *109*, 231–246. [CrossRef]
28. ASI (Austrian Standards Institute). *ÖNORM EN 15442—Solid Recovered Fuels—Methods for Sampling*; ASI: Vienna, Austria, 2011.
29. Viczek, S. Origins, Distribution, and Fate of Contaminants and Ash Constituents in Waste for SRF Production and Co-Processing. Ph.D. Thesis, Montanuniversitaet Leoben, Leoben, Austria, 2021.
30. Khodier, K.; Viczek, S.A.; Curtis, A.; Aldrian, A.; O'Leary, P.; Lehner, M.; Sarc, R. Sampling and Analysis of Coarsely Shredded Mixed Commercial Waste. Part I: Procedure, Particle Size and Sorting Analysis. *Int. J. Environ. Sci. Technol.* **2019**, *17*, 959–972. [CrossRef]
31. Viczek, S.A.; Kandlbauer, L.; Khodier, K.; Aldrian, A.; Sarc, R. Sampling and analysis of coarsely shredded mixed commercial waste. Part II: Particle size-dependent element determination. *Int. J. Environ. Sci. Technol.* **2021**, under review.
32. Khodier, K. Empirical Modelling for the Optimized Operation and Real-Time Control of Coarse Shredders for Mixed Commercial Waste. Ph.D. Thesis, Montanuniversitaet Leoben, Leoben, Austria, 2021.

33. European Union (EU). *Directive 2008/98/EC of the European Parliament and of the Council of 19 November 2008*; Waste and Repealing Certain Directives: Brussels, Belgium, 2008.
34. BMLFUW. Verordnung des Bundesministers für Land- und Forstwirtschaft, Umwelt und Wasserwirtschaft und des Bundesministers für Wirtschaft, Familie und Jugend über die Verbrennung von Abfällen (Abfallverbrennungsverordnung—AVV) [Waste Incineration Ordinance] BGBl. II Nr. 389/2002, as Amended BGBl. II Nr. 135/2013. Available online: https://www.ris.bka.gv.at/GeltendeFassung.wxe?Abfrage=Bundesnormen&Gesetzesnummer=20002239 (accessed on 1 March 2021).
35. Lorber, K.E.; Sarc, R.; Aldrian, A. Design and quality assurance for solid recovered fuel. *Waste Manag. Res.* **2012**, *30*, 370–380. [CrossRef]
36. Viczek, S.A.; Aldrian, A.; Pomberger, R.; Sarc, R. Origins and carriers of Sb, As, Cd, Cl, Cr, Co, Pb, Hg, and Ni in mixed solid waste—A literature-based evaluation. *Waste Manag.* **2020**, *103*, 87–112. [CrossRef] [PubMed]
37. Viczek, S.A.; Khodier, K.; Kandlbauer, L.; Aldrian, A.; Redhammer, G.; Tippelt, G.; Sarc, R. The particle size-dependent distribution of chemical elements in mixed commercial waste and implications for enhancing SRF quality. *Sci. Total Environ.* **2021**, *776*, 145343. [CrossRef]
38. Curtis, A.; Adam, J.; Pomberger, R.; Sarc, R. Grain size-related Characterization of Various Non-hazardous Municipal and Commercial Waste for Solid Recovered Fuel (SRF) Production. *Detritus* **2019**, *7*, 55–67. [CrossRef]
39. Viczek, S.A.; Lorber, K.E.; Pomberger, R.; Sarc, R. Production of contaminant-depleted solid recovered fuel from mixed commercial waste. *Fuel* **2021**, *294*, 120414. [CrossRef]
40. Viczek, S.A.; Sarc, R.; Pomberger, R. *Grain Size Dependent Distribution of Contaminants in Coarse-Shredded Commercial Waste*; Thiel, S., Thomé-Kozmiensky, E., Winter, F., Juchelková, D., Eds.; IRRC, Waste Management: Neuruppin, Germany, 2019; ISBN 978-3-944310-48-0.
41. Khodier, K.; Feyerer, C.; Möllnitz, S.; Curtis, A.; Sarc, R. Efficient derivation of significant results from mechanical processing experiments with mixed solid waste: Coarse-shredding of commercial waste. *Waste Manag.* **2020**, *121*, 164–174. [CrossRef]
42. Möllnitz, S.; Küppers, B.; Curtis, A.; Khodier, K.; Sarc, R. Influence of Pre-Screening on Down-Stream Processing for the Production of Plastic Enriched Fractions for Recycling from Mixed Commercial and Municipal Waste. *Waste Manag.* **2020**, *119*, 365–373. [CrossRef] [PubMed]
43. Letcher, T.M. *Plastic Waste and Recycling—Environmental Impact, Societal Issues, Prevention, and Solutions*; Elsevier: London, UK, 2020.
44. Friedrich, K.; Möllnitz, S.; Holzschuster, S.; Pomberger, R.; Vollprecht, D.; Sarc, R. Benchmark Analysis for Plastic Recyclates in Austrian Waste Management. *Detritus* **2020**, *9*, 105–112. [CrossRef]
45. Möllnitz, S.; Feuchter, M.; Duretek, I.; Schmidt, G.; Sarc, R. Verification of the processability of different polymer fractions recovered from mixed wastes and determination of material properties. *Polym. Recycl. Spec. Issue Polym. Waste Recycl. Manag.* **2021**, *13*, 457. [CrossRef]
46. Möllnitz, S.; Bauer, M.; Schwabel, D.; Schmidt, G.; Sarc, R. Wet-mechanical processing of plastic-rich 2D-fractions from mixed wastes for recycling purposes. *Waste Manag. Res.* **2021**, 1–13. [CrossRef]
47. Bauer, M.; Lehner, M.; Schwabl, D.; Flachberger, H.; Kranzinger, L.; Pomberger, R.; Hofer, W. Sink–float density separation of post-consumer plastics for feedstock recycling. *J. Mater. Cycles Waste Manag.* **2018**, *20*, 1781–1791. [CrossRef]
48. Bauer, M.; Lehner, M. Post-consumer plastics in Austria and their potential for chemical recycling. In Proceedings of the ISWA World Congress 2013 Vienna, Vienna, Austria, 7–11 October 2013; International Solid Waste Association: Vienna, Austria, 2013.
49. CEN. *CENprEN 15359:2010 Solid Recovered Fuels—Specifications and Classes*; CEN: Brussels, Belgium, 2010.
50. BMLFUW. *Richtlinie für Ersatzbrennstoffe [Guideline for Waste Fuels]*; BMLFUW: Vienna, Austria, 2008.
51. Sarc, R. Herstellung, Qualität und Qualitätssicherung von Ersatzbrennstoffen zur Erreichung der 100%-Igen Thermischen Substitution in der Zementindustrie [Design, Quality and Quality Assurance of Solid Recovered Fuel (SRF) for Achieving 100% Thermal Substitution in Cement Industry]. Ph.D. Thesis, Montanuniversitaet Leoben, Leoben, Austria, 2015.
52. GCCA (Global Cement and Concrete Association). Getting the Numbers Right. 2020. Available online: https://gccassociation.org/gnr/Excel/GNR%20-%20Totals_&_Averages%20-%20Light%20Report%202018.xls (accessed on 1 March 2021).
53. Viczek, S.A.; Aldrian, A.; Pomberger, R.; Sarc, R. Determination of the material-recyclable share of SRF during co-processing in the cement industry. *Resour. Conserv. Recycl.* **2020**, *156*, 104696. [CrossRef]
54. Hökfors, B.; Boström, D.; Viggh, E.; Backman, R. On the phase chemistry of Portland cement clinker. *Adv. Cement Res.* **2015**, *27*, 50–60. [CrossRef]
55. Aldrian, A.; Viczek, S.A.; Pomberger, R.; Sarc, R. Methods for identifying the material-recyclable share of SRF during co-processing in the cement industry. *Methods X* **2020**, *7*, 100837. [CrossRef]
56. Viczek, S.A.; Aldrian, A.; Pomberger, R.; Sarc, R. Origins of mineral constituents in SRF and implications for recycling during co-processing in the cement industry. *Waste Manag.* **2021**, *126*, 423–432. [CrossRef]
57. Sarc, R.; Hermann, R. *Unternehmensbefragung zum Thema Abfallwirtschaft 4.0 [Company Survey to the Topic Waste Management 4.0]*; Pomberger, R., Adam, J., Aldrian, A., Curtis, A., Friedrich, K., Kranzinger, L., Küppers, B., Lorber, K.E., Möllnitz, S., Neuhold, S., et al., Eds.; Recy&DepoTech: Leoben, Austria, 2018.
58. Khodier, K.; Sarc, R. Distribution-independent empirical modeling of particle size distributions—Coarse-shredding of mixed commercial waste. *Processes* **2021**, *9*, 414. [CrossRef]

59. Kandlbauer, L.; Khodier, K.; Ninevski, D.; Sarc, R. Sensor-based Particle Size Determination of Shredded Mixed Commercial Waste based on two-dimensional Images. *Waste Manag.* **2021**, *120*, 784–794. [CrossRef] [PubMed]
60. Weißenbach, T.; Sarc, R. Investigation of particle-specific characteristics of non-hazardous, fine shredded mixed waste. *Waste Manag.* **2021**, *119*, 162–171. [CrossRef]
61. Pomberger, R.; Küppers, B. Entwicklungen in der Sensorgestützten Sortiertechnik [Developments in Sensor-Based Sorting Technology]. In Proceedings of the Österreichische Abfallwirtschaftstagung 2017—Die Digitalisierung der Abfallwirtschaft, Graz, Austria, 10–11 May 2017.
62. Flamme, S.; Hams, S.; Zorn, M. *Sensortechnologien in der Kreislaufwirtschaft [Sensor Technologies in the Recycling Economy]*; Pomberger, R., Adam, J., Aldrian, A., Curtis, A., Friedrich, K., Kranzinger, L., Küppers, B., Lorber, K.E., Möllnitz, S., Neuhold, S., et al., Eds.; Recy&DepoTech: Leoben, Austria, 2018.
63. Weißenbach, T.; Sarc, R. Investigation of particle-specific characteristics of coarse-shredded waste. *Waste Manag.* **2021**, *119*, 162–171. [CrossRef] [PubMed]
64. Curtis, A.; Küppers, B.; Möllnitz, S.; Khodier, K.; Sarc, R. Real time material flow monitoring in mechanical waste processing and the relevance of fluctuations. *Waste Manag.* **2021**, *120*, 687–697. [CrossRef] [PubMed]
65. Curtis, A.; Sarc, R. Real-Time Monitoring of Volume Flow, Mass Flow and Shredder Power Consumption in Mixed Solid Waste Processing. *Waste Manag.* **2021**, under review.

processes MDPI

Article

Distribution-Independent Empirical Modeling of Particle Size Distributions—Coarse-Shredding of Mixed Commercial Waste

Karim Khodier [1] **and Renato Sarc** [2,*]

[1] Department of Environmental and Energy Process Engineering, Chair of Process Technology and Industrial Environmental Protection, Montanuniversitaet Leoben, Franz-Josef-Straße 18, 8700 Leoben, Austria; karim.khodier@unileoben.ac.at
[2] Department of Environmental and Energy Process Engineering, Chair of Waste Processing Technology and Waste Management, Montanuniversitaet Leoben, Franz-Josef-Straße 18, 8700 Leoben, Austria
* Correspondence: renato.sarc@unileoben.ac.at

Abstract: Particle size distributions (PSDs) belong to the most critical properties of particulate materials. They influence process behavior and product qualities. Standard methods for describing them are either too detailed for straightforward interpretation (i.e., lists of individual particles), hide too much information (summary values), or are distribution-dependent, limiting their applicability to distributions produced by a small number of processes. In this work the distribution-independent approach of modeling isometric log-ratio-transformed shares of an arbitrary number of discrete particle size classes is presented. It allows using standard empirical modeling techniques, and the mathematically proper calculation of confidence and prediction regions. The method is demonstrated on coarse-shredding of mixed commercial waste from Styria in Austria, resulting in a significant model for the influence of shredding parameters on produced particle sizes (with classes: >80 mm, 30–80 mm, 0–30 mm). It identifies the cutting tool geometry as significant, with a p-value $< 10^{-5}$, while evaluating the gap width and shaft rotation speed as non-significant. In conclusion, the results question typically chosen operation parameters in practice, and the applied method has proven to be valuable addition to the mathematical toolbox of process engineers.

Keywords: particle size distribution; compositional data analysis; simplex; isometric log-ratios; multivariate multiple linear regression; mechanical processing; waste treatment; commercial waste; shredder

Citation: Khodier, K.; Sarc, R. Distribution-Independent Empirical Modeling of Particle Size Distributions—Coarse-Shredding of Mixed Commercial Waste. *Processes* **2021**, *9*, 414. https://doi.org/10.3390/pr9030414

Academic Editor: Aneta Magdziarz

Received: 8 February 2021
Accepted: 22 February 2021
Published: 25 February 2021

1. Introduction

The size distribution belongs to the most critical properties of solid particulate materials, and particularly mixed solid waste, for example: The quality classes of solid recovered fuels (SRF) demand specific maximum particle sizes [1]. The particle size distribution (PSD) of the organic fraction of municipal waste impacts its anaerobic digestion [2]. The particle sizes of municipal solid waste influence the yields of dry gas, char, and tar in fixed bed reactor pyrolysis [3]. And the PSD influences the mass throughput of robotic sorters, which are limited by picks per hour [4]; hence, smaller particle sizes (and the corresponding smaller weights) decrease the possible mass throughput.

Concerning mixed commercial waste, besides the PSD's relevance as a technical quality criterion for processing products (e.g., SRF [5]), and its influence on the performance of reactors and processing machines (e.g., wind sifters [6]), different types of materials also concentrate in different particle size ranges (e.g., sorting analysis by Khodier et al. [7] and the size distribution of different plastic types according to Möllnitz et al. [8]). Hence, beyond influencing the shares of a plant's throughput that pass specific machines (due to material flow separation by screens), the PSD also determines the kinds of materials that pass through these machines.

Therefore, beneficial PSDs increase the effectiveness, as well as economic and ecologic efficiency of mixed solid waste treatment. Consequently, the PSD of mixed solid waste is deliberately influenced during mechanical processing, which is usually the first treatment stage for this kind of material, mainly through a combination of shredding and sieving [1–3,7,9–13].

1.1. Describing Particle Size Distributions

Strictly speaking, the PSD of a collective is described as a list of the individual particles' sizes. But the representation as such a list is not suitable for analyzing and comparing PSDs. Consequently, various more useful methods for describing PSDs exist, which were summarized by Polke et al. [14]: Collectives of particles are often described through average equivalent diameters. An example is the Sauter diameter d_S, which gives information on the specific surface of the collective (Equation (1), where V_t is the total volume of all particles, and A_t is the total surface area of all particles).

$$d_S = 6V_t/A_t \tag{1}$$

Often, information on the width of the distribution is also essential. Consequently, measures of this width are frequently provided. Examples are the sample standard deviation S (Equation (2), where d_i is the size of the ith particle, $\overline{d}$ is the arithmetic average size, and N is the number of particles), or distribution-independent measures of the width, as shown, for example, in Equation (3) (where W is the width, and d_b is the bth percentile particle size).

$$S = \sqrt{\frac{\sum_{i=1}^{N}\left(d_i - \overline{d}\right)^2}{N-1}} \tag{2}$$

$$W = \frac{d_{75} - d_{25}}{d_{50}} \tag{3}$$

Considering the described influence of the PSD on the path individual particles take through a plant, it is essential in mechanical waste processing to have more detailed knowledge on it than just summary values. Hence, the results of a PSD analysis are often reported graphically: the frequency density is shown in a histogram [15], where particle sizes are summarized into particle size classes (PSCs)—which is also the level of information obtained from sieve analyses (Figure 1a). Another representation, which is more suitable for comparing PSDs, is the sum distribution (Figure 1b) [14].

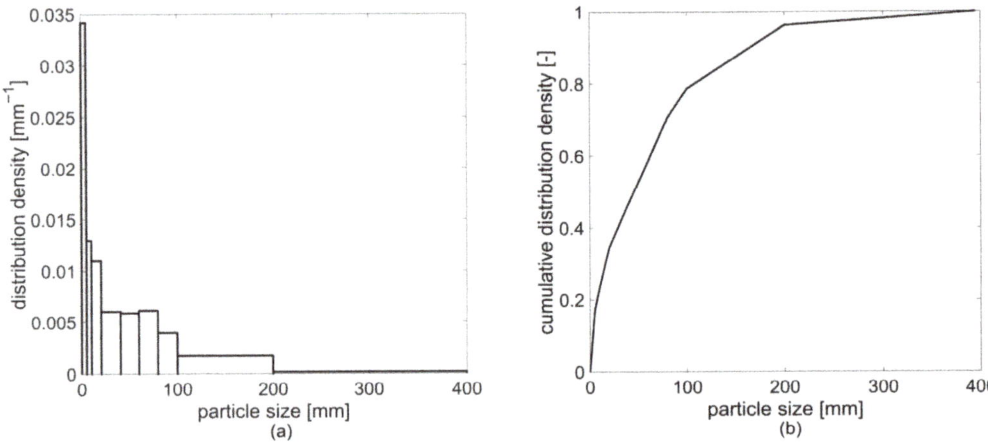

Figure 1. Representation of the overall particle size distribution (PSD) of a mixed commercial waste according to Reference [7]: (**a**) frequency density; (**b**) cumulative frequency density.

These graphical representations correspond to an empirical distribution [15]. As the sample size approaches infinity and the class width approaches zero, the histogram's representation becomes a continuous function: the probability density function (PDF). This PDF can also be approximated from analyses, where only summary information on discrete PSCs is available (e.g., sieve analyses), for example, through cubic splines [16] or Kernel density estimation [17].

Sometimes, the PDF can be approximately described by an analytical expression. In such cases, reporting the momentums of such an analytical distribution is sufficient to describe the PSD. So, for example, reporting the arithmetic mean particle size $\bar{d}$ of the sample and its standard deviation S, usually implies the underlying assumption of a normal distribution, according to Equation (4), where $q(d)$ is the probability density or frequency density for particles of size d, μ is the arithmetic average of the population, which is estimated through $\bar{d}$, and σ is the population's standard deviation, which is estimated through S [18].

$$q(d) = \frac{1}{\sqrt{2\pi\sigma^2}} \cdot \exp\left(-\frac{(d-\mu)^2}{2\sigma^2}\right) \tag{4}$$

Three further analytical PDFs, are reported as being relevant to the description of PSDs [14]: The log-normal distribution describes materials, where the logarithm of the particle size follows a normal distribution. It is a positively skewed distribution, which—in contrast to the normal distribution—only includes positive and, therefore, meaningful particle sizes. It is shown in Equation (5), where μ and σ are estimated by the arithmetic average and the sample standard deviation of the logarithm of the particle sizes.

$$q(d) = \begin{cases} 0 & \text{if } d < 0 \\ \frac{1}{d\sqrt{2\pi\sigma^2}} \cdot \exp\left(-\frac{(\log(d)-\mu)^2}{2\sigma^2}\right) & \text{otherwise} \end{cases} \tag{5}$$

The Gates-Gaudin-Schuhmann (GGS) distribution [19] is an empirical approximation that is often suitable for describing products of coarse comminution processes. Its parameters are the maximum particle size d_{max} and the uniformity parameter m_u. Its PDF is shown in Equation (6).

$$q(d) = \begin{cases} 0 & \text{if } d < 0 \text{ or } d > d_{max} \\ \frac{m_u}{d_{max}} \left(\frac{d}{d_{max}}\right)^{m_u - 1} & \text{otherwise} \end{cases} \tag{6}$$

The Rosin-Rammler-Sperling-Bennet (RRSB) distribution [20] is also an empirical approximation. Its first parameter d' is equal to the particle size, where the cumulative frequency density reaches the value $1 - 1/e \approx 0.632$, and its second parameter n_u is a uniformity parameter. Its PDF is shown in Equation (7). The RRSB distribution is often used for describing products of fine comminution and dusts. Examples of the discussed distributions are shown in Figure 2.

$$q(d) = \begin{cases} 0 & \text{if } d < 0 \\ \frac{n_u}{d'} \left(\frac{d}{d'}\right)^{n_u - 1} \exp\left[-\left(\frac{d}{d'}\right)^{n_u}\right] & \text{otherwise} \end{cases} \tag{7}$$

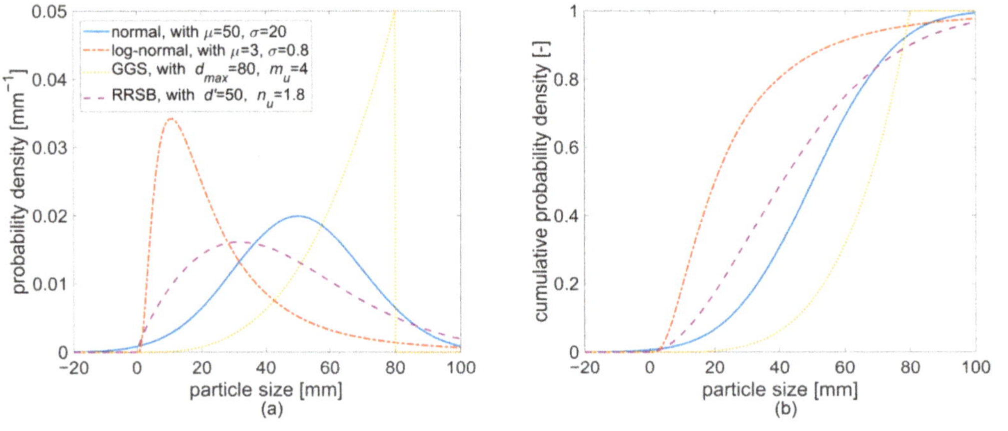

Figure 2. Frequency density (**a**) and cumulative frequency density (**b**) of a normal, log-normal, Gates-Gaudin-Schuhmann (GGS), and Rosin-Rammler-Sperling-Bennet (RRSB) distribution.

1.2. Modeling Particle Size Distributions

Products' PSDs result from their original condition and the kinds and parameters of machines that process them. Hence, to beneficially influence PSDs through process design and the choice and parametrization of machines, modeling and predicting them is desirable.

The most sophisticated and advantageous models are physical models, which provide an in-depth understanding of the phenomena that influence PSDs. Simulations based on such models are usually implemented using the discrete element method (DEM). Examples in literature are the works of Sinnott and Cleary [21] on the particle flows and breakage in impact crushers, Lee et al. [22] on breakage and liberation behavior of recycled aggregates from impact-breakage of concrete waste, and Dong et al. [23] on particle flow and separation on vibrating screens.

While DEM-based models improve process understanding, they also require high amounts of computational resources and detailed models and data on the processing machines and the materials to be comminuted, which limits their applicability in practice in many cases [24]: no published models, for example, incorporate the variability of materials, geometries and particle interactions for real mixed solid waste.

When physical models cannot be used, empirical regression models can deliver basic insights on machine and parameter influences on PSDs. For ensuring the reliability of the results, it is essential to involve statistical analyses for finding and interpreting the models. For scalar target values, the procedure has been thoroughly described by Khodier et al. [24], based on the example of the parametrization-dependent energy demand and throughput behavior of coarse shredders for mixed commercial waste.

PSDs are non-scalar: they are either described as PDFs, which are continuous functions, or as compositions of PSCs and, therefore, as multivariate vectors. Analytical PDFs, as those presented in Section 1.1., are defined by a set of momentums, which can also be treated as multivariate vectors. Consequently, modeling PSDs requires extensions of the methods used by Khodier et al. [24] to multivariate dependent variables.

The most widely applied variation of regression modeling is linear regression. It relates one or more dependent variables to one or more independent variables based on a set of linear regression coefficients. Its most general form—which is relevant to this work—is multivariate multiple linear regression, which involves multiple independent variables and multivariate dependent variables. Its model Equation is shown in Equation (8) [25]. The matrix $\mathbf{Y}$ (Equation (9)) is the $P \times R$ matrix of P observations of the R-variate vector of the dependent variable, with elements $y_{p,r}$. $\boldsymbol{\varepsilon}$ (Equation (12)) is also a $P \times R$ matrix and shows the corresponding model residuals $\varepsilon_{p,r}$. $\boldsymbol{X}$ (Equation (10)) is a $P \times (Q+1)$ matrix of the settings $x_{p,q}$ of the Q independent variables corresponding to the P observations. Its first column (which is indexed with 0) is a column of ones, corresponding to constant terms in linear regressions. And the matrix $\boldsymbol{\beta}$ (Equation (11)) contains the $(Q+1) \times R$ regression coefficients $\beta_{q,r}$.

$$\boldsymbol{Y} = \boldsymbol{X} \cdot \boldsymbol{\beta} + \boldsymbol{\varepsilon} \tag{8}$$

$$\boldsymbol{Y} = \begin{bmatrix} y_{1,1} & y_{1,2} & \cdots & y_{1,R} \\ y_{2,1} & y_{2,2} & \cdots & y_{2,R} \\ \vdots & \vdots & \ddots & \vdots \\ y_{P,1} & y_{P,2} & \cdots & y_{P,R} \end{bmatrix} \tag{9}$$

$$\boldsymbol{X} = \begin{bmatrix} x_{1,0} & x_{1,1} & x_{1,2} & \cdots & x_{1,Q} \\ x_{2,0} & x_{2,1} & x_{2,2} & \cdots & x_{2,Q} \\ \vdots & \vdots & \vdots & \ddots & \vdots \\ x_{P,0} & x_{P,1} & x_{P,2} & \cdots & x_{P,Q} \end{bmatrix} = \begin{bmatrix} 1 & x_{1,1} & x_{1,2} & \cdots & x_{1,Q} \\ 1 & x_{2,1} & x_{2,2} & \cdots & x_{2,Q} \\ \vdots & \vdots & \vdots & \ddots & \vdots \\ 1 & x_{P,1} & x_{P,2} & \cdots & x_{P,Q} \end{bmatrix} \tag{10}$$

$$\boldsymbol{\beta} = \begin{bmatrix} \beta_{0,1} & \beta_{0,2} & \cdots & \beta_{0,R} \\ \beta_{1,1} & \beta_{1,2} & \cdots & \beta_{1,R} \\ \vdots & \vdots & \ddots & \vdots \\ \beta_{Q,1} & \beta_{Q,2} & \cdots & \beta_{Q,R} \end{bmatrix} \tag{11}$$

$$\boldsymbol{\varepsilon} = \begin{bmatrix} \varepsilon_{1,1} & \varepsilon_{1,2} & \cdots & \varepsilon_{1,R} \\ \varepsilon_{2,1} & \varepsilon_{2,2} & \cdots & \varepsilon_{2,R} \\ \vdots & \vdots & \ddots & \vdots \\ \varepsilon_{P,1} & \varepsilon_{P,2} & \cdots & \varepsilon_{P,R} \end{bmatrix} \tag{12}$$

The resulting model is obtained by minimizing the sum of squares of the residuals $\varepsilon_{p,r}$. It is shown in Equation (13), where $\hat{\boldsymbol{Y}}$ is the matrix of the $P \times R$ model predictions $\hat{y}_{p,r}$ for the dependent variable, corresponding to the observations in $\boldsymbol{Y}$, and $\hat{\boldsymbol{\beta}}$ is the matrix of the $(Q+1) \times R$ least-squares estimates $\hat{\beta}_{q,r}$ of the regression coefficients in $\boldsymbol{\beta}$.

$$\hat{\boldsymbol{Y}} = \boldsymbol{X} \cdot \hat{\boldsymbol{\beta}} \tag{13}$$

For linear regression models, it is necessary to describe the dependent variable as a vector of a fixed length. In the case of analytical PDFs, this is the vector of the momentums. So, the immediate result of a model for these momentums is the vector of their expected values, and the corresponding confidence bands. From there, the PDF and its confidence region can be calculated.

Theoretically, $\hat{\boldsymbol{\beta}}$ can also be calculated from Q univariate linear regressions. The resulting values $\hat{\beta}_{q,r}$ are identical. But multivariate methods, involving multivariate linear regression and multivariate analysis of variance (MANOVA, e.g., Reference [26]), are preferable: they allow a more accurate calculation of confidence regions, considering correlations between the momentums. Moreover, the evaluation of parameter's significance should be based on multivariate considerations to find coherent models for the resulting—interdependent—distribution of the particle sizes.

Analytical PDFs are practical, as they allow a detailed description of PSDs with a small number of variables—the momentums. However, only few processes produce PSDs that follow analytical functions, according to Polke et al. [14]. Consequently, methods for distribution-independent modeling of PSDs are also needed, which is especially true for the mechanical processing of mixed solid waste, considering the variety of processing machines used there: e.g., shredders, screens, magnetic separators, and sensor-based sorters [10].

Empirical distributions allow the distribution-independent description of PSDs: particle sizes are amalgamated to D discrete PSCs. As a result, the PSD is described as a composition—a D-dimensional vector of the shares of those PSCs.

Mathematically, the compositional nature of the vector of PSCs has wide-reaching consequences: D-dimensional compositions are constrained to a vector space called the D-dimensional simplex $\boldsymbol{S}^{(D)}$, which is a sub-space of the D-dimensional real space $\boldsymbol{R}^{(D)}$ [27]. Compositions, being simplicial vectors, have specific common properties: Their D parts are not linearly independent. This results from a summation constraint: their parts sum up to a fixed constant, e.g., 1 or 100%. Furthermore, parts may only have positive values or zero. So, more precisely, $\boldsymbol{S}^{(D)}$ is a sub-space of $\boldsymbol{R}_0^{+(D)}$. The most well-known representation of a simplex is the ternary diagram (see, e.g., Reference [28]), which shows the three-dimensional simplex.

When modeling simplicial vectors, the constraints of the simplex and the interdependence of the compositional parts must be considered. While Khodier et al. [29] report the empirical observation that the constraints are automatically fulfilled for the prediction values $\hat{\boldsymbol{Y}}$, if all observations in $\boldsymbol{Y}$ are valid compositions, this does not apply to corresponding confidence regions. Hence, different methods were developed in the past decades to handle and model simplicial data.

The most widely applied approach for handling compositions is transforming data using log-ratios, as suggested by Aitchison [30]. There are many kinds of such log-ratios, but the state of the art approach in the compositional data community is the application of so-called isometric log-ratios (ilr) (as proposed by Egozcue [31]), according to Reference [32]. These are bijective projections of the D-dimensional Simplex onto a $(D-1)$-dimensional real-space with an orthonormal basis [27]. The resulting ilr-coordinates are unconstrained, linearly independent coordinates. Hence, standard statistics can be applied in the projected log-ratio space, as demonstrated, for example, by Edjabou et al. [33] for waste composition analysis. Consequently, the use of ilr-transformations allows the application of standard multivariate multiple linear regression to predict PSDs, described as empirical distributions.

In this work, the approach of modeling influences on empirical PSDs using multivariate multiple linear regression and ilrs is applied on particle size data, using the programming language R. The data was obtained within Khodier et al.'s [24] industry-scale coarse-shredding experiments with real mixed commercial waste. Based on this example, this work aims at presenting the method to the process engineering and waste processing communities, enabling the distribution-independent empirical modeling of particle size distributions, while preserving all relevant information. Moreover, the method's suitability for PSD modeling and its limitations and potential pitfalls in the interpretation of the results are discussed. And, finally, insights on the influences of coarse shredders' parameters on the PSDs of mixed commercial waste are presented, complementing findings of Khodier et al. [24] on their effects on shredders' throughput behavior and energy demand.

2. Materials and Methods

2.1. Experimental Design and Setup

The choice of the experimental design and the shredding experiment setup have been explained in detail by Khodier et al. [24]. Hence, they are only summarized in the following. The extension of the experiment with material sampling and particle size analysis are described in detail.

2.1.1. Experimental Design

The shredding experiment examines the influence of three independent variables on the throughput behavior and energy demand of a Terminator 5000 SD, which is a single-shaft shredder from the Austrian company Komptech GmbH (Frohnleiten, Austria)—a research partner in the funded project ReWaste 4.0. These independent variables are the radial gap width w, the shaft rotation speed s, and the cutting tool geometry c. The factor range of w was defined from 0% to 100% of the maximum gap width, with discrete levels at a step size of 10%. The minimum of the factor range of s was chosen at 60% and the maximum at 100% of the maximum shaft rotation speed of 31 rpm, again width discrete 10% steps. Concerning c, three different geometries are examined, called "F", "XXF", and "V" (for more details, cf. Reference [24] and Figure A1 and Table A1 in the Appendix A).

The factors are coded for the design and analysis of the experiment: for the numerical factors s and w, their range is adjusted to -1 to 1, representing the minimum and maximum factor settings. Concerning the nominal factor c, it is represented by two variables c_1 and c_2, based on sum contrasts (cf. Reference [34]). The values of these variables that correspond to the cutting tool geometries are shown in Table 1.

Table 1. Contrast matrix for the cutting tool geometry.

Cutting Tool Geometry	c_1	c_2
F	1	0
XXF	−1	−1
V	0	1

The experimental design settings of the independent variables were chosen based on a statistical Design of Experiments (cf. Reference [35]). A 32 runs, completely randomized D-optimal design (cf. Reference [36]) was chosen, with no blocking (cf. Reference [37]) and with five replicate points and five lack-of-fit points. It is based on the reduced-cubic design model, shown in Equation (14), where $\hat{y}^{(r)}$ is the model prediction for the rth (univariate) response (=dependent variable), $\vec{x}$ is a vector of the factors' settings, and $K^{(r)}_{jkmn}$ is the model coefficient for the rth response and the factor or interaction (=multiplication of factors) $w^j s^k c_1^m c_2^n$.

$$\hat{y}^{(r)}\left(\vec{x}\right) = \sum_{j=0}^{2}\sum_{k=0}^{2-j}\sum_{m=0}^{1}\sum_{n=0}^{1-m}\left(K^{(r)}_{jkmn} w^j s^k c_1^m c_2^n\right) \tag{14}$$

Equation (14) can easily be extended to multivariate responses; therefore, the design is also valid for multivariate multiple linear regression. To extend it, $\hat{y}^{(r)}$ equals $\hat{y}_{p,r}$ in Equation (13): the rth univariate response becomes the rth dimension of the multivariate response, and p indexes the corresponding vector of factor settings $\vec{x}$, which is simply a row of $\mathbf{X}$. Each factor or interaction $w^j s^k c_1^m c_2^n$ is represented by a column of $\mathbf{X}$, and each coefficient $K^{(r)}_{jkmn}$ corresponds to a coefficient $\hat{\beta}_{q,r}$.

2.1.2. Setup of the Shredding Experiment

The flowchart of the experiment is shown in Figure A2 in the Appendix A: The feed material is waste, declared as mixed commercial waste (cf. Reference [24] for more details), collected in Styria in Austria in October 2019. It is fed into the shredder's feeding bunker using a wheel loader. From the shredder's output belt, the material is passed to a digital material flow monitoring system (DMFMS), consisting of a belt-scale and optical sensors (cf. Reference [11]). The material leaving the DMFMS is collected on a product heap. Each experimental run has a total duration of one hour.

2.1.3. Sampling

For analyzing the PSD of the shredded waste, samples must be taken. The design of the sampling process in this experiment is based on Pierre Gy's Theory of Sampling (TOS), as described in the Danish standard DS 3077 [38] and the work of Khodier et al. [7] on its application on coarsely shredded mixed commercial waste.

According to TOS, the fundamental sampling principle must be considered: each particle must have the same probability of ending up in the sample. The most beneficial sampling situation is a one-dimensional sampling [39], e.g., taking the samples from a falling stream. Furthermore, a sample should spatially cover the whole lot, which is achieved by composite sampling: the sample is a composite of so-called increments (the sampled material from one individual sampling step).

In this work, the sample for each experimental run consisted of 40 such increments. They were taken from a falling stream, swiveling the output conveyor belt of the DMFMS back and forth over a sampling box, elevated by a forklift (see Figure 3), every 3 minutes, taking two increments. The inner dimensions of the box were 115 × 915 × 565 (length × width × height in mm). For ensuring that all desired material ends up in the box, the height of the back-side wall of the box was increased by placing 1.5 m-long wood boards inside. The box was changed after every ten increments. Based on the share of the samples in the total processed mass, each increment covered approximately 0.61 seconds of throughput. For intermediate storage, the samples were finally transferred to 1 m^3 big bags.

Figure 3. Sampling setup.

2.1.4. Particle Size Analysis

The PSDs of the samples were analyzed using a Komptech (Frohnleiten, AT) Nemus 2700 drum screen and five screening drums that have square holes with side lengths of 80, 60, 40, 20, and 10 mm. The geometries of the drums are shown in Figure A3 and Table A2 in the Appendix A.

First, one big bag of a sample was evenly distributed in the feeding bunker of the screen, which has a length of 4033 mm in the direction of the material flow, and a width of 1035 mm. Then, the drum was started with a rotation speed of 11.5 rpm, and the conveyor belt of the feeding bunker was operated at 0.026 m/s. The drum screen was only stopped after all material had passed. The produced fine fraction was then screened with the subsequent finer drum. The scale used for measuring the masses of the fraction has an uncertainty of 10 g.

2.2. Analysis of the Results

The analyses of the results, which are explained in this section, were performed in R version 4.0.2, based on the work of van den Boogaart and Tolosana-Delgado [40]. The implementation is attached as an HTML export of a jupyter notebook (see Supplementary Material).

For the analyses in this work, the six particle size fractions from the particle size analysis are aggregated to three PSCs for easier visualization. To ensure the relevance of the findings for mechanical waste processing, the PSCs were chosen, based on the particle size limits of SRF premium quality (30 mm) and SRF medium quality (80 mm) [1]. Since none of the used screen drums had a mesh width of 30 mm, equal shares of the screening fraction 20–40 mm are assigned to the particle size fractions 0–30 mm and 30–80 mm, which corresponds to linear interpolation.

2.2.1. Isometric Log-Ratios

The ilr transformation (denoted as a function ilr()) is an isometry of the vector spaces $\boldsymbol{S^D}$ and $\boldsymbol{R^{D-1}}$ [27], which means that the distance between two compositions $\vec{y}^{(1)}$ and $\vec{y}^{(2)}$ is preserved in the transformation. For interpretation, it is essential to understand that the distance referred to is not the Euclidian distance Δ_E (Equation (15), where y_i is the ith element of a R-dimensional composition $\vec{y}$), but rather the Aitchison distance Δ_A (Equation (16), where $\text{ilr}_i()$ is the function for the ith ilr dimension) (cf. Reference [27]). Consequently, the preserved distance is of a relative, multiplicative nature and not an absolute, additive nature. Hence, the least-squares minimization of the model residuals in ε, when calculating the coefficients' estimates $\hat{\boldsymbol{\beta}}$, is also based on the Aitchison distance when applying the ilr transformation. Considering this is particularly important when the order of magnitude of the parts' shares differs significantly since small absolute differences become very significant on a relative scale for very small shares (cf. Reference [41]).

$$\Delta_E\left(\vec{y}^{(1)},\vec{y}^{(2)}\right) = \sqrt{\sum_{i=1}^{R}\left(y_i^{(1)} - y_i^{(2)}\right)^2} \tag{15}$$

$$\Delta_A\left(\vec{y}^{(1)},\vec{y}^{(2)}\right) = \sqrt{\frac{1}{2R}\sum_{i=1}^{R}\sum_{j=1}^{R}\left[\ln\left(\frac{y_i^{(1)}}{y_j^{(1)}}\right) - \ln\left(\frac{y_i^{(2)}}{y_j^{(2)}}\right)\right]^2} = \sqrt{\sum_{i=1}^{R-1}\left[\text{ilr}_i\left(\vec{y}^{(1)}\right) - \text{ilr}_i\left(\vec{y}^{(2)}\right)\right]^2} \tag{16}$$

Furthermore, the ilr transformation is not defined if any part has a value of zero. While there are different approaches to handling such values, they complicate the application of the transformation [27].

In the following, ilr-transformed coordinates are marked with "∗" so that $\boldsymbol{Y}^* = \text{ilr}(\boldsymbol{Y})$. The back-transformation function is denoted "$\text{ilr}^{-1}()$". For calculating ilr coordinates, the compositional parts are sequentially grouped: first, each component is assigned to a group +1, or −1. For subsequent ilr coordinates, the elements of one group are again assigned to new groups +1 or −1, and the parts of the other group are assigned to group 0. Each ilr coordinate is then calculated according to Equation (17), where $a_{i,r}$ is the scaling factor for the rth compositional part and the ith ilr coordinate. The calculation of $a_{i,r}$ is shown in Equation (18), where t is the number of parts in group +1, and v is the number of parts in group −1 [31].

$$\text{ilr}_i\left(\vec{y}\right) = \sum_{r=1}^{R}\left[a_{i,r}\ln(y_r)\right] \tag{17}$$

$$a_{i,r} = \begin{cases} +\frac{1}{t}\sqrt{\frac{tv}{t+v}} & \text{for parts of group } +1 \\ -\frac{1}{v}\sqrt{\frac{tv}{t+v}} & \text{for parts of group } -1 \\ 0 & \text{for parts of group } 0 \end{cases} \tag{18}$$

Greenacre [42] documents concerns regarding the interpretation of ilr-transformed data since it incorporates the geometric means of compositional parts. Consequently, in this work, results are interpreted based on graphical representations of back-transformed data. Hence, the exact choice of groups is arbitrary, and the standard grouping of the "ilr()" function of the "compositions" package version 2.0-1 in R [43] is used. In this work, $r = 1$ stands for the fraction > 80 mm, $r = 2$ for the fraction 30–80 mm, and $r = 3$ for the fraction 0–30 mm (see Supplementary Material). Resulting from the standard grouping, the ilr-transformed representation $\vec{y}^*$ of a particle size composition $\vec{y}$ (corresponding to a row of **Y**) is calculated according to Equation (19).

$$\vec{y}^* = \text{ilr}\left(\vec{y}\right) = \begin{pmatrix} \ln\left[\left(\frac{y_2}{y_1}\right)^{\sqrt{0.5}}\right] \\ \ln\left[\frac{y_3{}^{\sqrt{2/3}}}{(y_1 y_2)^{\sqrt{1/6}}}\right] \end{pmatrix} \quad (19)$$

2.2.2. Model Reduction: MANOVA

To obtain a final, reliable model, the factors and interactions in Equation (14) must be checked on their significance, eliminating non-significant ones, but retaining model hierarchy (cf. Reference [24]). For univariate dependent variables, this is done through F-tests in an analysis of variance (ANOVA) (cf. Reference [35]). The multivariate character of the compositional dependent variable in this work requires a multivariate extension of the ANOVA: the MANOVA. Different from the ANOVA, there is more than one definition of the F-statistic in the MANOVA. The most commonly used definitions are the Pillai-Barlett trace, Wilk's lambda, the Hotelling-Lawley trace, and Roy's largest eigenvalue statistic, according to Hand and Taylor [26]. These are also the ones implemented in the "regr" package, version 1.1 [44], used for the MANOVA in this work.

For the analyses at hand, the Pillai-Barlett trace was chosen, based on the recommendation of Olson [45] as cited in Reference [26]. Model reduction is performed applying backward selection: the least significant term, which can be removed without violating model hierarchy, is eliminated, as long as removable non-significant factors or interactions are present (cf. Reference [40]). Analogous to Reference [24], 0.1 is chosen as the threshold so that factors and interactions with an empirical significance (p-value) higher than that threshold are discarded. The relevant p-values are calculated using the "drop1()" function from the "regr" package (version 1.1).

For evaluating the final model, three performance values are calculated: The coefficient of determination R^2 calculates how much of the variance of the data is explained by the model. The adjusted coefficient of determination R^2_{adj} is a measure similar to R^2. But it is adjusted by the terms in the model and thereby evaluates the model's efficiency [35]. Both are calculated using the "R2()" function in R.

The prediction coefficient of determination R^2_{pred} determines the share of variance, which is explained by models fitted without considering the very point which is evaluated. High differences between R^2_{pred} and R^2_{adj} indicate overfitting. R^2_{pred} is calculated according to Equation (20), where $PRESS$ is the prediction residual sum of squares and SS_{tot} is the total sum of squares [46]. $PRESS$ is calculated, using the "PRESS()" function from the "MPV" package version 1.56 [47]. And SS_{tot} is calculated, according to Equation (21). $y^*_{p,r}$ is the pth observation of the rth of $(R-1)$ coordinates of the ilr-transformed dependent variable. And $\overline{y}^*_r$ is the arithmetic mean of the observations of the rth ilr coordinate (see Equation (22)).

$$R^2_{pred} = 1 - \frac{PRESS}{SS_{tot}} \quad (20)$$

$$SS_{tot} = \sum_{p=1}^{P}\sum_{r=1}^{R-1}\left(y^*_{p,r} - \overline{y}^*_r\right)^2 \quad (21)$$

$$\overline{y}^*_r = \frac{1}{P}\sum_{p=1}^{P} y^*_{p,r} \quad (22)$$

2.2.3. Analysis of the Residuals

The tests in the MANOVA require multivariate normality of the (ilr-transformed) residuals ε^*. Hence, to validate the final model, the distribution of the residuals must be examined. Each coordinate of variables, which follow a multivariate normal distribution, also follows a univariate normal distribution [48]. Hence, a quantile-quantile plot for each coordinate is examined as a first visual step. Since the individual coordinates' univariate normality is a necessary, but not sufficient condition for multivariate normality, multivariate tests are also applied, particularly Mardia's Skewness and Mardia's Kurtosis [49]. Both tests are part of the "mvn()" function in "MVN" package version 5.8 [50], which also tests the univariate normality of the individual coordinates using the Shapiro-Wilk test.

2.2.4. Confidence and Prediction

The resulting model Equation shows the most likely prediction value. Two regions express the uncertainty of this prediction: the confidence region and the prediction region (cf. Reference [40]). The confidence region covers the uncertainty of the model parameter estimation. The region reflects likely average PSC distributions (on an ilr scale) for specific parameter settings, for extended operation times, on a chosen confidence level. The prediction region adds the residual variability around the expected value. Hence, it shows likely PSC distributions for one hour of operation (since this is the experimental duration the data is based on).

Due to the required multivariate character of the ilr-transformed residuals, the resulting confidence and prediction regions are equipotential-ellipses (resulting from the PDF of the multivariate normal distribution) on an ilr-scale. Van den Boogaart and Tolosana Delgado [40] provide R-code for calculating these regions and visualizing their back-transformed representation in ternary diagrams. This code is used in this work (see Supplementary Materials).

3. Results and Discussion

3.1. Data and Model

The experimental design and the resulting shares of the PSCs are shown in the Supplementary Materials. Their order corresponds to the order in which the experimental runs were performed. Originally, a completely randomized order was planned. As reported by Khodier et al. [24], due to an unintentional change of the motor rotation speed of the V cutting tool during the experiments, three runs had to be repeated. Since the tight timescale did not allow re-randomizing all remaining runs, considering the time consumed when switching shredders, the randomness is slightly impaired.

A significant model was found based on the experimental data. It is visualized in Figure 4 and shown in Equation (23), where $\vec{y}$ is a vector showing the model predictions of the mass shares of the three PSCs. Its first element corresponds to the coarse fraction (>80 mm), its second element to the medium fraction (30–80 mm), and the third element shows the fine fraction (0–30 mm). As shown in the Equation, the only significant factor (at a threshold of $p = 0.1$) is the cutting tool geometry c.

$$\vec{y} = \mathrm{ilr}^{-1}\left[\begin{pmatrix} 0.137 \\ 0.047 \end{pmatrix} - \begin{pmatrix} 0.307 \\ 0.178 \end{pmatrix} c_1 + \begin{pmatrix} 0.344 \\ 0.145 \end{pmatrix} c_2\right] = \begin{cases} \begin{pmatrix} 0.39 & 0.31 & 0.30 \end{pmatrix}^T & \text{for } c = \text{"F"} \\ \begin{pmatrix} 0.30 & 0.35 & 0.35 \end{pmatrix}^T & \text{for } c = \text{"XXF"} \\ \begin{pmatrix} 0.21 & 0.42 & 0.37 \end{pmatrix}^T & \text{for } c = \text{"V"} \end{cases} \quad (23)$$

The analysis of the ilr-residuals proves a multivariate normal distribution, with p-values of 0.67 and 0.89 for the Shapiro-Wilk test on univariate normality of each coordinate and p-values of 0.89 and 0.71 for Mardia's Skewness and Mardia's Kurtosis, respectively.

R^2 for the model is 0.57 and R^2_{adj} is 0.54. These values are noticeably lower than those of the models for the throughput behavior and energy demand (see Reference [24]), which range from 0.73 to 0.87 for R^2 and from 0.67 to 0.81 for R^2_{adj}.

The lower coefficients of determination are reasonable since material fluctuations are likely to influence the material quality more than the process behavior, and sampling adds random noise (cf. Reference [7]). Considering the high expectable noise in processing experiments with real waste, and especially in shredding experiments (cf. Reference [24]), the model performance is satisfactory.

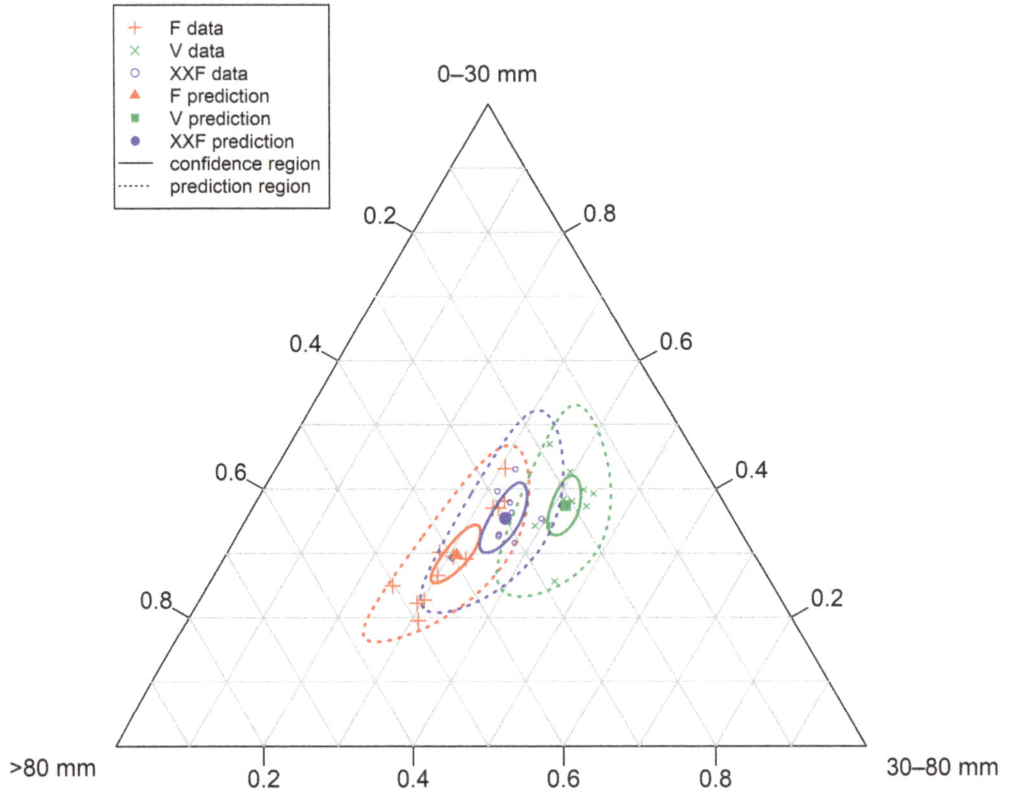

Figure 4. Prediction values and confidence and prediction regions for the particle size class distributions resulting from different cutting tool geometries.

3.2. Discussion of the Method

The data from the particle size analyses were amalgamated to three discrete PSCs and then freed from the restrictions of the simplex by applying an ilr-transformation. The resulting model, the multivariate normal distribution of its residuals, and the consequent successful calculation of confidence and prediction regions prove the potential of this approach in terms of distribution-independent, mathematically correct empirical modeling of particle size distributions.

It is another potential application of ilr-transformations in the context of waste management, besides the application on waste compositions, in terms of sorting fractions (cf. Reference [33]). The method is a notable extension to the toolbox of chemical and process engineers in general, besides familiar approaches, like equivalent diameters or certain analytical distributions (see Section 1.1).

The discussed restriction of the method concerning zero-values is not an issue for the present data. When data include zeros, the potential impact of zero-handling techniques must be considered. These include amalgamations of compositional parts or zero-replacement (cf. Reference [42]).

Furthermore, the impact of the relative scale of ilr-transformed data must be kept in mind. Concerning the aim of the present work, it cannot be ignored, since the absolute values of the PCSs' shares result in absolute waste masses, which are of interest. But the effect of the relative scale is low when the shares of all parts are of similar orders of magnitude, as is the case in this work (see Equation (23)). For other cases, calculations of weighted variances (cf. Reference [42]), for example, can be used to counteract the impact of the relative scale, if necessary. The relative scale can also be beneficial in other cases (as explained by Pawlowsky-Glahn et al. [27]), for example, where the share of trace elements in a chemical composition has a high impact.

The discussed issue of interpreting ilrs, is solved by graphical representation in this work. While this is a straightforward approach for three or fewer fractions, the question of how to represent more fractions arises. Graphic solutions include sets of ternary diagrams of two specific fractions and amalgamations of the others (cf. Reference [40]) and area plots (for the expected values, but not for confidence and prediction regions). Some other approaches include: The application of easier interpretable log-ratios (e.g., additive log-ratios or log-ratios incorporating amalgamations, cf. Reference [42]) at the cost of the exact representation of the variance structure in the data, or modeling non-transformed data, while incorporating potential issues with the restrictions of the simplex, particularly for calculating confidence and prediction regions.

Finally, the results in Figure 4 confirm the decision of Khodier et al. [24] to perform a Design of Experiments-based investigation, with multiple runs: the 95% prediction regions overlap even for the F and V unit. Hence, conclusions that contradict the findings in this work could be drawn when comparing only single runs of one hour.

3.3. Discussion of the Modeling Results

For the gap width and shaft rotation speed, no significant impacts on the produced PSDs were identified. In conclusion, either no such effects exist, or they are too small, compared to the residual variance in the data, to be detected based on the data at hand (resulting in so-called type II or β error). According to Biemann [51], no matter how small, any effect becomes significant if the amount of data is big enough. Hence, the order of magnitude of potential, non-identified effects is of interest. For the chosen limit p-value of 0.1 in the MANOVA, linear effects are preserved in the model if the 90% confidence regions of their extreme settings (which get smaller with more data) do not overlap. And potential effects are likely not to exceed the maximum distance of the borders of these regions.

Consequently, the confidence regions for the minimum and maximum settings of w and s at factor setting 0 for the other variables were plotted (see Figure 5), based on a linear model (see Equation (24)), to give a visual impression of potential type II errors. The residuals of these models also follow a multivariate normal distribution. Figure 5 shows, that w influences the allocation to PSCs of 0 to about 13% of the product material, and s influences 0 to about 9%, for average settings of the other factors.

$$\hat{\vec{y}} = \mathrm{ilr}^{-1}\left[\begin{pmatrix} K^{(1)}_{0000} \\ K^{(2)}_{0000} \end{pmatrix} + \begin{pmatrix} K^{(1)}_{1000} \\ K^{(2)}_{1000} \end{pmatrix} w + \begin{pmatrix} K^{(1)}_{0100} \\ K^{(2)}_{0100} \end{pmatrix} s + \begin{pmatrix} K^{(1)}_{0010} \\ K^{(2)}_{0010} \end{pmatrix} c_1 + \begin{pmatrix} K^{(1)}_{0001} \\ K^{(2)}_{0001} \end{pmatrix} c_2\right] \quad (24)$$

Differences in the PSD, caused by the shaft rotation speed, would most likely be based on differences in the breakage of brittle materials in the waste, caused by the impact speed of the shaft's teeth. Since the used machines are slow running shredders, the non-existence of such effects appears feasible. Furthermore, since potential impacts of this kind are most likely relatively small, the low share of brittle materials in mixed commercial waste (cf. Reference [7]) complicates their identification.

Concerning the gap width, considering the geometry of the cutting tools, a slight increase of the coarse fraction (>80 mm) would be reasonable due to falling through of uncomminuted particles between the teeth of the counter comb. But the non-significance of potential effects makes sense, considering that the breakage situation, according to Feyerer [52], does not change with the gap width for the F and XXF geometry and only slightly for the V geometry.

The influence of the cutting tool geometry is highly significant, with a factor p-value $<10^{-5}$. As Figure 4 shows, it is largest, comparing the PSDs produced by the F and the V geometry. But the F and XXF geometries also differ significantly on a 95% confidence level. The results confirm expectations, considering the geometries (cf. Figure A1 and Table A1 in the Appendix A): The smaller axial gap between the counter comb teeth of the XXF geometry leads to a finer product, compared to the F unit. To be more precise, the share of the coarse fraction decreases in favor of the two other fractions.

Concerning the V geometry, the gap between the counter comb teeth is smaller than the gap of the F geometry. It is larger than the XXF geometry's gap close to the shaft but gets much smaller with increasing distance. This smaller gap, combined with a comb system (cf. Reference [24]), leads to the finest product among the examined geometries—again, mainly in terms of an even lower share of coarse material compared to the XXF geometry.

Relating the results to the findings of Khodier et al. [24], the choice of a cutting tool geometry depends on the requirements of the process: The V tool produces the finest material of the three but at the cost of higher energy demand and smaller but steadier throughput.

Concerning the gap width and the shaft rotation speed, the standard operation with minimum gap widths and maximum shaft rotation speeds must be questioned: increasing the gap width is beneficial for the throughput and energy demand and hardly affects the throughput steadiness, according to Reference [24]. The shares of the PSCs chosen in this work (typical for SRF production) are not or only a little affected, with a maximum-likelihood influence on only about 3% of the material.

For the shaft rotation speed, the maximum likelihood influence of this non-significant factor on the PSD only concerns about 2% of the material, while the mass flow and energy demand show an optimum at about 84% and 80% of the maximum shaft rotation speed, respectively.

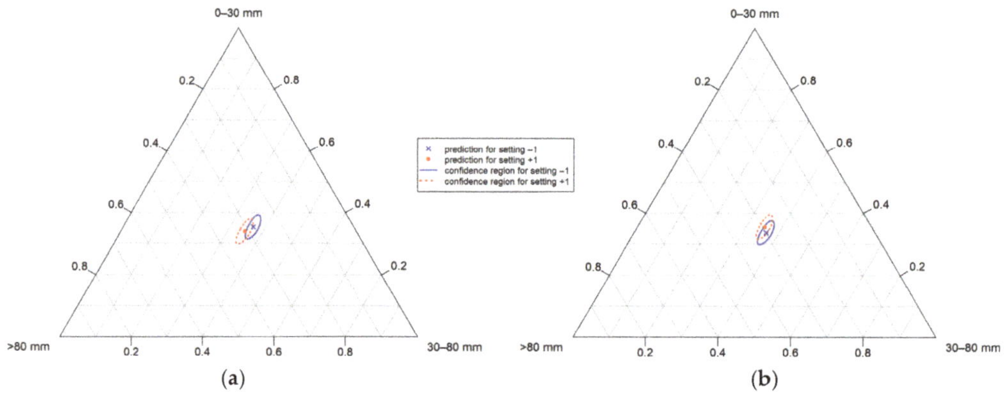

Figure 5. Ninety percent confidence regions for the minimum (−1) and maximum (+1) settings of w (**a**) and s (**b**), at factor setting 0 for the corresponding other factors of w, s, c_1 and c_2, based on the linear model from Equation (24).

4. Conclusions

Multivariate multiple linear regression modeling was applied in this work on ilr-transformed PSC data from a Design of Experiments-based 32 runs coarse-shredding experiment with mixed commercial waste. A significant model, with an R^2 of 0.57 was found, identifying the cutting tool geometry as a highly significant influence on the PSD.

The gap width and shaft rotation speed were found not to be significant, with maximum-likelihood influences on 3% and 2% on the material, respectively. If the discussed potential type II errors are rated as economically relevant or other particle size classes are of interest, further data should be generated and analyzed. Otherwise, the PSD can be treated as invariant to these factors when optimizing the throughput and energy demand. Consequently, based on the new insights from this work, a much more efficient operation of mechanical waste processing plants can be reached. The influence of the cutting tool, on the contrary, is highly significant. Its choice depends on process requirements.

What was not investigated are selective influences on specific material fractions, e.g., metals or wood. The investigation of the PSDs of such fractions may give more detailed insights and lead to different conclusions on process parametrization. It is subject to further research, which can make use of the presented modeling methods.

Using the presented method requires an in-depth understanding of the implications of applying the ilr transformation. It is essential to avoid wrong interpretations, caused by the introduced relative scale or by zero-replacement practices, where necessary. Nonetheless, establishing such understanding is rewarding: in conclusion, the method has proven to be suitable for the distribution-independent modeling of PSDs. Hence, it is a valuable addition to the toolkit of engineers, dealing with particulate materials.

Supplementary Materials: The following is available online at https://www.mdpi.com/2227-9717/9/3/414/s1, Supplementary Material: an HTML print of the jupyter notebook which contains the used code.

Author Contributions: Conceptualization, K.K.; methodology, K.K.; software, K.K.; formal analysis, K.K.; investigation, K.K.; data curation, K.K.; writing—original draft preparation, K.K.; writing—review and editing, K.K. and R.S.; visualization, K.K.; supervision, R.S.; project administration, K.K. and R.S.; funding acquisition, R.S. All authors have read and agreed to the published version of the manuscript.

Funding: The Center of Competence for Recycling and Recovery of Waste 4.0 (acronym ReWaste4.0) (contract number 860 884) under the scope of the COMET—Competence Centers for Excellent Technologies—financially supported by BMK, BMDW, and the federal state of Styria, managed by the FFG.

Institutional Review Board Statement: Not applicable.

Informed Consent Statement: Not applicable.

Data Availability Statement: The data is contained within the article and Supplementary Materials.

Conflicts of Interest: The authors declare no conflict of interest.

Abbreviations

$a_{i,r}$	scaling factor for the rth compositional part and the ith ilr coordinate
ANOVA	analysis of variance
A_t	total surface area of all particles
c	cutting tool geometry
c_1, c_2	coded representation of the cutting tool geometry
d	particle size
$\overline{d}$	arithmetic average particle size
d'	characteristic particle size of the RRSB distribution
d_b	b^{th} percentile particle size
d_i	size of the ith particle
d_{max}	maximum particle size
d_S	Sauter diameter
D	number of particle size classes
DEM	discrete element method
DMFMS	digital material flow monitoring system
GGS	Gates-Gaudin-Schuhmann
ilr	isometric log-ratios
ilr_i	i^{th} ilr dimension
j	factor exponent
k	factor exponent

$K^{(r)}_{jkmn}$	model constant for the factor or interaction $w^j s^k c_1^m c_2^n$ and the response r
m	factor exponent
m_u	uniformity parameter of the GGS distribution
MANOVA	multivariate analysis of variance
n	factor exponent
n_u	uniformity parameter of the RRSB distribution
N	number of particles
p	empirical significance
P	number of observations
PDF	probability density function
$PRESS$	prediction residual sum of squares
PSC	particle size class
PSD	particle size distribution
$q(d)$	frequency density for particles of size d
Q	number of independent variables
R	number of dimensions of the dependent variable
R^2	coefficient of determination
R^2_{adj}	adjusted coefficient of determination
R^2_{pred}	prediction coefficient of determination
$\boldsymbol{R}^{(D)}$	D-dimensional real space
$\boldsymbol{R}_0^{+(D)}$	D-dimensional positive real space, including 0
RRSB	Rosin-Rammler-Sperling-Bennet
s	shaft rotation speed
S	sample standard deviation
$\boldsymbol{S}^{(D)}$	D-dimensional simplex
SRF	solid recovered fuel
SS_{tot}	total sum of squares
t	number of parts in group +1
TOS	Theory of Sampling
v	number of parts in group -1
V_t	total volume of all particles
w	radial gap width
W	width of a distribution
$x_{p,q}$	p^{th} setting of the qth independent variable
$\boldsymbol{X}$	matrix of settings of the independent variables
$\vec{y}$	compositional vector
$\vec{y}^*$	ilr-transformed compositional vector
$\hat{\vec{y}}$	model prediction of the shares of the particle size classes
y_i	i^{th} element of $\vec{y}$
$y_{p,r}$	p^{th} observation of the rth dimension of the dependent variable
$y^*_{p,r}$	p^{th} observation of the rth dimension of the ilr-transformed dependent variable
$\hat{y}_{p,r}$	model prediction for $y_{p,r}$
$\hat{y}^{(r)}$	model prediction of the response r
$\overline{y^*_r}$	arithmetic mean of the rth ilr coordinate
$\boldsymbol{Y}$	matrix of the dependent variable
$\boldsymbol{Y}^*$	matrix of the ilr-transformed dependent variable
$\hat{\boldsymbol{Y}}$	matrix of model predictions of the dependent variable
$\boldsymbol{\beta}$	matrix of the regression coefficients
$\hat{\boldsymbol{\beta}}$	matrix of least squares estimates of the regression coefficients
$\beta_{q,r}$	q^{th} regression coefficient for the rth dimension of the dependent variable
$\hat{\beta}_{q,r}$	least squares estimate of $\beta_{q,r}$
Δ_A	Aitchison distance
Δ_E	Euclidian distance
$\boldsymbol{\varepsilon}$	matrix of the model residuals
$\boldsymbol{\varepsilon}^*$	matrix of the ilr-transformed residuals
$\varepsilon_{p,r}$	model residual corresponding to $y_{p,r}$
μ	arithmetic average of a population
σ	standard deviation of a population

Appendix A

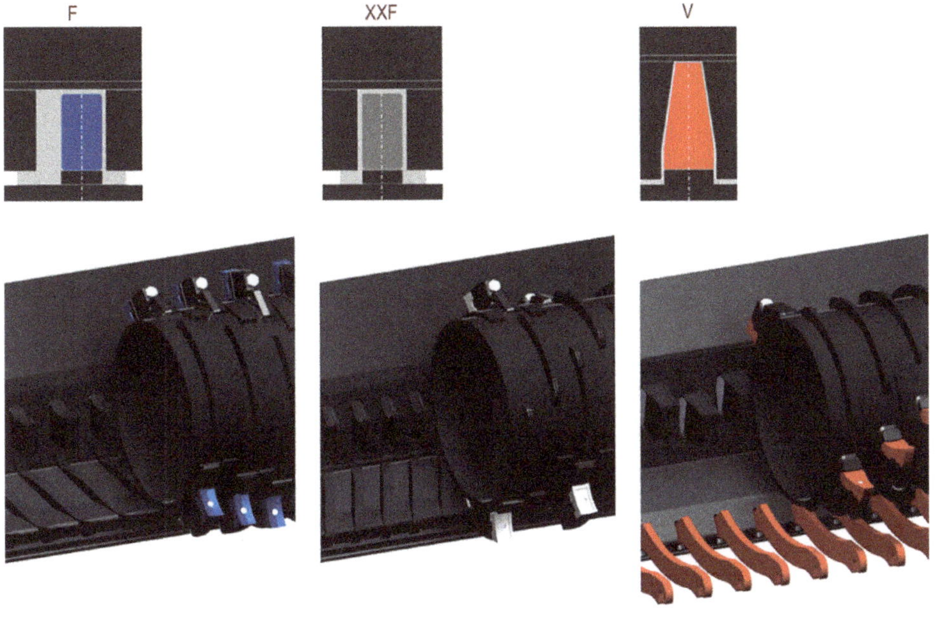

Figure A1. Cutting tool geometries [24].

Table A1. Technical data of the cutting tool geometries [24].

Type	F	XXF	V
number of cutting teeth (shaft) [pcs.]	32	22	32
position of cutting teeth (shaft) [-]	double helix	chevron	chevron
width of cutting teeth (shaft) [mm]	70	70	42/85 *
height of cutting teeth (shaft) [mm]	124	124	183
width of cutting teeth (counter comb) [mm]	64	54	81/100 *
height of cutting teeth (counter comb) [mm]	142	136	202
cutting circle [mm]	1070	1070	1170
length of shredding-shaft [mm]		3000	
right side cutting gap (axial) [mm]	3.5	2	3
left side cutting gap (axial) [mm]	39	2	3
minimum cutting gap (radial) [mm]		0	
maximum cutting gap (radial) [mm]	33	35	30/38 *
comb-system [-]	no	no	yes

* bottom/top of the teeth.

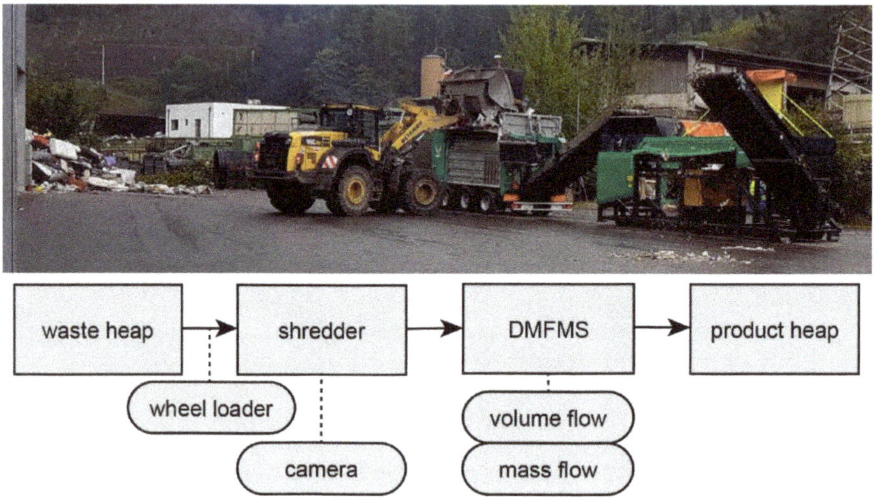

Figure A2. Experimental setup: photo and flow chart [24].

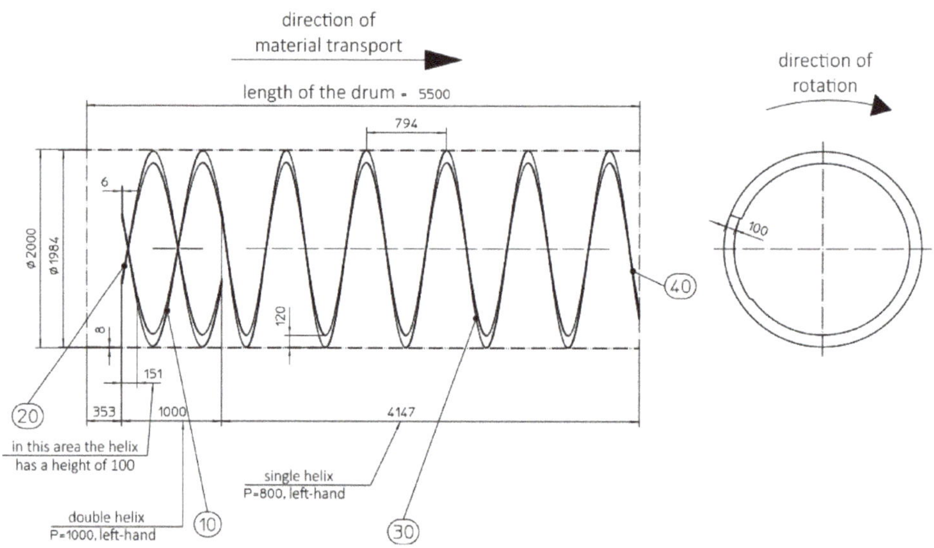

Figure A3. Screening drum geometry.

Table A2. Data of the screening drums.

side length of the square-shaped holes (mm)	80	60	40	20	10
total hole area (m^2)	16, 61	17, 06	17, 14	17, 96	14, 55

References

1. Sarc, R.; Lorber, K.E.; Pomberger, R. Manufacturing of Solid Recovered Fuels (SRF) for Energy Recovery Processes. In *Waste Management*; Thomé-Kozmiensky, K.J., Thiel, S., Eds.; TK Verlag Karl Thomé-Kozmiensky: Neuruppin, Germany, 2016; pp. 401–416.
2. Zhang, Y.; Banks, C.J. Impact of different particle size distributions on anaerobic digestion of the organic fraction of municipal solid waste. *Waste Manag.* **2013**, *33*, 297–307. [CrossRef]

3. Luo, S.; Xiao, B.; Hu, Z.; Liu, S.; Guan, Y.; Cai, L. Influence of particle size on pyrolysis and gasification performance of municipal solid waste in a fixed bed reactor. *Bioresour. Technol.* **2010**, *101*, 6517–6520. [CrossRef] [PubMed]
4. Sarc, R.; Curtis, A.; Kandlbauer, L.; Khodier, K.; Lorber, K.E.; Pomberger, R. Digitalisation and intelligent robotics in value chain of circular economy oriented waste management—A review. *Waste Manag.* **2019**, *95*, 476–492. [CrossRef]
5. Sarc, R.; Seidler, I.M.; Kandlbauer, L.; Lorber, K.E.; Pomberger, R. Design, quality and quality assurance of solid recovered fuels for the substitution of fossil feedstock in the cement industry—Update 2019. *Waste Manag. Res.* **2019**, *37*, 885–897. [CrossRef]
6. Leschonski, K. Windsichten [wind sifting]. In *Handbuch der Mechanischen Verfahrenstechnik [Handbook of Mechanical Process Engineering]*; Schubert, H., Ed.; John Wiley & Sons: Hoboken, NJ, USA, 2012; pp. 584–612. ISBN 3-527-30577-7.
7. Khodier, K.; Viczek, S.A.; Curtis, A.; Aldrian, A.; O'Leary, P.; Lehner, M.; Sarc, R. Sampling and analysis of coarsely shredded mixed commercial waste. Part I: Procedure, particle size and sorting analysis. *Int. J. Environ. Sci. Technol.* **2020**, *17*, 959–972. [CrossRef]
8. Möllnitz, S.; Khodier, K.; Pomberger, R.; Sarc, R. Grain size dependent distribution of different plastic types in coarse shredded mixed commercial and municipal waste. *Waste Manag.* **2020**, *103*, 388–398. [CrossRef]
9. Feil, A.; Coskun, E.; Bosling, M.; Kaufeld, S.; Pretz, T. Improvement of the recycling of plastics in lightweight packaging treatment plants by a process control concept. *Waste Manag. Res.* **2019**, *37*, 120–126. [CrossRef] [PubMed]
10. Gundupalli, S.P.; Hait, S.; Thakur, A. A review on automated sorting of source-separated municipal solid waste for recycling. *Waste Manag.* **2017**, *60*, 56–74. [CrossRef] [PubMed]
11. Curtis, A.; Küppers, B.; Möllnitz, S.; Khodier, K.; Sarc, R. Digital material flow monitoring in waste processing—The relevance of material and throughput fluctuations. *Waste Manag.* **2021**, *120*, 687–697. [CrossRef] [PubMed]
12. Möllnitz, S.; Küppers, B.; Curtis, A.; Khodier, K.; Sarc, R. Influence of pre-screening on down-stream processing for the production of plastic enriched fractions for recycling from mixed commercial and municipal waste. *Waste Manag.* **2021**, *119*. [CrossRef]
13. Müller, W.; Bockreis, A. Mechanical-Biological Waste Treatment and Utilization of Solid Recovered Fuels—State of the Art. In *Waste Management*; Thomé-Kozmiensky, K.J., Thiel, S., Eds.; TK Verlag Karl Thomé-Kozmiensky: Neuruppin, Germany, 2015; pp. 321–338.
14. Polke, R.; Schäfer, M.; Scholz, N. Charakterisierung disperser Systeme [Characterization of disperse systems]. In *Handbuch der Mechanischen Verfahrenstechnik [Handbook of Mechanical Process Engineering]*; Schubert, H., Ed.; John Wiley & Sons: Hoboken, NJ, USA, 2012; pp. 7–129. ISBN 3-527-30577-7.
15. Wasserman, L. *All of Statistics: A Concise Course in Statistical Inference*; Springer: New York, NY, USA, 2013; ISBN 978-0-387-21736-9.
16. Micula, G.; Micula, S. *Handbook of Splines*; Springer Netherlands: Dordrecht, The Netherlands, 1999; ISBN 978-94-010-6244-2.
17. Scott, D.W. *Multivariate Density Estimation*; Wiley: New York, NY, USA, 1992; ISBN 9780471547709.
18. Heumann, C.; Michael Schomaker, S. *Introduction to Statistics and Data Analysis: With Exercises, Solutions and Applications in R*; Springer: Berlin/Heidelberg, Germany, 2017; ISBN 978-3319461601.
19. German Institute for Standardization. *DIN 66143:1974-03, Darstellung von Korn-(Teilchen-)Größenverteilungen; Potenznetz [Graphical Representation of Particle Size Distributions; Power-Function Grid]*; Beuth Verlag GmbH: Berlin, Germany, 1974.
20. German Institute for Standardization. *DIN 66145:1976-04, Darstellung von Korn-(Teilchen-)Größenverteilungen; RRSB-Netz [Graphical Representation of Particle Size Distributions; RRSB-Grid]*; Beuth Verlag GmbH: Berlin, Germany, 1976.
21. Sinnott, M.D.; Cleary, P.W. Simulation of particle flows and breakage in crushers using DEM: Part 2—Impact crushers. *Miner. Eng.* **2015**, *74*, 163–177. [CrossRef]
22. Lee, H.; Kwon, J.H.; Kim, K.H.; Cho, H.C. Application of DEM model to breakage and liberation behaviour of recycled aggregates from impact-breakage of concrete waste. *Miner. Eng.* **2008**, *21*, 761–765. [CrossRef]
23. Dong, K.; Esfandiary, A.H.; Yu, A.B. Discrete particle simulation of particle flow and separation on a vibrating screen: Effect of aperture shape. *Powder Technol.* **2017**, *314*, 195–202. [CrossRef]
24. Khodier, K.; Feyerer, C.; Möllnitz, S.; Curtis, A.; Sarc, R. Efficient derivation of significant results from mechanical processing experiments with mixed solid waste: Coarse-shredding of commercial waste. *Waste Manag.* **2021**, *121*, 164–174. [CrossRef]
25. Johnson, R.A.; Wichern, D.W. *Applied Multivariate Statistical Analysis*, 6th ed.; Pearson/Prentice Hall: Upper Saddle River, NJ, USA, 2007; ISBN 978-0-13-187715-3.
26. Hand, D.J.; Taylor, C.C. *Multivariate Analysis of Variance and Repeated Measures*; Springer Netherlands: Dordrecht, The Netherlands, 1987; ISBN 978-94-010-7913-6.
27. Pawlowsky-Glahn, V.; Egozcue, J.J.; Tolosana-Delgado, R. *Modeling and Analysis of Compositional Data*; John Wiley & Sons Inc.: Chichester, UK, 2015; ISBN 9781118443064.
28. Pomberger, R.; Sarc, R.; Lorber, K.E. Dynamic visualisation of municipal waste management performance in the EU using Ternary Diagram method. *Waste Manag.* **2017**, *61*, 558–571. [CrossRef]
29. Khodier, K.; Lehner, M.; Sarc, R. Multilinear modeling of particle size distributions. In *Proceedings of the 8th International Workshop on Compositional Data Analysis (CoDaWork2019): Terrassa, 3–8 June 2019*; Ortego, M.I., Ed.; Universitat Politècnica de Catalunya-BarcelonaTECH: Barcelona, Spain, 2019; pp. 82–85. ISBN 978-84-947240-1-5.
30. Aitchison, J. The Statistical Analysis of Compositional Data. *J. R. Stat. Soc. B* **1982**, *44*, 139–177. [CrossRef]
31. Egozcue, J.J. Isometric Logratio Tranformations for Compositional Data Analysis. *Math. Geol.* **2003**, *35*, 279–300. [CrossRef]
32. Weise, D.R.; Palarea-Albaladejo, J.; Johnson, T.J.; Jung, H. Analyzing Wildland Fire Smoke Emissions Data Using Compositional Data Techniques. *J. Geophys. Res. Atmos.* **2020**, *125*, 139. [CrossRef]

33. Edjabou, M.E.; Martín-Fernández, J.A.; Scheutz, C.; Astrup, T.F. Statistical analysis of solid waste composition data: Arithmetic mean, standard deviation and correlation coefficients. *Waste Manag.* **2017**, *69*, 13–23. [CrossRef]
34. Chambers, J.M.; Hastie, T.J. Statistical Models. In *Statistical Models in S*; Chambers, J.M., Hastie, T.J., Eds.; Chapman & Hall: London, UK, 1993; pp. 13–44. ISBN 978-0-534-16765-3.
35. Siebertz, K.; van Bebber, D.; Hochkirchen, T. *Statistische Versuchsplanung [Design of Experiments]*; Springer: Berlin/Heidelberg, Germany, 2010; ISBN 978-3-642-05492-1.
36. Stat-Ease Inc. Optimality Criteria. Available online: https://www.statease.com/docs/v11/contents/advanced-topics/optimality-criteria/ (accessed on 26 May 2020).
37. Dean, A.; Voss, D.; Draguljić, D. *Design and Analysis of Experiments*, 2nd ed.; Springer: Cham, Switzerland, 2017; ISBN 978-3-319-52250-0.
38. Danish Standards Foundation. *DS 3077 Representative Sampling—Horizontal Standard*; Danish Standards Foundation: Charlottenlund, Denmark, 2013.
39. Esbensen, K.H.; Wagner, C. Theory of sampling (TOS) versus measurement uncertainty (MU)—A call for integration. *TrAC Trends Anal. Chem.* **2014**, *57*, 93–106. [CrossRef]
40. van den Boogaart, K.G.; Tolosana-Delgado, R. *Analyzing Compositional Data with R*; Springer: Dordrecht, The Netherlands, 2013; ISBN 978-3-642-36809-7.
41. Khodier, K.; Lehner, M.; Sarc, R. Empirical modeling of compositions in chemical engineering. In *Proceedings of the 16th Minisymposium Verfahrenstechnik and 7th Partikelforum (TU Wien, Sept. 21/22, 2020)*; Jordan, C., Ed.; TU Wien: Vienna, Austria, 2020; pp. 118–121. ISBN 978-3-903337-01-5.
42. Greenacre, M. *Compositional Data Analysis in Practice*; CRC Press: Boca Raton, FL, USA, 2019; ISBN 978-1-138-31643-0.
43. van den Boogart, K.G.; Tolosana-Delgado, R. Package 'Compositions' (Version 2.0-1). Available online: https://cran.r-project.org/web/packages/compositions/compositions.pdf (accessed on 13 January 2021).
44. Stahel, W.A. Package Regr for an Augmented Regression Analysis. Available online: https://rdrr.io/rforge/regr/f/inst/doc/regr-description.pdf (accessed on 14 January 2021).
45. Olson, C.L. On choosing a test statistic in multivariate analysis of variance. *Psychol. Bull.* **1976**, *83*, 579–586. [CrossRef]
46. Pareto, A. Predictive R-Squared According to Tom Hopper. Available online: https://rpubs.com/RatherBit/102428 (accessed on 14 January 2021).
47. Braun, W.J.; MacQueen, S. Package 'MPV' (Version 1.56). Available online: https://cran.r-project.org/web/packages/MPV/MPV.pdf (accessed on 14 January 2021).
48. Wang, C.-C. A MATLAB package for multivariate normality test. *J. Stat. Comput.* **2014**, *85*, 166–188. [CrossRef]
49. Mardia, K.V. Measures of multivariate skewness and kurtosis with applications. *Biometrika* **1970**, *57*, 519–530. [CrossRef]
50. Korkmaz, S.; Goksuluk, D.; Zararsiz, G. Package 'MVN' (Version 5.8). Available online: https://cran.r-project.org/web/packages/MVN/MVN.pdf (accessed on 14 January 2021).
51. Biemann, T. Logik und Kritik des Hypothesentests [Logic and criticism of the hypothesis test]. In *Methodik der Empirischen Forschung [Methodology of Empirical Research]*; Albers, S., Klapper, D., Konradt, U., Walter, A., Wolf, J., Eds.; Springer Fachmedien: Wiesbaden, Germany, 2007; ISBN 978-3-8349-0469-0.
52. Feyerer, C. Interaktion des Belastungskollektives und der Werkzeuggeometrie Eines Langsamlaufenden Einwellenzerkleinerers [Interaction of the Load Collective and Tool Geometry of a Low-Speed Single-Shaft Shredder]. Master's Thesis, Montanuniversitaet Leoben, Leoben, Austria, 2020.

processes

MDPI

Review

The EU Training Network for Resource Recovery through Enhanced Landfill Mining—A Review

Daniel Vollprecht [1,*], Lieven Machiels [2] and Peter Tom Jones [3]

1 Chair of Waste Processing Technology and Waste Management, Montanuniversität Leoben, 8700 Leoben, Austria
2 Department of Chemistry, KU Leuven, 3001 Leuven, Belgium; lieven.machiels@kuleuven.be
3 Department of Materials Engineering, KU Leuven, 3001 Leuven, Belgium; peter.jones@kuleuven.be
* Correspondence: daniel.vollprecht@unileoben.ac.at; Tel.: +43-3842-402-5110

Abstract: The "European Union Training Network for Resource Recovery Through Enhanced Landfill Mining (NEW-MINE)" was a European research project conducted between 2016 and 2020 to investigate the exploration of and resource recovery from landfills as well as the processing of the excavated waste and the valorization of the obtained waste fractions using thermochemical processes. This project yielded more than 40 publications ranging from geophysics via mechanical process engineering to ceramics, which have not yet been discussed coherently in a review publication. This article summarizes and links the NEW-MINE publications and discusses their practical applicability in waste management systems. Within the NEW-MINE project in a first step concentrates of specific materials (e.g., metals, combustibles, inert materials) were produced which might be used as secondary raw materials. In a second step, recycled products (e.g., inorganic polymers, functional glass-ceramics) were produced from these concentrates at the lab scale. However, even if secondary raw materials or recycled products could be produced at a large scale, it remains unclear if they can compete with primary raw materials or products from primary raw materials. Given the ambitions of transition towards a more circular economy, economic incentives are required to make secondary raw materials or recycled products from enhanced landfill mining (ELFM) competitive in the market.

Keywords: enhanced landfill mining; NEW-MINE; waste treatment

Citation: Vollprecht, D.; Machiels, L.; Jones, P.T. The EU Training Network for Resource Recovery through Enhanced Landfill Mining—A Review. *Processes* **2021**, *9*, 394. https://doi.org/10.3390/pr9020394

Received: 21 December 2020
Accepted: 18 February 2021
Published: 22 February 2021

1. Introduction

Although landfill mining, "the process for extracting minerals or other solid natural resources from waste materials that have previously been disposed of by burying them in the ground" [1], has been investigated since 1953 [2], the interest in this topic only really started in the 1980s in the USA [3] and the 1990s in Europe [4]. The main reasons for Landfill Mining towards at the end of the 20th century comprised landfill remediation and landfill volume recovery [5]. Between 2017 and 2015, several national (research) projects such as the "Closing the Circle" project (Flanders/Belgium) [6], the LAMIS project (Austria) [7–10], the MINERVE project (Wallonia/Belgium) [11] and the TönsLM project (Germany) [12] were conducted, which focused more on material and energy recovery. These projects covered different process steps ranging from geophysical exploration via exploration to dry and wet mechanical processing.

In summary, these studies demonstrated that using state-of-the-art technologies can yield concentrates of metals, inorganic-nonmetallic materials and combustibles. The metal concentrates can already be sold at the market, since standard pyrometallurgical recycling routes exist, which makes this fraction a key economic driver for landfill mining [13]. However, the inorganic-nonmetallic and combustible fractions are more challenging, while the remaining fine fraction represents the final sink for contaminants [14] that have to be removed from the circular economy [15]. In detail, recycling of inorganic-nonmetallic materials even from fresh construction and demolition waste is mainly as aggregate [16], which is a low-value application that is, furthermore, associated with a quality decrease due to the negative impact on the attached cement paste on concrete properties [17]. Similarly, energy recovery from calorific fractions of the waste may either occur by classical waste incineration processes [18] or after mechanical processing to refuse derived fuel (RDF) in co-incineration plants [19]. Nevertheless, even in the latter case the waste supplier has to pay for the material instead of obtaining revenues. Finally, regarding the fine fractions it has to be mentioned that 78% of excavated soils, most of them not contaminated, are landfilled in Austria [20] due to low landfilling costs and low material values. Therefore, it seems unlikely that contaminated soil-like materials from landfills would be recycled.

For these reasons the question arose if innovative technology might be an option to increase the economic feasibility of landfill mining projects. Consequently, in 2013 the term "enhanced landfill mining (ELFM)" was defined as "the safe conditioning, excavation and integrated valorization of (historic and/or future) landfilled waste streams as both materials (waste-to-material, WtM) and energy (waste-to-energy, WtE), using innovative transformation technologies and respecting the most stringent social and ecological criteria" [6]. To investigate the potential of such "innovative transformation technologies" and to cluster the existing European expertise in the field of landfill mining, the European Enhanced Landfill Mining Consortium (EURELCO) was founded in 2014 [21]. From this consortium the "EU Training Network for Resource Recovery through Enhanced Landfill Mining (NEW-MINE)" was developed and received funding from the European Union in 2016. For a period of four years, 15 early stage researchers (ESRs) investigated the exploration of and resource recovery from landfills as well as the processing of the excavated waste and the valorization of the obtained waste fractions using thermochemical processes (Figure 1).

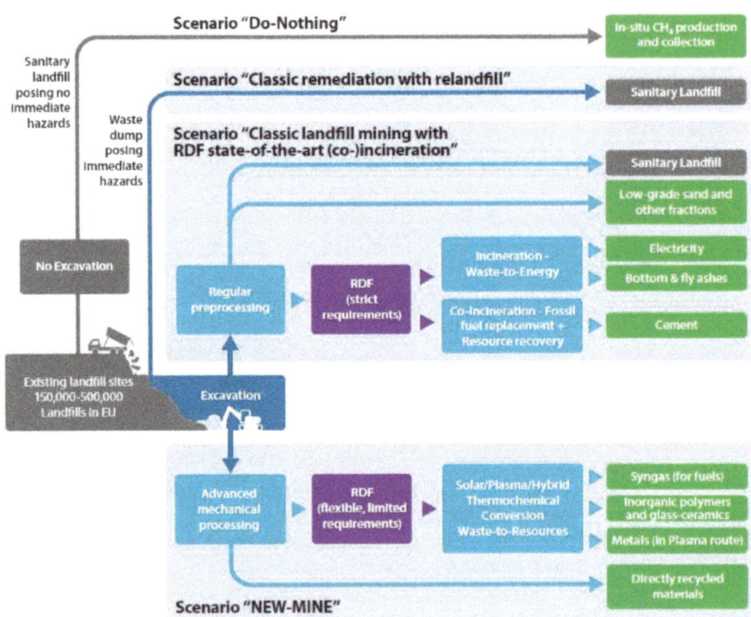

Figure 1. Approach of the NEW-MINE project (www.new-mine.eu accessed on 16 November 2020).

The NEW-MINE project yielded a significant output of scientific publications, ranging from geophysics via mechanical process engineering to ceramics. Although there is already a review article that links the first NEW-MINE results with previous investigations [22], there is no comprehensive review assessing the entire research output of NEW-MINE with respect to its practical applicability in waste management systems, which is the aim of this review article.

This article is structured in the following way: Firstly, the methods for literature review and discussion of the project results are given. Secondly, the project results are summarized referring to the individual publications. In this context also the state-of-the-art before the project and the publications referring to the NEW-MINE publications are shortly summarized. Thirdly, in the Discussion section, the practical applicability of the NEW-MINE results in waste management systems is discussed. Finally, a summary of the project results and the challenges regarding their practical application is given.

2. Materials and Methods

Based on the publications listed on the NEW-MINE website (www.new-mine.eu accessed on 16 November 2020), additional publications of the authors were identified using Google Scholar and searching for the names of the NEW-MINE researchers. In a second step papers citing the NEW-MINE publications were identified and set into the context of NEW-MINE. The focus was on peer-reviewed publications, but certain conference proceedings were selected in those cases when the contained information was not found in a peer-reviewed article. Additionally, the search term "NEW-MINE" was used to find publications referring to this project. In the Google Scholar searches between dozens and few hundred articles appeared which were checked for consistency to the review topic. Publications by authors with similar names who were obviously other persons than the NEW-MINE researchers were excluded as well as publications referring to "new mines", but not to the NEW-MINE project. The entire literature research was conducted in November and December 2020. Finally, the results of the NEW-MINE project were checked for practical applicability based on the experience of the authors in waste management and on personal communications with colleagues over the last years. In this context, some theoretical reflections on resource classification and varying terminology in interdisciplinary research were added.

3. Results

3.1. Innovative Landfill Exploration & Mechanical Processing

3.1.1. Landfill Exploration

Landfills have been investigated long before the NEW-MINE project with respect to environmental problems arising from the landfilled waste [23–25]. With respect to the recovery of resources from landfills, the task for geophysical exploration is different: The landfill as a potential anthropogenic deposit shall be explored with respect to the quality, quantity and "bonitaet" (=geological factors influencing the mining) [26] of the contained secondary raw materials [27]. In contrast to pollutant-related exploration tasks, the resource-related exploration task aims not for the identification of individual objects or leakages but for the estimation of the average material composition. Before the onset of the NEW-MINE project only few studies focused on these kinds of exploration, e.g., for a distinction between landfilled foundry sands and iron-rich materials using magnetic methods [28]. Parallel to the NEW-MINE project, but not referring to it, another research group distinguished between metal enrichments and plastic enrichments in a landfill using a combination of electromagnetic and magnetic methods [29].

The publications within the NEW-MINE project include both applied studies focusing on the exploration of landfills with respect to the material composition and fundamental studies focusing on data processing of geophysical sensors.

The first applied study [30] investigated the suitability of electromagnetic induction measurements and ground penetrating radar to characterize the geometry and electric properties of waste layers. It was appreciated by scientists of two other EU project on ELFM, the SMARTGROUND project who cited it with respect to electromagnetic measurements in their study on ELFM in the UK [31], and the RAWFILL project who cited it with respect to geophysical exploration in general [32]. The second applied study [33] used magnetic total-field measurements to estimate the bulk magnetic susceptibility of the Hollabrunn landfill, Austria, by inverse modelling and validated the resulting susceptibility by manual sorting of drill-core samples. In parallel, the magnetic susceptibility of reference materials was determined by laboratory analyses and the resulting bulk susceptibility of the landfilled waste mixture was computed using a weighted mean. The differences between the bulk susceptibility derived from inverse modelling of the field data (0.06 to 0.11 SI) and the values obtained from computing from the values from the reference materials (0.01 to 0.05 SI) highlight the challenges to determine the iron content in landfills from magnetic data. This publication has not yet been cited, although two publications released in 2020 have addressed magnetic exploration in the context of landfill mining [34,35].

Consistent with the overall character of the NEW-MINE project and the requirements of the Marie Skłodowska Curie Actions, the share of fundamental research was dominant which is also reflected in the higher number of publications regarding the basic aspects of geophysical exploration. For example, a probabilistic inversion of electromagnetic data using the so-called Kalman ensemble generator, was introduced [36]. This approach has already been already cited twice by other authors [37,38] and was further extended to a joint inversion of direct current resistivity and small-loop electromagnetic data [39], which was published in a mathematical journal, *Algorithms*, highlighting the interdisciplinarity of the NEW-MINE project. Finally, offset parameters were incorporated into the probabilistic inversion framework to estimate systematic errors in electromagnetic measurements [40].

In summary, the role of geophysics in post-NEW-MINE projects (e.g., RAWFILL) is higher than in pre-NEW-MINE projects (e.g., TönsLM), and post-NEW-MINE projects refer to the results of NEW-MINE.

3.1.2. Mechanical Processing

Mechanical processing of landfilled waste was the main focus in pre-NEW-MINE projects with dry methods being investigated in the LAMIS project [7] and wet methods in the TönsLM project [12].

Based on a comprehensive study on state-of-the-art processing of excavated waste from the Halbenrain landfill, Austria, using an existing mechanical-biological treatment (MBT) plant [41], a novel processing concept using a ballistic separator as first processing step for untreated excavated landfilled waste was developed and tested at the Mont-Saint-Guibert landfill, Belgium [42]. The NEW-MINE approach to use state-of-the-art MBT was addressed outside NEW-MINE in the context of the problematic character of the fine fraction [43], its biological treatment to increase the value of the fine fraction [44], and regarding landfill mining in general [45].

A second key innovation in this field reached by NEW-MINE was the demonstration of the feasibility of near-infrared (NIR) sorting for the beneficiation of heavy fractions from dry-mechanical processing [46]. The authors could enrich the inert materials in these fractions from values between 85.6 to 98.8 wt% in the input to a range between 97.7 and 99.6 wt% in the output, by ejecting the plastics: this represents an important step towards a possible recycling of the inert fraction. Another finding, i.e., the impact of attached defilements on sorting performance, was subsequently considered in Austria's largest waste management project, ReWaste 4.0 [47]. The applicability of results from mechanical processing of ELFM materials to "fresh" waste highlights the relevance of NEW-MINE results in this field. Within NEW-MINE, the influence of surface roughness and surface moisture of plastics on NIR sorting was investigated [48], and the findings were used outside NEW-MINE for material flow characterization in sensor-based sorting systems using an instrumented particle [49] and for a study of the influence of material alterations and machine impairment [50]. NEW-MINE learnings on the effects of throughput rate and input composition on sensor-based sorting [51] were subsequently used within the above mentioned ReWaste 4.0 project [47,52].

Finally, the challenge of the fine fraction, which has been identified in pre-NEW-MINE projects, e.g., in the LAMIS project [53], was firstly summarized in two review publications [54,55]. A post-NEW-MINE study on contaminants in landfills [56], despite using a misleading soil definition [57], added further information on this topic. Other publications [44,58] also addressed the observation that the fine fraction makes up the majority of the landfilled material, but used correct terminology. Further post-NEW-MINE studies on leachate characteristics [59] and composition of the fine fraction [60] added further information to these fields. The need to further process the fine fraction to minimize re-landfilling was mentioned by a publication on geophysical exploration of landfills [61]. The problem stated within NEW-MINE that currently no valorization options for the fine fraction exist because of the possibility of contamination of this fraction [43] was later considered by an Estonian [62] and by an Indian study [63]. The same statement was also extracted from the second review [55] but with the suggestion to tackle this problem by further processing [44]. In contrast, a Hungarian group extracted the contradicting statement that fine fractions can be considered as a relevant source of metals, calorific fractions, inert fractions and soil-like material recovery [64]. The explanation that the amount of the fine fraction increases over time due to humification processes was mentioned in an Indian study on the application of Fourier-transformed infrared spectroscopy for its characterization [65]. The need for a deeper understanding of the physico-chemical properties of the fine fractions including their distribution across the different grain size fractions was also applied to shredder fine fractions [66]. Within the field of landfill mining, the general observations regarding the fine fractions were considered by authors of the SMARTGROUND project [67].

Then, the fine fraction < 90 mm obtained by ballistic separation was characterized [68] and further separated into a light fraction, a heavy fraction, ferrous and non-ferrous metals and a fine fraction < 4.5 mm using windsifting, magnetic separation, eddy-current separation and screens [69]. The heavy fraction was investigated with respect to the utilization as construction material and the light fraction with respect to the use as refuse derived fuel (RDF) [70]. The fine fraction < 4.5 mm was further processed and characterized with respect to the relationship between mineralogy and leachability indicating the stable mineralogical bonding of many contaminants (e.g., Pb as metal and Pb-Ca phosphate and Zn as Fe-Zn alloy, ZnS and $ZnSO_4$) [71].

In summary, novel research results obtained within NEW-MINE, especially using sensor-based sorting, can be applied to areas beyond ELFM.

3.2. *Thermochemical Conversion and Solar Energy Storage*

3.2.1. Thermochemical Conversion

Most pre-NEW-MINE landfill mining projects covered the value chain from exploration via excavation and mechanical treatment to the production of concentrates of secondary raw materials, i.e., metals for pyrometallurgical recycling, mineral wastes for the production of construction materials and combustibles for energy recovery. These concentrates are still wastes (and not products) but state-of-the-art recyling and recovery routes do exist, which is why the actual recycling or recovery was not part of most of these projects. With respect to the combustibles these recovery routes include among others:

- Municipal solid waste incineration plants (MSWI plants)
- Refuse derived fuel power plants (RDF power plants)
- Cement works (co-incineration)
- Coal-fired power plants (co-incineration)
- Industrial power plants (co-incineration) [72].

Consequently, previous projects investigated the combustibles with respect to the (co-) incineration in existing facilities considering not only calorific properties but also the content of heavy metals [7] as these restrict the valorization options of RDF by co-incineration [19]. For those cases in which the combustibles fulfil the requirements for co-incineration in the cement industry, the inorganic constituents of the RDF are incorporated into the mineral phases of the clinker contributing to its desired hydraulic properties, which justifies to address the share of RDF used on a material level as "recycling index" [73]. In contrast, for those cases in which contaminant concentrations exceed limit values for

RDF, energy recovery in MSWI plants is the state-of-the-art option. MSWI yields bottom ashes whose recycling is restricted by limit values for total concentrations and/or leachable concentrations of contaminants [74,75]. The chemical composition and leachability, as well as the element-specific limit values significantly vary between individual EU member states. Table 1 presents the ranges of concentrations relative to ranges of limit values for material used in road construction, showing that the total contents of Cd, Cu, Cr, Pb and Zn and the leachable contents of Pb are most problematic.

Table 1. Ranges of total (mg/kg) [76] and leachable (mg/L) [76,77] contents and respective limit values (CH, DK, SWE, FIN, FRA, AUT, D, NL, BEL for total and DK, ESP, SWE, FIN, FRA; AUT, GER; NED, BEL for leachable contents) for recycling of MSWI bottom ashes.

Total Content	Concentrations	Limit Values
Cd	1–177	0.2–10
Cu	738–17620	40–500
Cr	115–852	40–500
Ni	38–850	35–500
Pb	197–6441	20–1250
Zn	1142–9370	120–1250
Leachable Content	**Concentrations**	**Limit Values**
Cl	259–416	80–5000
Cr	0.01	0.006–0.5
Ni	<0.05	0.01–0.35
Sb	0.016–0.023	0.006–0.2
SO_4	15–106	70–6500
Pb	1.8–6	0.02–0.82

Following the pathway of classic incineration for those calorific fractions exceeding the threshold values for co-incineration would, therefore, result in the need to re-landfill significant proportions of the resulting bottom ashes. Disposal of MSWI bottom ash at increasing scarcity of landfill volume and the need for landfill aftercare for up to more than 100 years [78] is associated with significant costs, which will represent a massive financial and environmental burden for governments and the public in the future.

This is why within NEW-MINE, despite sharp criticism [79], gasification and pyrolysis were investigated as alternative thermochemical conversion processes within NEW-MINE with the background idea to use the produced syngas and to couple these processes to plasma technology to convert the resulting ashes directly into a vitreous slag, which might be favorable compared to MSWI bottom ashes with respect to mechanical and leaching properties. As it was not possible within the project to test plasma technology, vitreous slag from previous tests was used for further experiments, but also MSWI bottom ash was melted to obtain a similar slag.

Within NEW-MINE, the light fraction obtained by mechanical processing of excavated waste from the Mont-Saint-Guibert landfill was subjected to pyrolysis (N_2 atmosphere, 900 °C, 1 h) and the resulting char was subjected to steam gasification using thermogravimetry analysis (TGA) to study the reaction kinetics [80]. This publication was cited as one of the few studies on gasification kinetics of MSW chars [81] and regarding the observed increase in activation of the char by gasification [82]. Furthermore, the study was cited with respect to the energy savings, low NO_x emissions and pollutant reduction of steam gasification compared to conventional thermochemical conversion technologies [83]. In a second study, a sample of the fine fraction (<10 mm) and two samples of combustibles (with different share of defilements) were used for a similar experimental setup [84]. The latter publication was cited in a review on co-incineration of organic waste and coal [85] and with respect to the higher production of CO and CO_2 in pyrolysis when using finer waste fractions in a publication on the prediction of gaseous products from RDF pyrolysis [86]. Two further results originated from this NEW-MINE publication: (1) pyrolysis of landfilled waste yields less oil and gas than pyrolysis of fresh waste and (2) gasification of char from pyrolysis of landfilled waste yields waste fuel with a higher reactivity compared to the same process using fresh waste, which is explained by the catalytic effect of the metals present in landfilled waste [87]. Further studies citing this publication are on pyrolysis of pre-treated trommel fines [88]. In the third study on this topic, it was shown that steam co-gasification of landfilled waste with biochar increases H_2 yields [89]. This publication was mentioned in a review article on waste-to-energy technologies [90] and with respect to the use of the syngas from gasification for ammonia synthesis [91].

The second issue addressed within NEW-MINE regarding gasification was the cracking of tars from the produced syngas which would restrict its utilization options. At first, a review on the state of the art regarding the role of plasma in syngas tar cracking [92] was published, which was cited outside of NEW-MINE by two other review articles [93,94] mentioning the efficiency, but also the short lifetime and high costs of this technology, as well as a novel warm gas tar cleaning processing called OLGA. In a next step, naphthalene was used as a model tar molecule and Corona plasma-aided thermal cracking was demonstrated successfully for its removal at 800 °C [95]. This study was cited with respect to tar generation and conversion kinetics [96], regarding the required high temperatures of 1100 °C which represent a disadvantage compared to catalytic reforming [97], and as an example for thermal tar cracking [98]. Finally, two CaO-rich catalysts doped with Sr were synthesized from mussel shells and successfully tested for tar cracking in the syngas [99].

Another NEW-MINE publication focused on further aspects of the pyrolysis process. Lab-scale pyrolysis tests at 400 to 700 °C with the above-mentioned light fraction of the Mont-Saint-Guibert landfill revealed the enrichment of polycyclic aromatic hydrocarbons (PAH) in the condensable pyrolysis products, highlighting the need to further treat this output fraction [100]. Although the main focus of thermochemical conversion technologies within NEW-MINE was on the treatment of the combustibles, thermal treatment (400/450 °C, 30 min) was also used to assess the quality of the nonferrous metals obtained from mechanical processing of the excavated waste from Mont-Saint-Guibert landfill [101].

3.2.2. Solar Energy Storage

In contrast to waste incineration and co-incineration, which produce energy and even contribute to 3.7% of the German end energy consumption [72], gasification requires external energy. The use of energy from renewable sources for this purpose is environmentally favorable, but requires energy storage. Consequently, within NEW-MINE, MgO-stabilized SrO was synthesized and successfully tested for heat storage via the $SrO/SrCO_3$ cycle [102]. This research was considered outside NEW-MINE with respect to radiation propagation in a heat exchanger transforming solar radiation into high-temperature heat [103], thermochemical energy storage in the $CaO/CaCO_3$ [104] and $MnAl_2O_4/MnAl_2O_{4-x}$ [105] system and two reviews on thermochemical energy storage [106,107]. The most recent NEW-MINE publication in this field deals with thermochemical heat storage via the CuO/Cu_2O redox cycle and describes the synthesis of granules with a gravimetric energy storage density in the range of 470 to 615 kJ kg^{-1} [108].

In summary, NEW-MINE publications on thermochemical conversion and solar energy storage focused on the organic constituents of gasification chars and of oxide/carbonate systems for solar energy storage and created a novel state-of-the-art. However, the fate of inorganic pollutants during gasification was not addressed and the intended production of vitreous slags via plasma processing that might immobilize these pollutants was not realized in the project.

3.3. *Benefication of Products from Thermochemical Conversion*

3.3.1. Melting and Vitrification

Since the intended plasma gasification of ELFM materials was not realized within NEW-MINE, three groups of input materials were selected for further processing, i.e., (1) MSWI bottom ashes, (2) vitreous slag from plasma gasification ("Plasmastone") from a pre-NEW-MINE project and (3) synthetic vitreous slags mimicking the plasmastone composition. The aims of the beneficiation of these materials comprised the recovery of metals and the beneficiation of the nonmetallic fractions for specific applications in civil engineering beyond the state-of-the-art applications of MSWI bottom ashes as low-strength aggregates [109]. Thermal treatment of MSWI bottom ashes within NEW-MINE aimed for the recovery of metals and beneficiation of the mineral fraction. For this purpose, two heating technologies were investigated.

Firstly, a submerged arc furnace (SAF) and an electric resistance furnace (ERF) were used to melt MSWI bottom ash obtained from a Dutch MSWI plant at 1500 and 1400 °C, respectively. These experiments yielded an Cu-Fe alloy and a vitreous CaO-Al_2O_3-SiO_2 slag, which might be used for the production of glass ceramics or inorganic binders [110] and were linked to the application of SAF technology in other fields [111].

Secondly, microwave heating was used to study the dielectric properties from the above mentioned MSWI bottom ash from a Dutch MSWI plant and revealed that MSWI bottom ash absorbs microwaves with low losses until 320 °C but above this temperature pyrolysis enhances the dielectric loss [112]. This study was cited with respect to microwave cladding, an emerging surface modification technique [113]. In a next step, flash microwave vitrification was realized at 2.45 GHz within 1.5 min. The advantage of this technology for vitrification is that the cooling rate is naturally so fast, due to the inherent cold environment, that no specific efforts are required for quenching [114].

3.3.2. Alkali Activation

The next group of studies within NEW-MINE dealt with the further benefication of vitrified MSWI bottom ashes. In one study, the vitrified bottom ash was subjected to alkali activation (1 M NaOH and 2.5 M NaOH and sinter crystallization (800 °C or 900 °C), thereby yielding porous (70 vol%) glass ceramics with a compressive strength of 3 MPa, which might be used as an alternative to lightweight concrete [115]. Another group of studies addressed the production of inorganic polymers by alkali activation. Compared to geopolymers, which are the main subgroup of alkali activated materials, inorganic polymers differ with respect to their chemical composition.

A synthetic plasmastone analogue was used to study the impact of solid-to-liquid and K_2O/SiO_2 ratio and the type of activation solution on the resulting inorganic polymers. Adjustment of these process parameters yielded inorganic parameters with a compressive strength of up to 119 MPa [116]. This study was cited in a review article as an example to valorize vitrified MSWI bottom ash [117]. Based on these findings, strategies to increase the volumetric stability of inorganic polymers were developed [118]. This work was cited in the context of alternative cementitious materials for radioactive waste encapsulation, taking into account that strengthening of inorganic polymers was observed due to radiation-induced iron oxidation [119]. One approach to increase the volumetric stability is the addition of CaO-rich materials, which was demonstrated in another study [120].

3.3.3. Sintering of Glass-Ceramics

The last group of studies refers to the sintering of vitreous slags from plasma processing of MSW or from MSWI vitrification to glass ceramics, which might be used as building materials such as aggregates and panels. The most-cited publication of NEW-MINE [121] dealt with the production of porous glass ceramics from the vitreous slag obtained by plasma processing of MSW. It was demonstrated how increasing the firing temperature of glass ceramics can trap chromium in stable pyroxene group minerals and adding borosilicate glass can decrease the leaching of Cr, Mo and V as boron acts as a network former. This paper was cited in publications on the preparation of glass-containing foams from geopolymers [122] and vitrified MSWI bottom ash [123] in which the formation of wollastonite and the freezing of the microstructural evolution were mentioned. Other papers cited this publication with respect to the recycling of glass waste into foam glass [124–129], porous waste glass for lead removal in wastewater treatment [130], lead stabilization through alkali activation and sintering of Pb-bearing sludge [131], utilization of waste glass for the production of sulphuric acid resistant concrete [132], mechanical and alkali activation of MSWI fly and bottom ashes for the production of low-range alkaline cement [133] and foam glass-ceramics [134], inorganic gel casting for manufacturing of boro-alumino-silicate glass foams [135], porous glass-ceramics derived from MgO-CuO-TiO_2-P_2O_5 glasses [136], alkali activation of coal and biomass fly ashes [137], nickel-based catalysts for steam reforming of naphthalene utilizing MSW gasification slag as support [138], production of porous glass ceramics from titanium mine tailings and waste glass [139], porous bioactive glass microspheres [140], Al-SiO_2 composites [141], glass-ceramic foams from alkali-activated vitrified MSWI bottom ash and waste glasses [142]. Another study used vitrified MSWI bottom ash as input material to obtain similar porous glass ceramics [143] and was cited by some of the publications that also cited the first study.

In a next step, dense glass with good mechanical and environmental properties (low leaching due to stable incorporation of heavy metals in stable hedenbergite, wollastonite and iron oxides) were obtained from a mixture of "Plasmastone", recycled soda-lime glass and kaolin clay by cold pressing and fast heat treatment (1000 °C, 40 °C/min) [144]. This publication was referred to as a case study for the influence of crystallization time on glass ceramics [145].

The crucial linkage between mineralogy and technical and environmental performance was also investigated in a further study in which low temperatures (800 °C) were sufficient to immobilize heavy metals in a stable matrix using spent borosilicate glasses and additional electromagnetic shielding functionalities were obtained by magnetite formation [146]. This approach was cited in a study on the substitution of feldspar by waste glass for porcelain production [147] and in a paper on cheap pore-generating agents for ceramics [148]. Finally, the promoting effect of firing in nitrogen atmosphere on the stabilization of pollutants and novel functionalities was investigated [149]; the paper was cited in studies on the recycling of iron-rich inorganic wastes into functional glass-ceramics [150] and on the production of glass ceramics from MSWI bottom ash and coal fly ash by melting and sintering [151].

In summary, the NEW-MINE publications on beneficiation of products of thermochemical conversion of (excavated) MSW, i.e., vitrified MSWI bottom ashes and vitreous slags of plasma gasification, yielded novel insights into the relation between (crystalline/vitreous) structure and (mechanical/leaching) properties of waste-derived glass ceramics and inorganic polymers. Although these specific approaches to address MSWI residues might not be realized in practical waste management, the underlying mechanisms are also relevant for the mobility of heavy metals in other mineral wastes and allow the eco-design of their properties and, consequently, enhanced recyclability.

3.4. *Multi-Criteria Assessment*

3.4.1. Life Cycle Assessment

Already before NEW-MINE, comprehensive methods for landfill mining projects have been developed. With respect to ecological and socio-economic criteria, this approach was based on utility analysis and then transferred into a utility-net present value chart [9]. In contrast, within NEW-MINE, for the ecological assessment life cycle assessment (LCA) was used, although it was not applied to landfill mining, but to the impact of sanitary landfills themselves [152]: this approach yields important information regarding the alternative to ELFM, i.e., state-of-the-art landfill aftercare. This LCA publication was cited with respect to the site selection for landfills [153,154] and in a study which yielded the controversial statement that "recycling metals except gold had more negative environmental impacts than mining" [155]. Further citing studies deal with environmental impacts of combustion-based energy production [156], composting on closed landfill sites [157], bioreactor landfills [158], suppression of methane generation [159], modelling of landfill gas production [160], organic waste enrichment [161]. Within NEW-MINE, LCA was also applied to compare smelting of MSWI bottom ash with the state-of-the-art mineral processing approach and revealed that the energy demand for smelting processes overrules, with respect to global warming, the positive effect of secondary raw material production [162].

In summary, LCA demonstrated the environmental impact of all scenarios of landfill management. Landfill aftercare requires a high effort for gas and leachate treatment, while traditional landfill mining with MSWI yields mineral fractions of MSWI bottom ashes which cannot substitute for high-value primary raw materials due to the leaching of pollutants. In contrast ELFM, with the production of vitreous slags–either via plasma gasification of municipal solid waste (MSW) or via smelting of MSWI bottom ashes, yields more valuable secondary raw materials which are characterized by a lower leaching, but their production requires more energy.

3.4.2. Techno-Economic Assessment

Based on a comprehensive study on existing economic assessment methods [163] it has been suggested within NEW-MINE to follow learning-oriented approaches. Subsequently, an economic and environmental assessment of a possible ELFM project in Sweden has been conducted and revealed that none of the scenarios is economically profitable, but of all them result in net avoided emissions due to the recovery of metals [164]. In a next step, more than 500,000 landfill mining scenarios were evaluated with respect to their net present value and the influencing factors were identified via global sensitivity analysis. It was found that 80% of these scenarios show negative results and that revenues from avoided landfill management costs are more important than revenues from resource recovery [165]. This study was cited in a review paper on the impact of landfills on the environment [166] in a study on the environmental and economic assessment of ELFM in Tehran [167] and with respect to geophysical exploration of landfills [35].

In summary, NEW-MINE results confirm that ELFM is ecologically favorable, but economically not profitable, which demonstrates the need for an integrated ecological and economic assessment [9,35]. This means, that for those cases in which this integrated assessment shows a positive effect of ELFM, economic incentives would be required to make secondary raw materials or recycled products from ELFM competitive in the market.

3.4.3. Sociological Assessment

Before NEW-MINE, landfill mining was investigated mainly from technological, environmental and economic perspectives. However, it is well known that large infrastructure projects often fail due to public resistance [168]. Therefore, within NEW-MINE, also stakeholders' perspectives on ELFM were investigated [169] and stakeholder needs were identified [170]. The latter study was cited in a study on the characterization of excavated waste [171]. Finally, stakeholder archetypes for ELFM were defined: The Engaged Citizen, the Entrepreneur, the Technology Enthusiast, the Visionary and the Skeptic [172]. In summary, the importance of the involvement of stakeholders in ELFM has been demonstrated and fundamentals for corresponding strategies have been laid.

4. Discussion

4.1. *Disciplinarity and Interdisciplinarity*

Within the NEW-MINE project the entire process chain, from landfill exploration via excavation and mechanical processing to thermochemical conversion and beneficiation of conversion products, has been investigated. This broad scope constituted a unique feature of the NEW-MINE project and, correspondingly, many publications do not address the landfill mining itself, but waste treatment technologies, which may also be used for waste from other sources than landfills. Some publications, e.g., dealing with energy storage, are only loosely linked to landfill mining. This broad scope enabled the interdisciplinary collaboration of researchers, for example between geophysicists and waste scientists [33], waste scientists and polymer scientists [48], experts in mechanical and thermal waste treatment [84,89,101] and in high and low temperature waste mineralogy [121,144]. However, the majority of the publications can be allocated to a single scientific discipline. With respect to the number of publications citing the NEW-MINE publications, no significant difference between disciplinary and interdisciplinary publications is visible. Although most disciplinary publications were cited by researchers from other disciplines (e.g., [165] by [35]), most of them were cited by colleagues from the authors' discipline.

Considering that the overall research question of the project was to investigate if innovative technologies increase the economic and environmental feasibility of landfill mining, more interdisciplinary cooperation would be required. For example, if the purity of an inert fraction obtained by sensor-based sorting is increased from 86 to 98% [46], what is then the impact on the economic, environmental and societal assessment of ELFM? This and several similar questions demonstrate the need for further interdisciplinary research in the field of ELFM for which NEW-MINE could have been an excellent starting point.

4.2. Recycling Waste from Landfills

Although a lot of reasons for landfill mining have been identified and excavation of landfills for environmental remediation has been practiced already for decades, it is the recovery of resources from landfills, which fostered the landfill mining projects in the last decade including the NEW-MINE project This is clearly indicated in the project's name, i.e., "EU Training Network for Resource Recovery Through Enhanced Landfill Mining". During the NEW-MINE project a lot of effort was devoted to treat the excavated waste by mechanical, thermochemical and chemical methods. The final aim of these treatments is to obtain either a secondary raw material, which may substitute for a primary raw material in an industrial process, or a product, which substitutes for a product with the same functionalities but is made from primary raw materials. Regarding the production of secondary raw materials, NEW-MINE demonstrated that e.g., the 2D fraction of ballistic separation can substitute for fossil fuels in co-incineration plants. In contrast, the use of the inert fraction as raw material for the production of concrete [70] and the use of the fine fraction as raw material for the production of compost soil [71] fails, due to the exceeded limit values for total and leachable contaminants.

Regarding the production of a product, it has been demonstrated that e.g., glass-ceramics [144] fulfil the mechanical and leaching requirements for building products. Apart from the problem that several countries, e.g., Austria, have also limit values for total contents, it has to be mentioned that these products have to compete with products with the same functionality but are made from primary raw materials. Consequently, comparative economic and environmental assessments are required to investigate the production costs and environmental impact of e.g., lightweight concrete made from primary raw materials (e.g., Airium™) and waste-derived glass-ceramics that compete for the same application. With respect to raw materials, the same applies: e.g., recycled aggregates compete with natural aggregates (with the problem of decreasing quality with increasing number of recycling cycles [173]), but also additional challenges occur: The RDF that can be produced by ballistic separation or windsifting has a negative value, i.e., the producer has to pay the cement plant to incinerate it. Therefore, within the Halbenrain case study in NEW-MINE, the lightweight fraction, which could have been delivered to an RDF production plant was re-landfilled, as this was the cheaper option. The reason for this paradox is that combustible waste must not be landfilled in Austria and incineration is an expensive process, considering also the taxes that have to be paid. Therefore, co-incineration plants can demand negative prices for RDF, as it is still cheaper than incineration. Only for the specific case study of ELFM, re-landfilling of combustibles is allowed and represents the cheapest option.

In summary, even if secondary raw materials or recycled products can be produced from ELFM materials, it remains unclear if they can compete with primary raw materials or products from primary raw materials. First observations for Austria suggest that this may often not be the case. Given the ambitions of transition towards a more circular economy, economic incentives are required to make secondary raw materials or recycled products from ELFM competitive in the market.

4.3. Waste Management Meets Mining Economics

Finally, as is already evident in the project name "NEW-MINE" in which terminology from mining economics ("resources") and waste management ("landfill") are used, it represents a good example to apply a figure illustrating the different terminologies from mining economics, waste management and process engineering [27] to MSW landfills and ELFM (Figure 2).

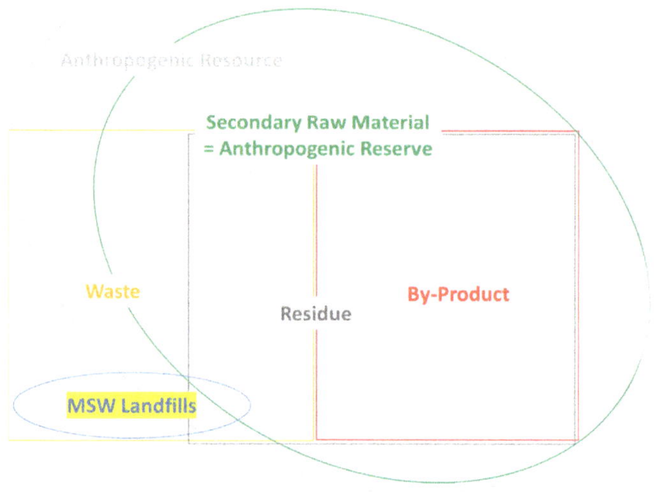

Figure 2. Location of MSW landfills in waste and resource terminology (Reprinted with permission from ref. [27]. Copyright 2020. Daniel Vollprecht).

Often, in the recycling business, the term "waste" is avoided due to its negative connotations. Instead, the terms "residue", "anthropogenic resource", "by-product" or "secondary raw material" are used, but they have a different meaning.

Waste is precisely defined in the European Waste Framework Directive [174] as "any substance or object which the holder discards or intends or is required to discard". Consequently, the material in an MSW landfill is waste, as the (former) holder discarded it. If the material is recycled, it reaches the end-of-waste at the moment when it substitutes for a primary raw material. Considering the technologies investigated within NEW-MINE, this means that the material remains a waste until the waste-derived glass ceramics or inorganic binders finally substitute for conventional construction ceramics or cement, respectively, and this only if they fulfil the technical and environmental requirements for the competing products made from primary raw materials.

A by-product is only a substance or object whose further use is certain [174], i.e., there cannot be a by-product in a landfill, since if the further use would have been certain, it could not have been disposed of. In contrast to "waste" and "by-product" which originate from waste management, the term "residue" stems from process engineering and refers to "the part that is left after the main part has gone or been taken away" [175]. In some MSW landfills also residues can be found, as industrial waste has not always been disposed of in separate landfills. Similar to "residue" also "secondary raw materials" are defined by the Great Soviet Encyclopaedia as "materials and articles which, after complete initial use (wear), may be used repeatedly in production as starting material" [176].

The terms "resource" (reasonable prospects for eventual economic extraction in the foreseeable future) and "reserve" originate from mining economics (current economic extraction possible) and have been introduced into waste management [13]. If current economic extraction is possible, the material will be a secondary raw material for a subsequent industrial process, i.e., "anthropogenic reserve" and "secondary raw material" are synonyms. With respect to ELFM, only a small fraction of the excavated materials within NEW-MINE, i.e., the metal fractions from magnetic and eddy current separation or from smelting processes, fall in this category. Some other fractions, e.g., the combustibles, might be classified as "anthropogenic resource" if RDF prices rise significantly above zero. However, the results for the fine fraction do not justify a classification as anthropogenic resource, as no reasonable prospects for their purification exist [71].

In summary, the quantity of resources that can be recovered from MSW landfills can be significant, if valorization options for the inert fraction and the combustibles become feasible. However, quality and "bonitaet" [26] of anthropogenic resources from MSW landfills represent significant challenges for future ELFM projects.

5. Conclusions

The NEW-MINE project (2016–2020) was a European research project investigating the entire process chain of ELFM. Within the NEW-MINE project in a first step concentrates of specific materials (e.g., metals, combustibles, inert materials) were produced which might be used as secondary raw materials. In a second step, recycled products (e.g., inorganic polymers, functional glass-ceramics) were produced from these concentrates at the lab scale. However, even if secondary raw materials or recycled products could be produced at a large scale, it remains unclear if they can compete with primary raw materials or products from primary raw materials. Given the ambitions of transition towards a more circular economy, economic incentives are required to make secondary raw materials or recycled products from ELFM competitive in the market.

Author Contributions: Conceptualization, D.V.; methodology, D.V.; investigation, D.V. and L.M.; writing—original draft preparation, D.V.; writing—review and editing, L.M., P.T.J.; visualization, L.M.; supervision, P.T.J., L.M.; project administration, P.T.J., L.M.; funding acquisition, P.T.J. All authors have read and agreed to the published version of the manuscript.

Funding: The NEW-MINE project has received funding from the European Union's EU Framework Program for Research and Innovation Horizon 2020 under Grant Agreement No 721185.

Acknowledgments: The authors thank the entire NEW-MINE team for the fruitful collaboration within the project.

Conflicts of Interest: The authors declare no conflict of interest. The funders had no role in the design of the study; in the collection, analyses, or interpretation of data; in the writing of the manuscript, or in the decision to publish the results.

References

1. Krook, J.; Svensson, N.; Eklund, M. Landfill mining: A critical review of two decades of research. *Waste Manag.* **2012**, *32*, 513–520. [CrossRef] [PubMed]
2. Savage, G.M.; Golueke, C.G.; von Stein, E.L. Landfill mining: Past and present. *Biocycle* **1993**, *34*, 58–61.
3. Cobb, C.E.; Ruckstuhl, K. Mining and reclaiming existing sanitary landfills. In Proceedings of the National Waste Processing Conference, Detroit, MI, USA, 17–20 May 1988; pp. 145–151.
4. Rettenberger, G.; Urban-Kiss, S.; Schneider, R.; Göschl, R.; Kremsl, W. Deponierückbau an der deponie burghof in vaihingen/enzhorrheim—Erfahrungen aus einem demonstrationsprojekt. *Korresp. Abwasser* **1995**, *2*, 196.
5. Fricke, K.; Münnich, K.; Heußner, C.; Schulte, B.; Wanka, S. Landfill mining—Ein beitrag der abfallwirtschaft für die ressourcensicherung. In Proceedings of the Berliner Recycling—Und Rohstoffkonferenz, Berlin, Germany, 26–27 March 2012; pp. 933–944.
6. Jones, P.T.; Geysen, D.; Tielemans, Y.; van Passel, S.; Pontikes, Y.; Blanpain, B.; Quaghebeur, M.; Hoekstra, N. Enhanced landfill mining in view of multiple resource recovery: A critical review. *J. Clean. Prod.* **2013**, *55*, 45–55. [CrossRef]
7. Wolfsberger, T.; Aldrian, A.; Sarc, R.; Hermann, R.; Höllen, D.; Budischowsky, A.; Zöscher, A.; Ragoßnig, A.; Pomberger, R. Landfill mining: Resource potential of Austrian landfills—Evaluation and quality assessment of recovered municipal solid waste by chemical analyses. *Waste Manag. Res.* **2015**, *33*, 962–974. [CrossRef] [PubMed]

8. Wolfsberger, T.; Nispel, J.; Sarc, R.; Aldrian, A.; Hermann, R.; Höllen, D.; Pomberger, A.; Ragossnig, A. Landfill mining: Development of a theoretical method for a preliminary estimate of the raw material potential of landfill sites. *Waste Manag. Res.* **2015**, *33*, 671–680. [CrossRef]
9. Hermann, R.; Wolfsberger, T.; Pomberger, R.; Sarc, R. Landfill mining: Developing a comprehensive assessment method. *Waste Manag. Res.* **2016**, *34*, 1157–1163. [CrossRef]
10. Wolfsberger, T.; Pinkel, M.; Polansek, S.; Sarc, R.; Pomberger, R. Landfill mining: Development of a cost simulation model. *Waste Manag. Res.* **2016**, *34*, 356–367. [CrossRef] [PubMed]
11. Dumont, G.; Pilawski, T.; Dzaomuho-Lenieregue, P.; Hiligsmann, S.; Delvigne, F.; Thonart, P.; Robert, T.; Nguyen, F.; Hermans, T. Gravimetric water distribution assessment from geoelectrical methods (ERT and EMI) in municipal solid waste landfill. *Waste Manag.* **2016**, *55*, 129–140. [CrossRef]
12. Wanka, S.; Münnich, K.; Fricke, K. Landfill mining—Wet mechanical treatment of fine MSW with a wet jigger. *Waste Manag.* **2016**, *59*, 316–323. [CrossRef]
13. Winterstetter, A.; Laner, D.; Rechberger, H.; Fellner, J. Framework for the evolution of anthropogenic resources: A landfill mining case study—Resource or reserve? *Resour. Conserv. Recycl.* **2015**, *96*, 19–30. [CrossRef]
14. Mönkäre, T.; Palmroth, M.; Rintala, J. Characterization of fine fraction mined from two Finnish landfills. *Waste Manag.* **2016**, *47A*, 34–39. [CrossRef] [PubMed]
15. Brunner, P. Final Sink: Prerequisite for a cycling society. In Proceedings of the ISWA Beacon 3rd International Conference on Final Sinks, Taipei, China, 23–26 August 2015.
16. Wu, H.; Zuo, J.; Zillante, G.; Wang, J.; Yuan, H. Status quo and future directions of construction and demolition waste research: A critical review. *J. Clean. Prod.* **2019**, *240*, 118163. [CrossRef]
17. Silva, R.V.; de Brito, J.; Dhir, R.K. Comparative analysis of existing prediction models on the creep behavior of recycled aggregate concrete. *Eng. Struct.* **2015**, *100*, 31–42. [CrossRef]
18. Makarichi, L.; Jutidamrongphan, W.; Techato, K. The evolution of waste-to-energy incineration: A review. *Renew. Sustain. Energy Rev.* **2018**, *91*, 812–821. [CrossRef]
19. Sarc, R.; Lorber, K.E. Production, quality and quality assurance of Refuse Derived Fuels (RDFs). *Waste Manag.* **2013**, *33*, 1825–1834. [CrossRef] [PubMed]
20. Vollprecht, D.; Riegler, C.; Ahr, F.; Stuhlpfarrer, S.; Wellacher, M. Sequential chemical extraction and mineralogical bonding of metals from Styrian soils. *Int. J. Environ. Sci. Technol.* **2020**, *17*, 3663–3676. [CrossRef]
21. European Enhanced Landfill Mining Consortium. Available online: www.eurelco.org (accessed on 16 November 2020).
22. Hernández Parrodi, J.C.; Lucas, H.I.; Gigantino, M.; Sauve, G.; Esguerra, J.L.; Einhäupl, P.; Vollprecht, D.; Pomberger, R.; Friedrich, B.; Van Acker, K.; et al. Integration of resource recovery into current waste management through (enhanced) landfill mining. *Detritus* **2019**, *8*, 141–156. [CrossRef]
23. Orlando, L.; Marchesi, E. Georadar as a tool to identify and characterise solid waste dump deposits. *J. Appl. Geophys.* **2001**, *48*, 163–174. [CrossRef]
24. Hermozilha, H.; Grangeia, C.; Senos Matias, M. An integrated 3D constant offset GPR and resistivity survey on a sealed landfill Ilhavo, NW Portugal. *J. Appl. Geophys.* **2010**, *70*, 58–71. [CrossRef]
25. Porsani, J.; Filho, W.; Elis, V.; Shimeles, F.; Dourado, J.; Moura, H. The use of GPR and VES in delineating a contamination plume in a landfill site: A case study in SE Brazil. *J. Appl. Geophys.* **2004**, *55*, 199–209. [CrossRef]
26. Fettweis, G.; Brandstätter, W.; Hruschka, F. Was ist Lagerstättenbonität? *Mitt. Osterr. Geol. Ges.* **1985**, *78*, 23–40.
27. Vollprecht, D. Exploration, Mobilization and Fixation of Constituents of Mineral Wastes, Landfills, Contaminated Sites and Waste Waters. Habilitation Thesis, Montanuniversität Leoben, Leoben, Austria, 17 April 2020.
28. Zanetti, M.; Godio, A. Recovery of foundry sands and iron fractions from an industrial waste landfill. *Resour. Conserv. Recycl.* **2006**, *48*, 396–411. [CrossRef]
29. Yannah, M.; Martens, K.; van Camp, M.; Walraevens, K. Geophysical exploration of an old dumpsite in the perspective of enhanced landfill mining in Kermt area, Belgium. *Bull. Eng. Geol. Environ.* **2019**, *78*, 55–67. [CrossRef]
30. Bobe, C.; Van De Vijver, E.; Van Meirvenne, M. Exploring the potential of electromagnetic surface measurements for the characterization of industrial landfills. In Proceedings of the 4th International Symposium on Enhanced Landfill Mining, Mechelen, Belgium, 5–6 February 2018; pp. 45–50.
31. Wagland, S.; Coulon, F.; Canopoli, L. Developing the case for enhanced landfill mining in the UK. *Detritus* **2019**, *5*, 105–110. [CrossRef]
32. Manrique, I.I.; Caterina, D.; Van De Vijver, E.; Dumont, G.; Nguyen, F. Assessment of geophysics as a characterization and monitoring tool in the dynamic landfill management (DLM) context: Opportunities and challenges. In Proceedings of the 17th International Waste Management and Landfill Symposium, Forte Village, Santa Margherita di Pula, Italy, 30 September–4 October 2019.
33. Vollprecht, D.; Bobe, C.; Stiegler, R.; Van de Vijver, E.; Wolfsberger, T.; Küppers, B.; Scholger, R. Relating magnetic properties of municipal solid waste constituents to iron content—Implications for enhanced landfill mining. *Detritus* **2019**, *8*, 31–46. [CrossRef]
34. Sandrin, A.; Keiding, J. Integrated geophysical study for landfill mining: A case study in Denmark. In Proceedings of the 82nd EAGE Annual Conference & Exhibition Workshop Programme, Amsterdam, The Netherlands, 8 December 2020; pp. 1–5. [CrossRef]

35. Sandrin, A.; Maricak, A.; Heincke, B.; Clausen, R.; Nielsen, L.; Keiding, J. Geophysics for urban mining and the first surveys in Denmark: Rationale, field activity and preliminary results. *GEUS Bull.* **2020**, *44*, 5240. [CrossRef]
36. Bobe, C.; Van de Vijver, E.; Keller, J.; Hanssens, D.; Van Meirvenne, M.; De Smedt, P. Probabilistic 1-D inversion of frequency-domain electromagnetic data using a kalman ensemble generator. *IEEE Trans. Geosci. Remote Sens.* **2020**, *58*, 3287–3297. [CrossRef]
37. Grana, D.; Liu, M.; Ayani, M. Prediction of CO_2 saturation spatial distribution using geostatistical inversion of time-lapse geophysical data. *IEEE Trans. Geosci. Remote Sens.* **2020**. [CrossRef]
38. Hanssens, D.; Waegeman, W.; Declercq, Y.; Dierckx, H.; Verschelde, H.; De Smedt, P. High-resolution surveying with small-loop requency-domain electromagnetic systems: Efficient survey design and adaptive processing. *IEEE Geosci. Remote Sens. Mag.* **2020**. [CrossRef]
39. Bobe, C.; Hanssens, D.; Hermans, T.; Van De Vijver, E. Efficient probabilistic joint inversion of direct current resistivity and small-loop electromagnetic data. *Algorithms* **2020**, *13*, 144. [CrossRef]
40. Bobe, C.; Van De Vijver, E. Offset errors in probabilistic inversion of small-loop frequency-domain electromagnetic data: A synthetic study on their influence on magnetic susceptibility estimation. In Proceedings of the International Workshop on Gravity, Electrical & Magnetic Methods and Their Applications, Xi'an, China, 19–22 May 2019; Society of Exploration Geophysicists: Tulsa, OK, USA, 2019; pp. 312–315. [CrossRef]
41. García López, C.; Küppers, B.; Clausen, A.; Pretz, T. Landfill mining: A case study regarding sampling, processing, and characterization of excavated waste from an Austrian Landfill. *Detritus* **2018**, *2*, 29–45. [CrossRef]
42. García López, C.; Ni, A.; Hernández Parrodi, J.C.; Pretz, T.; Raulf, K.; Küppers, B. Characterization of landfill mining material after ballistic separation to evaluate material and energy recovery potential. *Detritus* **2019**, *8*, 5–23. [CrossRef]
43. Datta, M.; Somani, M.; Ramana, G.V.; Sreekrishnan, T.R. Feasibility of re-using soil-like material obtained from mining of old MSW dumps as an earth-fill and as compost. *Process Saf. Environ. Prot.* **2021**, *147*, 477–487. [CrossRef]
44. Mönkäre, T.; Palmroth, M.R.; Sormunen, K.; Rintala, J. Scaling up the treatment of the fine fraction from landfill mining: Mass balance and cost structure. *Waste Manag.* **2019**, *87*, 464–471. [CrossRef]
45. Márquez, A.J.C.; Cassettari Filho, P.C.; Rutkowski, E.W.; de Lima Isaac, R. Landfill mining as a strategic tool towards global sustainable development. *J. Clean. Prod.* **2019**, *226*, 1102–1115. [CrossRef]
46. Küppers, B.; Hernández Parrodi, J.C.; García Lopez, C.; Pomberger, R.; Vollprecht, D. Potential of sensor-based sorting in enhanced landfill mining. *Detritus* **2019**, *8*, 24–30. [CrossRef]
47. Möllnitz, S.; Küppers, B.; Curtis, A.; Khodier, K.; Sarc, R. Influence of pre-screening on down-stream processing for the production of plastic enriched fractions for recycling from mixed commercial and municipal waste. *Waste Manag.* **2021**, *119*, 365–373. [CrossRef] [PubMed]
48. Küppers, B.; Schlögl, S.; Oreski, G.; Pomberger, R.; Vollprecht, D. Influence of surface roughness and surface moisture of plastics on sensor-based sorting in the near infrared range. *Waste Manag. Res.* **2019**, *8*, 843–850. [CrossRef] [PubMed]
49. Maier, G.; Pfaff, F.; Bittner, A.; Gruna, R.; Noack, B.; Kruggel-Emden, H.; Hanebeck, U.; Längle, T.; Beyerer, J. Characterizing material flow in sensor-based sorting systems using an instrumented particle. *Automatisierungstechnik* **2020**, *68*, 256–264. [CrossRef]
50. Küppers, B.; Schlögl, S.; Friedrich, K.; Lederle, L.; Pichler, C.; Freil, J.; Pomberger, R.; Vollprecht, D. Influence of material alterations and machine impairment on throughput related sensor-based sorting performance. *Waste Manag. Res.* **2020**, *39*, 197–198. [CrossRef]
51. Küppers, B.; Seidler, I.; Koinig, I.; Pomberger, R.; Vollprecht, D. Influence of throughput rate and input composition on sensor-based sorting efficiency. *Detritus* **2020**, *9*, 59–67. [CrossRef]
52. Curtis, A.; Küppers, B.; Möllnitz, S.; Khodier, K.; Sarc, R. Real time material flow monitoring in mechanical waste processing and the relevance of fluctuations. *Waste Manag.* **2020**, *120*, 687–697. [CrossRef]
53. Liebetegger, W. Landfill Mining—Charakterisierung der Fein—Und Heizwertreichen Fraktion. Master's Thesis, Montanuniversität Leoben, Leoben, Austria, 2015.
54. Parrodi, J.C.H.; Höllen, D.; Pomberger, R. Characterization of fine fractions from landfill mining: A review of previous investigations. *Detritus* **2018**, *2*, 46–62. [CrossRef]
55. Parrodi, J.C.H.; Höllen, D.; Pomberger, R. Potential and main technological challenges for material and energy recovery from fine fractions of landfill mining: A critical review. *Detritus* **2018**, *3*, 19–29. [CrossRef]
56. Hölzle, I. Contaminant patterns in soils from landfill mining. *Waste Manag.* **2019**, *83*, 151–160. [CrossRef]
57. Hartemink, A.E. The definition of soil since the early 1800s. *Adv. Agron.* **2016**, *137*, 73–126. [CrossRef]
58. Mutafela, R.N.; Marques, M.; Jani, Y.; Kriipsalu, M.; Hogland, W. Physico-chemical characteristics of fine fraction materials from an old crystal glass dumpsite in Sweden. *Chem. Ecol.* **2019**, *35*, 877–890. [CrossRef]
59. Somani, M.; Datta, M.; Ramana, G.V.; Sreekrishnan, T.R. Leachate characteristics of aged soil-like material from MSW dumps: Sustainability of landfill mining. *J. Hazard. Toxic Radioact. Waste* **2019**, *23*, 04019014. [CrossRef]
60. Mutafela, R.N.; Mantero, J.; Jani, Y.; Thomas, R.; Holm, E.; Hogland, W. Radiometrical and physico-chemical characterisation of contaminated glass waste from a glass dump in Sweden. *Chemosphere* **2020**, *241*, 124964. [CrossRef] [PubMed]
61. Mutafela, R.N.; Lopez, E.G.; Dahlin, T.; Kaczala, F.; Marques, M.; Jani, Y.; Hogland, W. Geophysical investigation of glass 'hotspots' in glass dumps as potential secondary raw material sources. *Waste Manag.* **2020**, *106*, 213–225. [CrossRef] [PubMed]
62. Pehme, K.M.; Orupõld, K.; Kuusemets, V.; Tamm, O.; Jani, Y.; Tamm, T.; Kriipsalu, M. Field study on the efficiency of a methane degradation layer composed of fine fraction soil from landfill mining. *Sustainability* **2020**, *12*, 6209. [CrossRef]

63. Singh, A.; Chandel, M.K. Effect of ageing on waste characteristics excavated from an Indian dumpsite and its potential valorisation. *Process Saf. Environ. Prot.* **2020**, *134*, 24–35. [CrossRef]
64. Mentes, D.; Sebe, E.; Kállay, A.A.; Póliska, C. The firing properties of the biofraction and RDF pellets. *Mater. Sci. Eng.* **2019**, *44*, 67–78.
65. Singh, A.; Chandel, M.K. Physicochemical and FTIR spectroscopic analysis of fine fraction from a municipal solid waste dumpsite for potential reclamation of materials. *Waste Manag. Res.* **2020**, *39*, 374–385. [CrossRef] [PubMed]
66. Gunaratne, T.; Krook, J.; Andersson, H.; Eklund, M. Guiding the future research on the valorization of shredder fine fractions: A review of four decades of research. *Detritus* **2020**, *9*, 150–164. [CrossRef]
67. Faitli, J.; Nagy, S.; Romenda, R.; Gombkötő, I.; Bokányi, L.; Barna, L. Assessment of a residual municipal solid waste landfill for prospective 'landfill mining'. *Waste Manag. Res.* **2019**, *37*, 1229–1239. [CrossRef] [PubMed]
68. Parrodi, J.C.H.; García Lopez, C.; Küppers, B.; Raulf, K.; Vollprecht, D.; Pretz, T.; Pomberger, R. Case study on enhanced landfill mining at Mont-Saint-Guibert landfill in Belgium: Characterization and potential of fine fractions. *Detritus* **2019**, *8*, 47–61. [CrossRef]
69. Parrodi, J.C.H.; Raulf, K.; Vollprecht, D.; Pretz, T.; Pomberger, R. Case study on enhanced landfill mining at Mont-Saint-Guibert landfill in Belgium: Mechanical processing of fine fractions for material and energy recovery. *Detritus* **2019**, *8*, 62–78. [CrossRef]
70. Parrodi, J.C.H.; Vollprecht, D.; Pomberger, R. Case study on enhanced landfill mining at Mont-Saint-Guibert landfill in Belgium: Physico-chemical characterization and valorization potential of combustibles and inert fractions recovered from fine fractions. *Detritus* **2020**, *10*, 44–61. [CrossRef]
71. Vollprecht, D.; Parrodi, J.C.H.; Lucas, H.I.; Pomberger, R. Case study on enhanced landfill mining at Mont-Saint-Guibert landfill in Belgium: Mechanical processing, physico-chemical and mineralogical characterization of fine fractions < 4.5 mm. *Detritus* **2020**, *10*, 26–43. [CrossRef]
72. Weber, K.; Quicker, P.; Hanewinkel, J.; Flamme, S. Status of waste-to-energy in Germany, Part I—Waste treatment facilities. *Waste Manag. Res.* **2020**, *38* (Suppl. 1), 23–44. [CrossRef]
73. Viczek, S.; Aldrian, A.; Pomberger, R.; Sarc, R. Determination of the material-recyclable share of SRF during co-processing in the cement industry. *Resour. Conserv. Recycl.* **2020**, *156*, 104696. [CrossRef]
74. Austrian Federal Ministry for Climate Action, Environment, Energy, Mobility, Innovation and Technology. Federal Waste Managment Plan 2017. Part 1. Available online: https://www.bmk.gv.at/themen/klima_umwelt/abfall/aws/bundes_awp/bawp.html (accessed on 24 November 2020).
75. Van Caneghem, J.; Verbinnen, B.; Cornelis, G.; de Wijs, J.; Mulder, R.; Billen, P.; Vandecasteele, C. Immobilization of antimony in waste-to-energy bottom ash by addition of calcium and iron containing additives. *Waste Manag.* **2016**, *54*, 162–168. [CrossRef]
76. Winter, B.; Szednyj, I.; Reiseinger, H.; Böhmer, S.; Janhsen, T. Abfallvermeidung und Verwertung: Aschen, Schlacken und Stäube in Österreich. Umweltbundesamt. 2005. Available online: https://www.umweltbundesamt.at/fileadmin/site/publikationen/rep0003.pdf (accessed on 24 November 2020).
77. Hans van der Sloot Consultancy; DH; ECN. HP Classification of European Incinerator Bottom Ash (IBA). Assessment of Hazardous Properties (HPs) of IBA; Executive Summary of Report Produced for CEWEP. Available online: https://www.cewep.eu/guidance-on-classification-of-iba/ (accessed on 15 December 2020).
78. Laner, D. Understanding and Evaluating Long-Term Environmental Risks from Landfills. Ph.D. Thesis, Technical Univeristy Vienna, Vienna, Austria, 2011. Available online: https://repositum.tuwien.at/handle/20.500.12708/9360 (accessed on 15 December 2020).
79. Quicker, P. Ein gespenst geht um in Europa. *Müll Abfall* **2019**, *5*, 221. Available online: https://muellundabfall.de/.download/_sid/MFYP-556488-cYzP/147572/mua_20190501.pdf (accessed on 2 December 2020).
80. Zaini, I.N.; Yang, W.; Jönsson, P.G. Steam gasification of solid recovered fuel char derived from landfill waste: A kinetic study. *Energy Procedia* **2017**, *142*, 723–729. [CrossRef]
81. Xu, F.; Wang, B.; Yang, D.; Qiao, Y.; Tian, Y. The steam gasification reactivity and kinetics of municipal solid waste chars derived from rapid pyrolysis. *Waste Manag.* **2018**, *80*, 64–72. [CrossRef] [PubMed]
82. Lokahita, B.; Samudro, G.; Huboyo, H.S.; Aziz, M.; Takahashi, F. Energy recovery potential from excavating municipal solid waste dumpsite in Indonesia. *Energy Procedia* **2019**, *158*, 243–248. [CrossRef]
83. Stąsiek, J.; Szkodo, M. Thermochemical conversion of biomass and municipal waste into useful energy using advanced HiTAG/HiTSG technology. *Energies* **2020**, *13*, 4218. [CrossRef]
84. Zaini, I.N.; López, C.G.; Pretz, T.; Yang, W.; Jönsson, P.G. Characterization of pyrolysis products of high-ash excavated-waste and its char gasification reactivity and kinetics under a steam atmosphere. *Waste Manag.* **2019**, *97*, 149–163. [CrossRef]
85. Liu, H.; Wang, Y.; Zhao, S.; Hu, H.; Cao, C.; Li, A.; Yu, Y.; Yao, H. Review on the current status of the co-combustion technology of organic solid waste (OSW) and coal in China. *Energy Fuels* **2020**, *34*, 15448–15487. [CrossRef]
86. Sieradzka, M.; Rajca, P.; Zajemska, M.; Mlonka-Mędrala, A.; Magdziarz, A. Prediction of gaseous products from refuse derived fuel pyrolysis using chemical modelling software Ansys Chemkin Pro. *J. Clean. Prod.* **2020**, *248*, 119277. [CrossRef]
87. Al-Salem, S.M.; Yang, Y.; Wang, J.; Leeke, G.A. Pyro-oil and wax recovery from reclaimed plastic waste in a continuous auger pyrolysis reactor. *Energies* **2020**, *13*, 2040. [CrossRef]
88. Eke, J.; Bridgwater, A.V.; Onwudili, J.A. Energy recovery by fast pyrolysis of pre-treated trommel fines derived from a UK-based MSW material recycling facility. *J. Energy Inst.* **2020**, *93*, 2006–2016. [CrossRef]

89. Zaini, I.N.; Gomez-Rueda, Y.; García López, C.; Ratnasari, D.K.; Helsen, L.; Pretz, T.; Jönsson, P.G.; Yang, W. Production of H_2-rich syngas from excavated landfill waste through steam co-gasification with biochar. *Energy* **2020**, *207*, 118208. [CrossRef]
90. Sharma, S.; Basu, S.; Shetti, N.P.; Kamali, M.; Walvekar, P.; Aminabhavi, T.M. Waste-to-energy Nexus: A sustainable development. *Environ. Pollut.* **2020**, *267*, 115501. [CrossRef]
91. Turk, K. Proizvodnja Amonijaka iz Odpadnih Plinov. Ph.D.Thesis, Maribor University, Maribor, Slovenia, 2020.
92. Gomez-Rueda, Y.; Helsen, L. The role of plasma in syngas tar cracking. *Biomass Convers. Biorefinery* **2020**, *10*, 857–871. [CrossRef]
93. Rozzi, E.; Minuto, F.D.; Lanzini, A.; Leone, P. Green synthetic fuels: Renewable routes for the conversion of non-fossil feedstocks into gaseous fuels and their end uses. *Energies* **2020**, *13*, 420. [CrossRef]
94. Shahabuddin, M.; Alam, M.T.; Krishna, B.B.; Bhaskar, T.; Perkins, G. A review of producing renewable aviation fuels from the gasification of biomass and residual wastes. *Bioresour. Technol.* **2020**, *312*, 123596. [CrossRef] [PubMed]
95. Gomez-Rueda, Y.; Zaini, I.N.; Yang, W.; Helsen, L. Thermal tar cracking enhanced by cold plasma—A study of naphthalene as tar surrogate. *Energy Convers. Manag.* **2020**, *208*, 112540. [CrossRef]
96. Du, Y.; Liu, H.; Ren, W. Numerical investigations of a fluidized bed biomass gasifier coupling detailed tar generation and conversion kinetics with particle-scale hydrodynamics. *Energy Fuels* **2020**, *34*, 8440–8451. [CrossRef]
97. Lu, G.; Bai, Y.; Ren, L.; Wang, J.; Song, X.; Yu, G. Role of phosphorus (P) additive in the performance of char-supported nickel (Ni) catalyst on tar reforming. *Energy Convers. Manag.* **2020**, *225*, 113471. [CrossRef]
98. Fauzi, M.A.; Setyono, P.; Pranolo, S.H. Environmental assessment of a small power plant based on palm kernel shell gasification. *AIP Conf. Proc.* **2020**, *2296*, 020038. [CrossRef]
99. Gomez-Rueda, Y.; Zaini, I.N.; Yang, W.; Helsen, L. Seashell waste-derived materials for secondary catalytic tar reduction in municipal solid waste gasification. *Biomass Bioenergy* **2020**, *143*, 105828. [CrossRef]
100. Jagodzińska, K.; Zaini, I.N.; Svanberg, R.; Yang, W.; Jönsson, P.G. Pyrolysis of excavated waste from landfill mining: Characterisation of the process products. *J. Clean. Prod.* **2020**, *279*, 123541. [CrossRef]
101. Lucas, H.I.; García Lopez, C.; Hernández Parrodi, J.C.; Vollprecht, D.; Raulf, K.; Pomberger, R.; Pretz, T.; Friedrich, B. Quality assessment of nonferrous metals recovered by means of landfill mining: A case study in Belgium. *Detritus* **2019**, *8*, 79–90. [CrossRef]
102. Gigantino, M.; Kiwic, D.; Steinfeld, A. Thermochemical energy storage via isothermal carbonation-calcination cycles of MgO-stabilized SrO in the range of 1000–1100 C. *Sol. Energy* **2019**, *188*, 720–729. [CrossRef]
103. Pratticò, L.; Bartali, R.; Crema, L.; Sciubba, E. Analysis of radiation propagation inside a hierarchical solar volumetric absorber. *Proceedings* **2020**, *58*, 27. [CrossRef]
104. Wild, M.; Steinfeld, A. Modelling of a high-temperature thermochemical storage reactor with radial flow across an annular packed bed using the $CaCO_3$-CaO cycle as a model reaction. In Proceedings of the ISES Solar World Congress, Santiago, Chile, 4–9 November 2019. [CrossRef]
105. Morabito, T.; Sau, S.; Tizzoni, A.C.; Spadoni, A.; Capocelli, M.; Corsaro, N.; D'Ottavi, C.; Licoccia, S.; Delise, T. Chemical CSP storage system based on a manganese aluminium spinel. *Sol. Energy* **2020**, *197*, 462–471. [CrossRef]
106. André, L.; Abanades, S. Recent advances in thermochemical energy storage via solid-gas reversible reactions at high temperature. *Energies* **2020**, *13*, 5859. [CrossRef]
107. Farulla, G.A.; Cellura, M.; Guarino, F.; Ferraro, M. A review of thermochemical energy storage systems for power grid support. *Appl. Sci.* **2020**, *10*, 3142. [CrossRef]
108. Gigantino, M.; Sas Brunser, S.; Steinfeld, A. High-temperature thermochemical heat storage via the CuO/Cu_2O redox cycle: From material synthesis to packed-bed reactor engineering and cyclic operation. *Energy Fuels* **2020**, *34*, 16772–16782. [CrossRef]
109. Dou, X.; Ren, F.; Nguyen, M.Q.; Ahamed, A.; Yin, K.; Chan, W.P.; Chang, V.W.C. Review of MSWI bottom ash utilization from perspectives of collective characterization, treatment and existing application. *Renew. Sustain. Energy Rev.* **2017**, *79*, 24–38. [CrossRef]
110. Lucas, H.; Friedrich, B. Thermodynamics of conditioning MSWI bottom ash using SAF for usage in mineral products. In Proceedings of the 6th International Slag Valorisation Symposium, Mechelen, Belgium, 2–4 April 2019; pp. 57–60. Available online: http://www.metallurgie.rwth-aachen.de/new/images/pages/publikationen/thermodynamics_id_7584.pdf (accessed on 27 November 2020).
111. Lucas, H.; Maier, J.; Friedrich, B. The use of submerged arc furnace (SAF) as a robust technology for upcycling waste into standard mineral products for construction industry. *Miner. Nebenprodukte Abfälle* **2020**, *7*, 272–287. Available online: https://www.vivis.de/wp-content/uploads/2020/11/272-287_Friedrich.pdf (accessed on 27 November 2020).
112. Flesoura, G.; Garcia-Banos, B.; Catala-Civera, J.M.; Vleugels, J.; Pontikes, Y. In-situ measurements of high-temperature dielectric properties of municipal solid waste incinerator bottom ash. *Ceram. Int.* **2019**, *45*, 18751–18759. [CrossRef]
113. Singh, B.; Zafar, S. Understanding time-temperature characteristics in microwave cladding. *Manuf. Lett.* **2020**, *25*, 75–80. [CrossRef]
114. Flesoura, G.; Dilissen, N.; Dimitrakis, G.; Vleugels, J.; Pontikes, Y. A new approach for the vitrification of municipal solid waste incinerator bottom ash by microwave irradiation. *J. Clean. Prod.* **2020**, *284*, 124787. [CrossRef]
115. Flesoura, G.; Rabelo Monich, P.; Alarcón, R.M.; Desideri, D.; Bernardo, E.; Vleugels, J.; Pontikes, Y. Porous glass-ceramics made from microwave vitrified municipal solid waste incinerator bottom ash. *Constr. Build. Mater.* **2020**, *270*, 121452. [CrossRef]

116. Ascensão, G.; Marchi, M.; Segata, M.; Faleschini, F.; Pontikes, Y. Reaction kinetics and structural analysis of alkali activated Fe–Si–Ca rich materials. *J. Clean. Prod.* **2020**, *246*, 119065. [CrossRef]
117. Kurda, R.; Silva, R.V.; de Brito, J. Incorporation of alkali-activated municipal solid waste incinerator bottom ash in mortar and concrete: A critical review. *Materials* **2020**, *13*, 3428. [CrossRef] [PubMed]
118. Ascensão, G.; Beersaerts, G.; Marchi, M.; Segata, M.; Faleschini, F.; Pontikes, Y. Shrinkage and mitigation strategies to improve the dimensional stability of CaO-FeO_x-Al_2O_3-SiO_2 inorganic polymers. *Materials* **2019**, *12*, 3679. [CrossRef] [PubMed]
119. Mast, B.; Cambriani, A.; Douvalis, A.P.; Pontikes, Y.; Schroeyers, W.; Vandoren, B.; Schreurs, S. The effect of high dose rate gamma irradiation on the curing of CaO-FeO_x-SiO_2 slag based inorganic polymers: Mechanical and microstructural analysis. *J. Nucl. Mater.* **2020**, *539*, 152237. [CrossRef]
120. Ascensão, G.; Marchi, M.; Segata, M.; Faleschini, F.; Pontikes, Y. Increasing the dimensional stability of CaO-FeO_x-Al_2O_3-SiO_2 alkali-activated materials: On the swelling potential of calcium oxide-rich admixtures. *Detritus* **2019**, *8*, 91–100. [CrossRef]
121. Monich, P.R.; Romero, A.R.; Höllen, D.; Bernardo, E. Porous glass-ceramics from alkali activation and sinter-crystallization of mixtures of waste glass and residues from plasma processing of municipal solid waste. *J. Clean. Prod.* **2018**, *188*, 871–878. [CrossRef]
122. Bai, C.; Li, H.; Bernardo, E.; Colombo, P. Waste-to-resource preparation of glass-containing foams from geopolymers. *Ceram. Int.* **2019**, *45*, 7196–7202. [CrossRef]
123. Romero, A.R.; Salvo, M.; Bernardo, E. Up-cycling of vitrified bottom ash from MSWI into glass-ceramic foams by means of 'inorganic gel casting' and sinter-crystallization. *Constr. Build. Mater.* **2018**, *192*, 133–140. [CrossRef]
124. Yatsenko, E.A.; Goltsman, B.M.; Smoliy, V.A.; Yatsenko, L.A. Perspective and experience of use of glass fraction of solid municipal waste in the production of silicate heat-insulating materials. In Proceedings of the 2018 IEEE International Conference of Management of Municipal Waste as an Important Factor of Sustainable Urban Development (WASTE), St. Petersburg, Russia, 4–6 October 2018; pp. 46–48. [CrossRef]
125. Yatsenko, E.A.; Goltsman, B.M.; Ryabova, A.V. Complex protection of pipelines using silicate materials based on local raw materials of the Far East. *Mater. Sci. Forum* **2019**, *945*, 46–52. [CrossRef]
126. Yatsenko, E.A.; Goltsman, B.M.; Ryabova, A.V.; Smoliy, V.A. Peculiarities of the use of siliceous raw materials of the Russian Far East in the integrated pipeline protection. In Proceedings of the International Conference on Advanced Functional Materials and Composites (ICAFMC 2018), Barcelona, Spain, 5–7 September 2018; EDP Sciences: Barcelona, Spain, 2018. [CrossRef]
127. Yatsenko, E.A.; Goltsman, B.M.; Smolii, V.A.; Goltsman, N.S.; Yatsenko, L.A. Study on the possibility of applying organic compounds as pore-forming agents for the synthesis of foam glass. *Glass Phys. Chem.* **2019**, *45*, 138–142. [CrossRef]
128. Ramteke, D.D.; Hujova, M.; Kraxner, J.; Galusek, D.; Romero, A.R.; Falcone, R.; Bernardo, E. Up-cycling of 'unrecyclable' glasses in glass-based foams by weak alkali-activation, gel casting and low-temperature sintering. *J. Clean. Prod.* **2021**, *278*, 123985. [CrossRef]
129. Yatsenko, L.A.; Yatsenko, E.A.; Goltsman, B.M. Development of a mathematical model of the interrelation between the technological parameters of the synthesis and properties of foamed glass materials. *Mater. Sci. Forum* **2020**, *992*, 922–928. [CrossRef]
130. Petrella, A.; Spasiano, D.; Race, M.; Rizzi, V.; Cosma, P.; Liuzzi, S.; De Vietro, N. Porous waste glass for lead removal in packed bed columns and reuse in cement conglomerates. *Materials* **2019**, *12*, 94. [CrossRef] [PubMed]
131. Abdel-Gawwad, H.A.; Mohammed, M.S.; Heikal, M. Ultra-lightweight porous materials fabrication and hazardous lead-stabilization through alkali-activation/sintering of different industrial solid wastes. *J. Clean. Prod.* **2020**, *244*, 118742. [CrossRef]
132. Bisht, K.; Kabeer, K.S.A.; Ramana, P.V. Gainful utilization of waste glass for production of sulphuric acid resistance concrete. *Constr. Build. Mater.* **2020**, *235*, 117486. [CrossRef]
133. Cristelo, N.; Segadães, L.; Coelho, J.; Chaves, B.; Sousa, N.R.; de Lurdes Lopes, M. Recycling municipal solid waste incineration slag and fly ash as precursors in low-range alkaline cements. *Waste Manag.* **2020**, *104*, 60–73. [CrossRef]
134. Liu, B.; Yang, Q.W.; Zhang, S.G. Integrated utilization of municipal solid waste incineration fly ash and bottom ash for preparation of foam glass-ceramics. *Rare Met.* **2019**, *38*, 914–921. [CrossRef]
135. Romero, A.R.; Tamburini, S.; Taveri, G.; Toušek, J.; Dlouhy, I.; Bernardo, E. Extension of the 'inorganic gel casting' process to the manufacturing of boro-alumino-silicate glass foams. *Materials* **2018**, *11*, 2545. [CrossRef]
136. Chen, F.; Zhang, W.; Liu, S. Porous glass-ceramics derived from MgO-CuO-TiO_2-P_2O_5 glasses with different additions of Fe_2O_3. *Ceram. Int.* **2020**, *46*, 6560–6566. [CrossRef]
137. Korat, L.; Ducman, V. Characterization of fly ash alkali activated foams obtained using sodium perborate monohydrate as a foaming agent at room and elevated temperatures. *Front. Mater.* **2020**, *7*, 308. [CrossRef]
138. Teoh, F.; Veksha, A.; Chia, V.W.; Udayanga, W.C.; Mohamed, D.K.B.; Giannis, A.; Lim, T.-T.; Lisak, G. Nickel-based catalysts for steam reforming of naphthalene utilizing gasification slag from municipal solid waste as a support. *Fuel* **2019**, *254*, 115561. [CrossRef]
139. Xi, C.; Zhou, J.; Zheng, F.; Gao, J.M.; Hu, P.; Li, Y.; Liu, J.L. Conversion of extracted titanium tailing and waste glass to value-added porous glass ceramic with improved performances. *J. Environ. Manag.* **2020**, *261*, 110197. [CrossRef]
140. Kraxner, J.; Michalek, M.; Romero, A.R.; Elsayed, H.; Bernardo, E.; Boccaccini, A.R.; Galusek, D. Porous bioactive glass microspheres prepared by flame synthesis process. *Mater. Lett.* **2019**, *256*, 126625. [CrossRef]
141. Kumaravel, S.; Alagumurthi, N. Material removal characteristics of Al-SiO_2 composite in WEDM. *Epitoanyag J. Silic. Based Compos. Mater.* **2020**, *72*, 20–24. [CrossRef]

142. Hujova, M.; Rabelo Monich, P.; Sedlacek, J.; Hnatko, M.; Kraxner, J.; Galusek, D.; Bernardo, E. Glass-ceramic foams from alkali-activated vitrified bottom ash and waste glasses. *Appl. Sci.* **2020**, *10*, 5714. [CrossRef]
143. Rabelo Monich, P.; Dogrul, F.; Lucas, H.; Friedrich, B.; Bernardo, E. Strong porous glass-ceramics from alkali activation and sinter crystallization of vitrified MSWI bottom ash. *Detritus* **2019**, *8*, 101–108. [CrossRef]
144. Rabelo Monich, P.; Vollprecht, D.; Bernardo, E. Dense glass-ceramics by fast sinter-crystallization of mixtures of waste-derived glasses. *Int. J. Appl. Ceram. Technol.* **2020**, *17*, 55–63. [CrossRef]
145. Li, B.; Guo, Y.; Fang, J. Effect of crystallization temperature on glass-ceramics derived from tailings waste. *J. Alloy. Compd.* **2020**, *838*, 155503. [CrossRef]
146. Rabelo Monich, P.; Desideri, D.; Bernardo, E. Low temperature upcycling of vitreous byproduct of the MSW plasma processing into multifunctional porous glass-ceramics. *Adv. Appl. Ceram.* **2019**, *118*, 366–371. [CrossRef]
147. Salman, M.M.; Nhabih, H.T. Assessment of the partial and total replacement of feldspar by waste glass on porcelain properties. *J. Ceram. Process. Res.* **2020**, *21*, 371–377. [CrossRef]
148. Salman, M.M.; Radhi, N.S.; Sabr, O.H.; Nhabih, H.T. Utilization of diverse cheap materials as pore generating agent to manufacture low-cost porous ceramic. *Cerâmica* **2020**, *66*, 179–185. [CrossRef]
149. Rabelo Monich, P.; Romero, A.R.; Desideri, D.; Bernardo, E. Waste-derived glass-ceramics fired in nitrogen: Stabilization and functionalization. *Constr. Build. Mater.* **2020**, *232*, 117265. [CrossRef]
150. Rincón Romero, A.; Desideri, D.; Boccaccini, A.R.; Bernardo, E. Up-cycling of iron-rich inorganic waste in functional glass-ceramics. *Minerals* **2020**, *10*, 959. [CrossRef]
151. Zhang, Z.; Wang, J.; Liu, L.; Ma, J.; Shen, B. Preparation of additive-free glass-ceramics from MSW incineration bottom ash and coal fly ash. *Constr. Build. Mater.* **2020**, *254*, 119345. [CrossRef]
152. Sauve, G.; Van Acker, K. The environmental impacts of municipal solid waste landfills in Europe: A life cycle assessment of proper reference cases to support decision making. *J. Environ. Manag.* **2020**, *261*, 110216. [CrossRef]
153. Chabok, M.; Asakereh, A.; Bahrami, H.; Jaafarzadeh, N.O. Selection of MSW landfill site by fuzzy-AHP approach combined with GIS: Case study in Ahvaz, Iran. *Environ. Monit. Assess.* **2020**, *192*, 1–15. [CrossRef]
154. Manyoma-Velásquez, P.C.; Vidal-Holguín, C.J.; Torres-Lozada, P. Methodology for locating regional landfills using multi-criteria decision analysis techniques. *Cogent Eng.* **2020**, *7*, 1776451. [CrossRef]
155. Pokhrel, P.; Lin, S.L.; Tsai, C.T. Environmental and economic performance analysis of recycling waste printed circuit boards using life cycle assessment. *J. Environ. Manag.* **2020**, *276*, 111276. [CrossRef]
156. Cho, H.H.; Strezov, V. A comparative review on the environmental impacts of combustion-based electricity generation technologies. *Energy Fuels* **2020**, *34*, 10486–10502. [CrossRef]
157. Vaverková, M.D.; Adamcová, D.; Winkler, J.; Koda, E.; Petrželová, L.; Maxianová, A. Alternative method of composting on a reclaimed municipal waste landfill in accordance with the circular economy: Benefits and risks. *Sci. Total Environ.* **2020**, *723*, 137971. [CrossRef] [PubMed]
158. Tan, W.; Wang, S.; Liu, N.; Xi, B. Tracing bacterial and fungal necromass dynamics of municipal sludge in landfill bioreactors using biomarker amino sugars. *Sci. Total Environ.* **2020**, *741*, 140513. [CrossRef]
159. Efremenko, E.; Senko, O.; Stepanov, N.; Mareev, N.; Volikov, A.; Perminova, I. Suppression of methane generation during methanogenesis by chemically modified humic compounds. *Antioxidants* **2020**, *9*, 1140. [CrossRef]
160. Manasaki, V.; Palogos, I.; Chourdakis, I.; Tsafantakis, K.; Gikas, P. Techno-economic assessment of landfill gas (LFG) to electric energy: Selection of the optimal technology through field-study and model simulation. *Chemosphere* **2020**, *41*, 128688. [CrossRef] [PubMed]
161. Raksasat, R.; Lim, J.W.; Kiatkittipong, W.; Kiatkittipong, K.; Ho, Y.C.; Lam, M.K.; Font-Palma, C.; Zaid, H.F.M.; Cheng, C.K. A review of organic waste enrichment for inducing palatability of black soldier fly larvae: Wastes to valuable resources. *Environ. Pollut.* **2020**, *267*, 115488. [CrossRef] [PubMed]
162. Sauve, G.; Van Acker, K. To mine or not to mine: A review of the effects of waste composition, time and long-term impacts of landfills in the decision making for ELFM. In Proceedings of the 4th International Symposium on Enhanced Landfill Mining, Mechelen, Belgium, 5–6 February 2018; pp. 379–385.
163. Esguerra, J.L.; Krook, J.; Svensson, N.; Van Passel, S. Assessing the economic potential of landfill mining: Review and recommendations. *Detritus* **2019**, *8*, 125–140. [CrossRef]
164. Esguerra, J.L.; Svensson, N.; Krook, J.; Van Passel, S.; Van Acker, K. The economic and environmental performance of a landfill mining project from the viewpoint of an industrial landfill owner. In Proceedings of the 4th International Symposium on Enhanced Landfill Mining, Mechelen, Belgium, 5–6 February 2018; pp. 389–395.
165. Laner, D.; Esguerra, J.L.; Krook, J.; Horttanainen, M.; Kriipsalu, M.; Rosendal, R.M.; Stanisavljević, N. Systematic assessment of critical factors for the economic performance of landfill mining in Europe: What drives the economy of landfill mining? *Waste Manag.* **2019**, *95*, 674–686. [CrossRef]
166. Vaverková, M.D. Landfill impacts on the environment. *Geosciences* **2019**, *9*, 431. [CrossRef]
167. Sabour, M.R.; Alam, E.; Hatami, A.M. Environmental and economic assessment of enhanced landfill mining in Tehran. *Environ. Sci. Pollut. Res.* **2020**, *27*, 34469–34483. [CrossRef] [PubMed]
168. Martinovsky, J. Repräsentative Demokratie in Österreich am Beispiel der Volksabstimmung über das Kernkraftwerk Zwentendorf. Diploma Thesis, Vienna University, Wien, Austria, 2012. [CrossRef]

169. Einhäupl, P.; Krook, J.; Svensson, N.; Van Acker, K.; Van Passel, S. Enhanced landfill mining at the Remo Site: Assessing stakeholders' perspectives for implementation. In Proceedings of the 4th International Symposium on Enhanced Landfill Mining, Mechelen, Belgium, 5–6 February 2018; pp. 367–377.
170. Einhäupl, P.; Krook, J.; Svensson, N.; Van Acker, K.; Van Passel, S. Eliciting stakeholder needs—An anticipatory approach assessing enhanced landfill mining. *Waste Manag.* **2019**, *98*, 113–125. [CrossRef]
171. Pecorini, I.; Iannelli, R. Characterization of excavated waste of different ages in view of multiple resource recovery in landfill mining. *Sustainability* **2020**, *12*, 1780. [CrossRef]
172. Einhäupl, P.; Van Acker, K.; Svensson, N.; Van Passel, S. Developing stakeholder archetypes for enhanced landfill mining. *Detritus* **2019**, *8*, 109–124. [CrossRef]
173. Müller, A. *Baustoffrecycling*; Springer: Wiesbaden, Germany, 2018. [CrossRef]
174. Directive 2008/98/EC of the European Parliament and of the Council of 19 November 2008 on Waste and Repealing Certain Directives. Available online: http://data.europa.eu/eli/dir/2008/98/oj (accessed on 24 November 2020).
175. *Cambridge Dictionary: Residue*; Cambridge University Press: Cambridge, UK; Available online: https://dictionary.cambridge.org/de/worterbuch/englisch/residue (accessed on 15 March 2020).
176. Prochorow, A. (Ed.) *The Great Soviet Encyclopedia. 1970–1979*; Available online: https://encyclopedia2.thefreedictionary.com/Secondary+Raw+Material (accessed on 17 November 2020).

processes

MDPI

Article

Advancing Plastic Recycling by Wet-Mechanical Processing of Mixed Waste Fractions

Daniel Schwabl [1,*], Markus Bauer [1] and Markus Lehner [2]

1 Circulyzer GmbH, Peter-Tunner-Straße 19, 8700 Leoben, Austria; office@circulyzer.at
2 Montanuniversitaet Leoben, Chair of Process Technology and Environmental Protection, Franz-Josef-Straße 18, 8700 Leoben, Austria; markus.lehner@unileoben.ac.at
* Correspondence: daniel.schwabl@circulyzer.at

Abstract: In this paper, an arc was drawn over ten years of research activities from three chairs of the Montanuniversitaet Leoben, as well as industrial partners. The superior objective of this research effort was to develop a wet-mechanical process for the recovery of polyolefin concentrates (90 wt% polyolefins) from mixed waste fraction for use in chemical recycling and to advance this new technology to commercial maturity. As a bridge technology, it would close the gap between state-of-the-art dry processing of mixed plastic waste materials and chemical plastic recycling via thermo-chemical conversion. The methods used were mainly tested in a lab-scale plant with a throughput capacity of 50 to 200 kg/h depending on the bulk density of the used feedstock. Further studies for the treatment and usage of the main products and by-products, as well as chemical analyses of them, were completed during the investigation. Within these series of tests, polyolefin concentrates, which satisfied the requirements for chemical recycling, could be recovered. With these data, a concept for an industrial pilot plant was developed and evaluated from an economic point of view. According to this evaluation, the realization of such an industrial pilot plant can be recommended.

Keywords: plastics recycling; chemical recycling; wet-mechanical processing; polyolefins; circular economy

Citation: Schwabl, D.; Bauer, M.; Lehner, M. Advancing Plastic Recycling by Wet-Mechanical Processing of Mixed Waste Fractions. *Processes* **2021**, *9*, 493. https://doi.org/10.3390/pr9030493

Academic Editors: Daniel Vollprecht and Renato Sarc

Received: 30 November 2020
Accepted: 5 March 2021
Published: 9 March 2021

1. Introduction

1.1. Plastic Recycling in Europe

The report "Plastics Europe 2019" [1] states that 4–6% of the global petroleum production was processed into polymer products, which amounted to 359 Mt in 2018. Of the global production of polymer products, 61.8 Mt or 17% was related to Europe (In [1] Europe is defined as the 27 member states of the European Union as well as Norway, Great Britain, and Switzerland). The European demand for polymer products was 51.2 Mt in 2018. These polymers can be classified according to their usage or chemical polymer structure, see Tables 1 and 2.

Table 1. European polymer demand by application in 2018 [1].

Application	Fraction (%)	Period of Application
Packaging	39.9	Days to weeks
Others (furniture, medical, etc.)	20.8	Months to years
Building and construction	19.8	Years to decades
Automotive	9.9	Years
Electrical and electronics	6.2	Years
Household, leisure and sport, and	4.1	Months to years
Agriculture	3.4	Weeks to months

Table 2. European polymer demand by chemical structure in 2018 [1].

Polymer	Fraction (%)	Exemplary Application
Polyethylene (PE)	29.7	Bags, containers, toys, houseware
Polypropylene (PP)	19.3	Food packaging, automotive, pipes
Others	19.0	-
Polyvinylchloride (PVC)	10.0	Cable insulation, pipes, coverings
Polyurethanes (PUR)	7.9	Insulation foams, pillows, mattresses
Polyethylene terephthalate (PET)	7.7	Bottles for drinks and other liquids
Polystyrene (PS)	6.4	Packaging, insulations, electronics

Polyethylene and polypropylene (the so-called polyolefins) make up approximately 50% of Europe's polymer demand. Mainly used as packaging with a short period of use, they represent a disproportionate amount of the European waste streams. In 2018 these polymer wastes amounted to 29.1 Mt [1], which were mostly collected as comparably heterogenous industrial or municipal mixed waste fractions. These mixed fractions are frequently unsuitable for mechanical recycling due to their heterogeneity and possible contaminations. Consequently, they are processed mechanically to be used as solid recovered fuels for energy recovery. This makes sense from an economic point of view because plastics have a comparable energy content to heating oil [2]. In countries with poorly developed waste management systems and without legislation demanding certain recycling quotas, substantial shares of plastic waste are still landfilled.

1.2. Plastic Recycling in Austria

In the last decade, some progress has been made to shift the treatment of mixed plastic wastes to a more circular economy approach. Whilst countries with poorly developed waste management systems are catching up, ruling countries like Austria, Germany, and Sweden have not been able to increase the proportion of polymer wastes in mechanical recycling. This trend can be seen in Table 3, which shows the data for Austria's plastic waste treatment.

Table 3. Treatment options of plastic waste fractions in Austria from 2006 to 2016 [1].

Treatment Option	Proportion of Polymer Wastes (%)					
	2006	2008	2010	2012	2014	2016
Mechanical recycling	23	28	26	24	26	27
Energy recovery	69	65	71	75	73	72
Landfilling	8	7	3	1	1	1

It seems there is a limit in current plastic waste treatment, especially regarding mechanical recycling. Different reasons for this have been discussed in numerous papers like "Mechanical and chemical recycling of solid waste" by K. Ragaert et al. [3]. The paper gives an overview of the most common types of solid plastic waste (SPW) by their origin (industry or consumer) as well as their possible further treatment according to significant properties like homogeneity, contamination, and composition. The overview emphasizes that most SPWs with known homogeneous composition and a low share of contamination are currently used for mechanical recycling. Such SPWs are predominantly so-called post-industrial (PI) plastics, such as fall-out products, waste from production (cuttings and trimmings), or residues from granulation [4]. In contrast, post-consumer (PC) plastics are more heavily mixed, degenerated by their usage, and contaminated with organic fractions like food waste or non-polymers like paper or metals [5]. Thus, their usage is limited to energy recovery.

When looking at the status of the Austrian waste management system, it can be noted that only about 13% of the SPWs in 2015 were of industrial or commercial origin [6] and would therefore be suitable for mechanical recycling without further treatment.

These facts translate to the current state of the art in industrial plants for waste plastic recycling and processing in Austria. According to Bauer [7] and Kranzinger [6], of the 2.46 Mt of waste streams containing PC and PI plastics in 2015, about 1.3 Mt or 54% were used for energy recovery in waste incineration plants like Spittelau in Vienna. The remaining 46% were processed by four general types of industrial plants to recover plastics for recycling or to be used as solid recovered fuels of different quality grades. The required criteria according for these quality grades according to the Austrian Waste Management Act are summarized in Table 4 [8]. As they mainly focus on calorific value, particle size, and contamination, the processing in these four types of waste treatment plants will also focus on improving these parameters.

Table 4. Criteria for different quality grades of solid recovered fuel according to the Austrian waste management act [8].

Criteria	Quality Grade of Solid Recovered Fuel					
	Power Plant	Low (Grate Combustion)	Middle (Fluidized-Bed Combustion)	High (Calzinator Combustion)	Premium (Primary Combustion)	Furnace Combustion
Calorific value (MJ/kg)	11–15	12–16	12–16	11–18	22–25	>25
Particle size (mm)	<50 *	<300	<20–100	<50–80	<10–30	<10 *
Oversize share (wt%)	0	<3	<2	<1	<1	0
Impurities share (wt%)	1	<3	<1	0	<1	0
Chlorine share (wt%)	1.5	<1	<0.8	<0.8	<0.8	<2
Ash share (wt%)	35	-	<20	-	<10	<10

* pelletized.

1.2.1. Packaging Waste Sorting Plants

Packaging waste sorting plants processed about 225,000 tons [6] or 9% of the Austrian waste streams containing PC and PI plastics in 2015. The comparably homogeneous feedstock is classified using screens or air separation to sort out big foil, which mainly consists of polyolefins. The smaller particles are then separated into 2D and 3D fractions, for instance, by ballistic separation. Flat and light particles like foils tend to accumulate in the 2D fraction whilst heavy plastics, metals, glass, or ceramics are discharged in the 3D fraction. The 2D particles are then separated by polymer type by a series of optical sorters. For the share intended for mechanical recycling (about 35% of feedstock), additional purification steps can be implemented; the remaining plastics are used as high to premium-grade solid recovered fuels. The 3D particles are treated by magnetic and eddy-current separation for metal recovery and then used as middle to low grade solid recovered fuels.

1.2.2. Mechanical Waste Splitting Plants

If mechanical waste splitting facilities for PI and PC plastics are added up, they accounted for 665,000 tons [6] or 27% of Austrian waste streams containing plastics in 2015. These plants are characterized by consecutive steps of comminution, classification, and metal recovery. When processing feedstocks that are rich in plastics and have lower moisture contents due to a low biogenic share, air classification and optical sorters are applied to recover middle to high-grade solid recovered fuels. Feedstocks containing higher contamination with biogenic material are treated by consecutive sieves and air classification to recover low to middle grad solid recovered fuels.

1.2.3. Mechanical-Biological Treatment Plants

Plastic waste streams with a significant biogenic share are treated in special mechanical-biological plants. These plants processed about 240,000 tons [6] or 10% of the Austrian waste streams containing plastics in 2015. Because of their high share of biogenic material like food, additional steps for biodegradation are added to the standard mechanical treatment. The waste material is fed into rotting tunnels, where it is moistened with water, ventilated with air, heated up to 40–50 °C, and slowly rotated for days to weeks. As a result, the biogenic share is reduced, and low- to middle-grade solid recovered fuels, rotting gas for energy recovery, and a stabilized fraction for landfilling can be recovered.

1.2.4. Solid Recovered Fuels Preparation Plants

These plants are mainly for processing pretreated fractions from packaging waste sorting plants or other fractions considered as high-grade solid recovered fuels. In 2015 these plants processed about 123,000 tons [6] or 5% of the Austrian waste streams containing plastics to generate high or premium-grade solid recovered fuels of consistent quality. Since these plants use a pretreated feedstock mainly consisting of plastics, only additional purification steps are conducted. Comminution, air classification, metal recovery by magnetic and eddy-current separation, and sometimes optical sorting to reduce the polyvinylchloride (PVC) share, and thus the chlorine load, are therefore implemented in these plants.

1.3. Background of Research Activities

To stimulate mechanical recycling of PI and PC plastics despite technical obstacles and most waste management systems being incineration centered, the European Union proposed a directive for a circular economy. According to this directive, the proportion of post-consumer plastics recycling shall be increased to 55% by 2025 [9]. To achieve this ambitious goal, maybe another technical approach, namely chemical recycling, could be a part of the solution.

A promising option to recycle more mixed waste plastics materially is feedstock or chemical recycling via thermal cracking or pyrolysis, i.e., connected to a refinery. Once liquefied by thermo-chemical conversion, the obtained hydrocarbon intermediates can be easily treated by diverse state-of-the-art refining processes according to their chemical and physical properties. At least they can be converted to ethylene and propylene, which can be further used to synthesize new polyethylene and polypropylene again. As a consequence, a plastic material cycle can be established which is capable of meeting the ever-growing challenge of material aging and accumulation of aggregates. However, decades of research and a multiplicity of failures indicate that plastic feedstock recycling in thermo-chemical conversion units needs adequately prepared feedstock of specified quality and high quantities.

Most of these failures were not due to technological limitations but rather for economic reasons. The feedstock of the required quality and quantity could not be obtained at prices that would enable a competitive process. Thus, a chemical recycling process—the so-called thermo-chemical conversion—with rather low-quality requirements for its feedstock (see Table 5) when comparing it to data of other known processes, according to Tukker [10] (see Table 6), was developed. This development was paired with research on a supplementary treatment process to ensure the long-term availability of the necessary feedstock.

Table 5. Feedstock requirements for thermo-chemical conversion [11].

Parameter	Requirement
Particle size (mm)	<30–40
Bulk density (kg/m^3)	50–100
Calorific vale (MJ/kg)	>30
Polyolefin share (wt%)	>90
Impurity share (wt%)	<10
Moisture (wt%)	<20
Halogen share (wt%)	<2

Table 6. Feedstock requirements for different chemical recycling processes [10].

Parameter	Feedstock Requirements				
	Texaco [1]	Polymer-Cracking [2]	BASF [3]	Veba Combi [4]	SVZ [5]
Mechanical Parameters					
Particle size (mm)	<10	20–1	-	<10	80–20
Fines share (<250 µm) (wt%)	<1	<1	-	-	-
Bulk density (kg/m^3)	>100	400–300	-	>300	-
Moisture (wt%)	<5	<1	-	<1	-
Ash share (wt%)	<6	5–2	-	-	<10
Plastics					
Total plastic share (wt%)	>90	95–90	-	>90	-
Polyolefin share (wt%)	-	80–70	-	-	-
PVC share (wt%)	<10	4–2	<5	<4	<2
Polystyrene share (wt%)	-	30–15	-	-	-
PET share (wt%)	-	5–3	-	-	-
Impurities					
Metal share (wt%)	<1	<1	-	<1	-
Paper share (wt%)	<10	-	-	-	-

[1] Texaco Gasification Process; [2] Polymer Cracking Process; [3] BASF (Badische Anilin- & Soda-Fabrik) Conversion Process; [4] Veba Combi Cracking Process; [5] SVZ (Sekundärrohstoff-Verwertungszentrum Schwarze Pumpe GmbH) Gasification Process.

This was the starting point for ten years of interdisciplinary research activities. The Chairs for Process Technology and Industrial Environmental Protection, Mineral Processing and Waste Processing Technology and Waste Management at Montanuniversitaet Leoben, as well as industrial partners, tried to develop, investigate, and optimize such a novel treatment process for mixed waste plastics. This treatment process should close the gap between the already existing dry mechanic processing and a thermo-chemical conversion in a refinery.

2. Materials and Methods

2.1. Feedstock

A broad variety of waste streams containing PC as well as PI plastics were taken into consideration as potential feedstock. According to the Austrian waste management plan 33, waste code numbers of such waste streams could be identified. To narrow them down, three criteria were set to determine whether a waste stream could potentially serve as feedstock:

1. The waste stream must be non-hazardous
2. The waste stream must have a share of polyolefins of at least 20 wt%
3. The waste stream must provide a mass flow of at least 20,000 t per year in Austria (except material from landfill mining)

With these criteria, 16 waste code numbers could be identified as promising feedstock, which were assorted into 6 feedstock groups: plastic foils, plastic containers, lightweight packaging, substitute fuels, rejects from paper recycling, and landfill mining. The latter feedstock was added due to its environmental importance (with an estimated 2.2 Mt in Austrian landfills [12]) even though no mass flow per year, as well as no share of polyolefins, could be specified. A detailed table of these feedstocks can be found in [12].

Samples of these 6 groups were collected in different Austrian waste processing facilities and examined for their suitability. By gathering additional data from waste companies, treatment facilities, and literature and combining them with the results of examinations, a review of the total polyolefin-rich waste streams emerging in the Austrian waste system could be achieved. According to Kranzinger [6], as can be seen in Table 7, more than half of the Austrian waste polyolefins were of municipal origin. Additionally, considerable amounts of industrial and commercial sources representing at least 430,000 tons per year in 2015 [6] could be used. (Potential feedstock from landfill mining was not taken into consideration for this estimation.) The most apparent difference of these sources is their polyolefin share which is comparatively high for industrial wastes. Of these 430,000 tons, only 60,000 (~15%) are mechanically recycled, according to this analysis.

Table 7. Available post-consumer polyolefins in Austria in 2015 [6].

Source	Total Amount (t/y)	Polyolefins (t/y)	Polyolefin Share (wt%)
Municipal	1,656,600	238,100	14
Commercial	600,000	84,000	14
Industrial	199,600	106,900	54

t/y: tons per year.

Samples of different waste streams (solid recovered fuels, mixed plastics, rejects from paper recycling, landfill mining, etc.) were prepared for treatment in the erected lab-scale plant by shredding them to a particle size of <20 mm because of the large number of input materials from different waste processing plants all over Austria three feedstocks were chosen as representatives and are further discussed in this paper.

Pictures of these three samples representing a feedstock with a high (A), middle (B), and low (C) share of polyolefins can be seen in Figure 1. Table 8 depicts their determined average mass fraction of polyolefins ($\overline{w}_{PO}$) and the eightieth percentile of the particle size distribution; particle size of 80% cumulative passing (p_{80}).

Figure 1. Waste fraction A, B, and C (left to right) [13] (Reproduced with permission from Schwabl D.; Proceedings of tenth "Recycling und Rohstoffe" Conference published by TK-Verlag, Berlin, 2017).

Table 8. Parameters of representing feedstock samples for high (A), middle (B), and low (C) share of polyolefins [13].

Waste Fraction	Origin	Composition	p_{80} (mm)	$\overline{w}_{PO}$ (wt%)
A	Lightweight packaging	2D and 3D plastics with superficial contamination	7.8	55.8
B	Middle-grade substitute fuels	2D and 3D plastics, polystyrene, wood, and paper textiles	9.8	34.4
C	Landfill mining	Soil, wood, paper, textiles, and plastics	10.9	11.8

p_{80}: particle size of 80% cumulative passing; $\overline{w}_{PO}$: average mass fraction of polyolefins.

2.2. *The Lab-Scale Plant*

A lab-scale plant was built to examine the possibility of recovering polyolefins from the determined feedstocks. The plant basically consisted of two core components, which each conducted a certain kind of density separation to recover a light (LF), middle (MF), and a heavy fraction (HF), as can be seen in Figure 2.

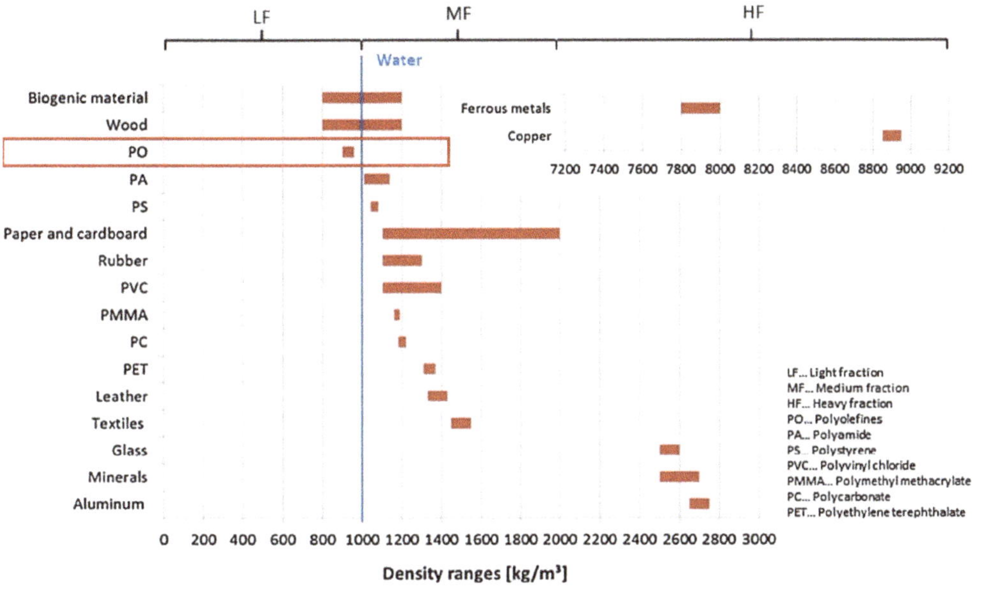

Figure 2. Density ranges of common components of mixed waste fractions and their respective product of the separation process [14].

2.2.1. Jig

The first separation step was conducted by a jig to partition heavy impurities like glass, metals, stones, and ceramics from light impurities and the contained plastics. The jig, as can be seen in Figure 3 on the left, which consisted of a stationary upper and movable lower tray. The former was equipped with a screen lining to collect the fed particles, and the latter executed an alternating up and down movement driven by an eccentric gear to generate a periodical upstream when filled with water. Consequently, the induced upstream scatters the particles in the bed and starts layering them according to their initial drop acceleration, which is mainly a function of particle size, shape, and density. In the case of considered post-consumer plastics, particle density is the dominant factor due to the plastics' huge density difference compared to other impurities. In the different density layers, the plastics concentrate in the upper and the impurities in the lower ones. Using a barrage, the plastics are transferred into the inter-fraction (IF) and the impurities into the heavy fraction (HF). Because the jig was constructed for mineral processing, it was suitable for abrasive materials and high throughputs. As a secondary effect, the particles of the IF are moistened and washed. This is beneficial to the second separation step due to the enhanced miscibility of these particles with the separation medium and the lower contamination of the used separation medium.

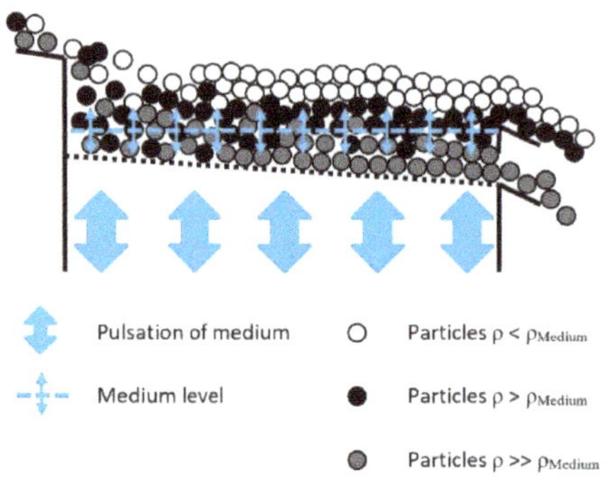

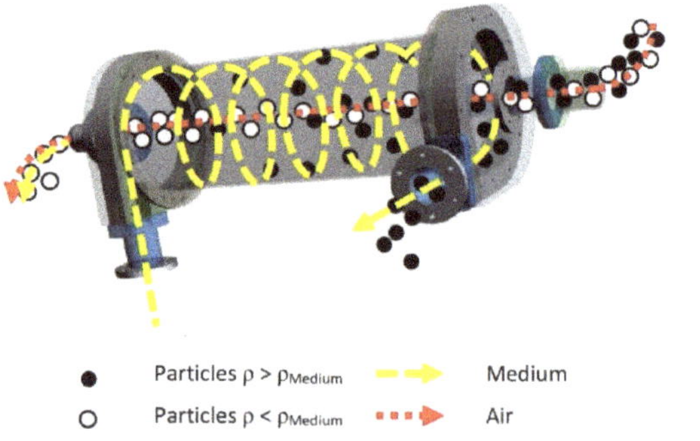

Figure 3. Function principle of the core components: Jig (**left**) and centrifugal force separator [13] (**right**) (Reproduced with permission from Schwabl D.; Proceedings of the tenth "Recycling und Rohstoffe" Conference published by TK-Verlag, Berlin, 2017).

2.2.2. Centrifugal Force Separator

In the second separation step, the polyolefins should be recovered from the IF, consisting basically of light impurities and mixed plastics. As can be seen in Figure 2, this can be done by density separation using water as a separation medium. A study of different density separation apparatuses [5] found a so-called centrifugal force separator (CFS) after the LarCoDeMS (Large Coal Dense Media Separation) -model, to be the most suitable for this step.

A CFS (as depicted in Figure 3, on the right) generally consists of an inclined cylinder with 4 tube-guided streams in and out of the separator. The feed material is separated by sink-/float-separation in a centrifugal field, thus having higher throughput than a sink-/float tank only using gravity. These centrifugal forces are induced by the separation medium, which is pumped into the separator tangentially at the lower end. Due to the shape of the CFS, the separation medium builds up a medium vortex with an air core and leaves the separator tangentially at the upper tangential outlet. The IF enters the separator axial from the upper central tube directly into the air core. The particles mix with the rotating separation medium, which induces centrifugal forces of about 15 times gravity. Particles with a higher density than the separation medium sink into the fluid and are dragged to the upper tangential outlet. There the particles exit the CFS as a middle fraction (MF) consisting of heavy plastics like polyethylene terephthalate (PET) or PVC and light impurities like paper and textiles. Particles with a density lower than the separation medium stay near the interface of the air core and the rotating fluid. Due to the inclination of the CFS, these particles are discharged with a small amount of the separation medium in the light fraction (LF), mostly consisting of polyolefins. The MF and LF are discharged onto separated compartments of a dewatering screen. The collected process water is stored in a basin and recirculated to the CFS.

2.2.3. Function Principle of the Lab-Scale Plant

The flow chart of the final construction of the lab-scale plant can be seen in Figure 4. As mentioned, the feed of mixed waste fractions must be shredded to a particle size of 20 mm or less to prevent plugging. The samples were fed to the jig via a vibrating feeder to guarantee a continuous material input stream and adjust its quantity. The feedstock sample was separated by density layering as mentioned above, and the HF and IF were discharged. Both fractions were dewatered afterward by the dewatering screen with a separate compartment for each fraction.

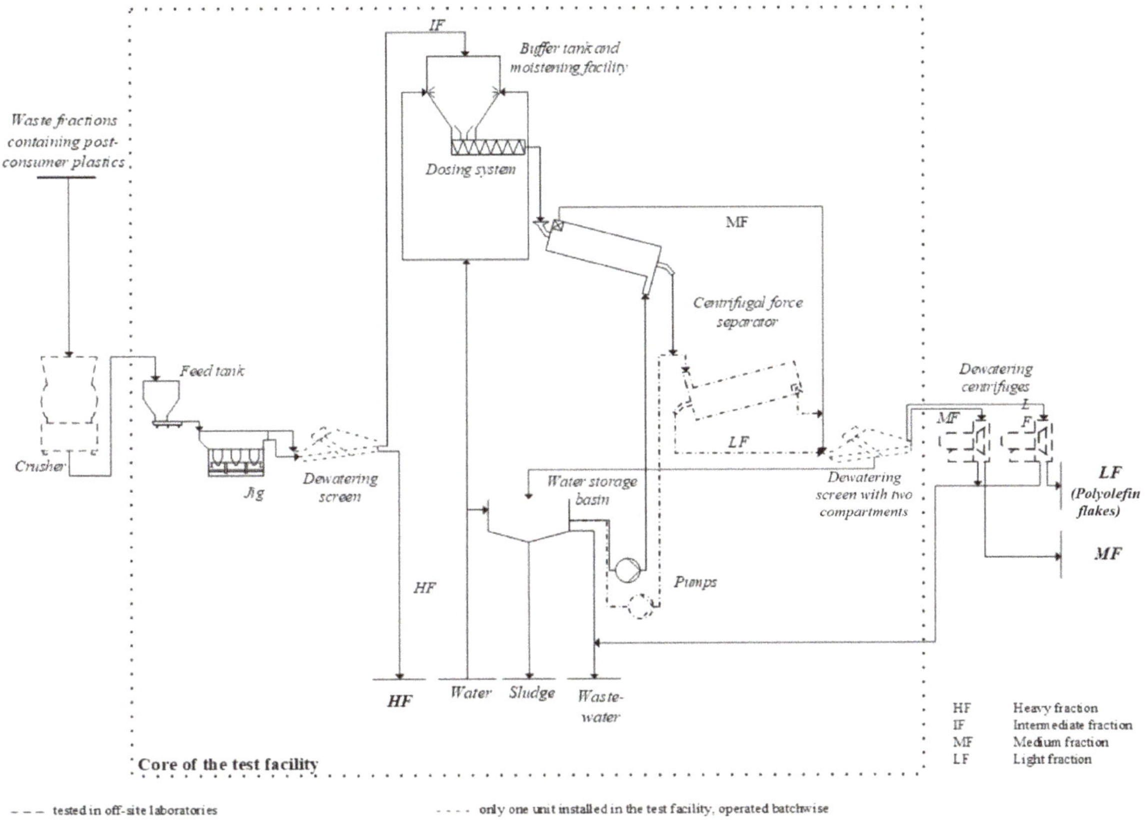

Figure 4. Flowchart of the lab-scale plant [14].

The lab-scale plant was only equipped with one water-circuit consisting of a pump with bypass control, a storage tank, and a dewatering screen. Thus, the jig and the CFS could not be operated in parallel. Consequently, after all the feedstocks had been processed by the jig, the IF was fed to the CFS. The CFS separated the MF and the LF, which were both again dewatered by the dewatering screen. If the desired share of polyolefins in the LF was not achieved, the generated fractions were processed multiple times in the CFS for further purification.

To summarize, the lab-scale plant can recover an HF fraction comprised of heavy impurities (glass, stones, ceramics, and metals), an MF (paper, rubber, and heavy plastics), and an LF (polyolefins) from different feedstock materials. The plant could process 50 to 200 kg of material per hour, depending on its bulk density. The operating parameters used for the jig and the CFS are summarized in Tables 9 and 10. These operating parameters were defined by preceding experiments [15,16].

Table 9. Operating parameters of the jig.

Parameter	Setting
Water pressure	1 bar
Water flow	16 m^3/h
Inclination of screen lining	4°
Lifting frequency	120 min^{-1}
Lifting height	20 mm

Table 10. Operating parameters of the centrifugal force separator (CFS).

Parameter	Setting
Water pressure	0.2 bar
Water flow	25 m^3/h
Inclination of CFS	40°

3. Results

3.1. Experimental Results

In the main test series regarding the recovery of polyolefins from different feedstocks, up to 50 kg of total input materials were processed in the lab-scale plant. The samples were processed as described in Section 2.2.3, with the exception that the LF was processed a second time with the CFS for further purification. All recovered fractions were examined by sink-/float-separation and drying at 80 °C to determine the share of polyolefins as well as the water content subsequently. Further samples of the main and by-products, as well as the process water, were examined for their pollution burden. Samples of the products were also analyzed chemically to figure out their calorific value. Further data from these analyses can be found in [14]. The data of the recovery of the polyolefins for the three representative samples are summarized in Table 11; a related flow chart can be seen in Figure 5. The results of the individual separation steps are shown in Tables 12–14.

Table 11. Mass fraction of feed (w_{Feed}), heavy (w_{HF}), middle (w_{MF}) and light fraction (w_{LF}) and corresponding polyolefin fraction(w_{PO}) of the test series at the lab-scale plant for feedstocks with a high (A), middle (B), and low (C) initial polyolefin share.

Waste Fraction	Feedstock [1]	Heavy Fraction		Middle Fraction		Light Fraction	
	w_{Feed} (wt%)	w_{HF} (wt%)	w_{PO} (wt%)	w_{MF} (wt%)	w_{PO} (wt%)	w_{LF} (wt%)	w_{PO} (wt%)
A	53.4	26.3	0.2	19.3	3.0	54.4	97.0
B	31.5	49.6	0.5	18.0	5.1	32.5	93.4
C	21.0	46.7	0.7	32.6	6.7	20.7	89.6

[1] w_{PO} differs from Table 7 because of the difference in sample size and examination.

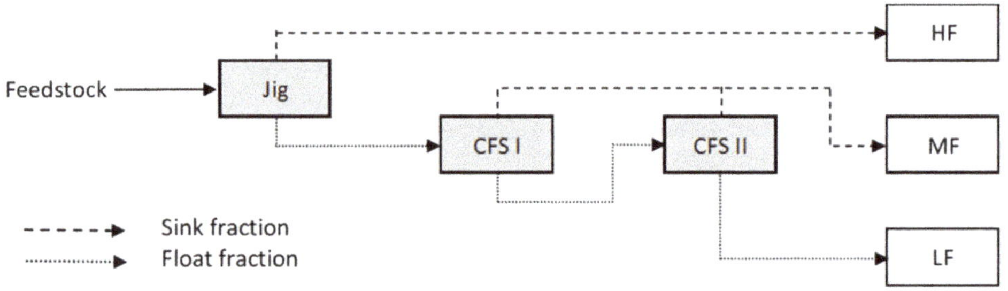

Figure 5. Flow chart of the conducted experiments; CFS I (centrifugal force separator's first pass), CFS II (centrifugal force separator's second pass), HF (heavy fraction), MF (middle fraction), and LF (light fraction).

Table 12. Mass fraction of the feed (w_{Feed}), heavy (w_{HF}) and inter fraction (w_{IF}) and corresponding polyolefin fraction(w_{PO}) of the jig for feedstocks with high (A), middle (B), and low (C) initial polyolefin share.

Waste Fraction	Feedstock [1]	Heavy Fraction		Inter Fraction	
	w_{Feed} (wt%)	w_{HF} (wt%)	w_{PO} (wt%)	w_{IF} (wt%)	w_{PO} (wt%)
A	53.4	26.3	0.2	73.7	72.4
B	31.5	49.6	0.5	50.4	62.0
C	21.0	46.7	0.7	53.3	38.8

[1] w_{PO} differs from Table 7 because of difference in sample size and examination.

Table 13. Mass fraction of the inter (w_{IF}), first middle ($w_{MF\,I}$) and first light fraction ($w_{LF\,I}$) and corresponding polyolefin fraction(w_{PO}) of the first CFS for feedstocks with high (A), middle (B), and low (C) initial polyolefin share.

Waste Fraction	Inter Fraction	Middle Fraction I		Light Fraction I	
	w_{IF} (wt%)	$w_{MF\,I}$ (wt%)	w_{PO} (wt%)	$w_{LF\,I}$ (wt%)	w_{PO} (wt%)
A	72.4	23.7	1.8	76.3	94.3
B	62.0	5.5	9.9	94.5	65.0
C	38.8	44.9	1.7	55.1	69.0

Table 14. Mass fraction of the first light fraction ($w_{LF\,I}$), second middle ($w_{MF\,II}$) and second light fraction ($w_{LF\,II}$) and corresponding polyolefin fraction(w_{PO}) of the second CFS for feedstocks with high (A), middle (B), and low (C) initial polyolefin share.

Waste Fraction	Light Fraction I	Middle Fraction II		Light Fraction II	
	$w_{LF\,I}$ (wt%)	$w_{MF\,II}$ (wt%)	w_{PO} (wt%)	$w_{LF\,II}$ (wt%)	w_{PO} (wt%)
A	94.3	3.5	13.2	96.7	97.0
B	65.0	31.9	4.3	68.2	93.3
C	69.0	29.5	19.8	70.5	89.6

3.2. Business Case

As the results of the test series conducted with the lab-scale plant were very promising, a business case for an industrial pilot plant was designed and evaluated from an economic point of view. Knowing that former processes and plants failed due to economic rather than technical reasons, it was deemed vital to also focus on this challenge.

This evaluation was based on the experimental as well as additional data from literature [17–20] and interviews with experts from the waste processing industry and the Montanuniversitaet Leoben. The plant was designed for a throughput of 95,000 tons per year of mixed plastic waste with an average polyolefin share of 25.5 wt%. Moreover, the estimation hypothesizes 3500 h of operation per year, producing 25,000 tons per year of a 90 wt% pure polyolefin flake. These and further basic parameters for the business case are depicted in Table 15.

Table 15. Material and production parameters for the business case.

	Feedstock	Polyolefin Flakes
Mass stream (moist)	95,000 t/y 27.1 t/h	25,000 t/y 6.2 t/h
Water content	15.0%	15.0%
Mass stream (dry)	82,500 t/y 23.6 t/h	21,800 t/y 6.2 t/h
Polyolefin fraction	25.5 wt%	90.0 wt%
Mass fraction	100.0 wt%	26.4 wt%

3.2.1. Plant Design

The basic design of the plant can be seen in Figure 6, which consisted of optional dry processing and wet-mechanical processing units. The dry processing unit was optional because, in many currently operating waste processing facilities, this part already exists (compare Section 1.2). Thus, the following discussion will focus only on the plant's wet-mechanical processing section.

The incoming feedstock from the dry processing facility was placed into a dosing bunker to cope with material throughput fluctuations and to supply the plant with a steady mass stream of input material. The feedstock was firstly treated by a flip-flop screen to separate small particles as well as fine biological and mineral impurities. This fine fraction must be separated prior to the wet processing steps to reduce the formation of sludge. Afterward, the feedstock was processed by sensor-based sorting devices to eject the PVC-plastics and decrease the chlorine burden. The latter processing step could be neglected if the feedstock is low in PVC. The data for the optical sorter was based on the findings in [21] can be seen in Table 16.

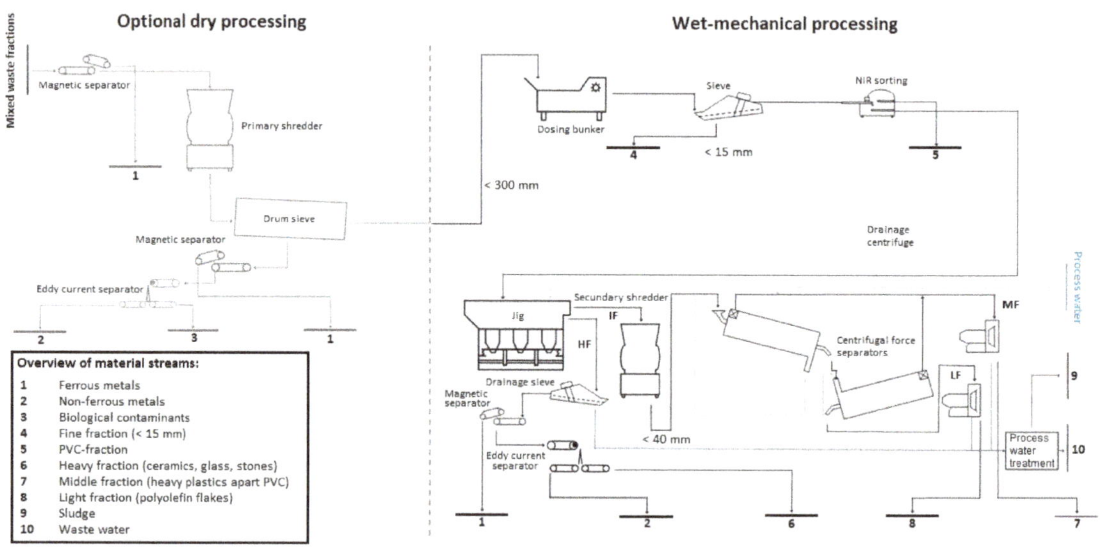

Figure 6. Flow chart of the designed industrial pilot plant [22] (Reproduced with permission from Schwabl D.; Proceedings of the fifteenth Recy and DepoTech-Conference; published by Chair of Waste Processing Technology and Waste Management (Montanuniversitaet Leoben; 2020).

Table 16. Operating parameters of the assumed NIR (Near Infrared) sorter.

Type	Redwave Multiplexer
Manufacturer	REDWAVE Waste GmbH
Wavelength	1000–2500 nm
Light stripes	64
Scanning frequency	50 Hz
Bandwidth	1 m

After these separation steps, the wet-mechanical processing section consisting of a jig and two CFS was conducted. The preprocessed material was fed to the jig and separated into the HF and IF as outlined in Section 2.2.1. The HF was dewatered and passed through an additional magnetic and eddy current separator to recover ferrous and non-ferrous metals. The IF can be comminuted by a secondary shredding step to adapt the particle size before entering the CFS if necessary. In the CFS, the IF was separated into the LF and HF by density separation, as described in Section 2.2.2. The LF and MF were discharged into centrifuges and dewatered for further use.

All accruing process waters from dewatering were collected and treated by removing sludge and small particles before being recirculated. The process water of the CFS, as well as the watering of the MF and LF, were recirculated to a separate water circuit from the process water of the jig and the dewatering of the HF. This was due to the higher contamination with sludge and fine particles, which could otherwise shift the required density of the CFS' separation medium. Despite these measures, about 10% of the process water had to be replaced every hour.

3.2.2. Economic Evaluation

Following the described design, a plant with a throughput of 95,000 tons of mixed waste material per year would require a capital expenditure of about 4 million € for the basic equipment (Table 17 shows their distribution to the different plant sections). As can be seen in Table 17, the dewatering and sensor-based sorting sections are the most expensive ones. To calculate the total capital expenditure, supplement factors were used to include costs for site area, buildings, pipes, measurements and controls, electrics, maintenance sites, construction, engineering, permits, and contingencies. According to M. Peters and K. Timmerhaus [20], these supplement factors sum up to 250% for such a combined dry and wet processing plant. Therefore the total capital expenditure could be estimated as 10 million € if planned in a green-field scenario without additional water treatment.

Table 17. Capital expenditure for equipment per plant section.

Plant Section	Share of Equipment Costs (%)
Feeding	1.8
Sieving	6.0
Optical sorting	31.3
Jig and metal recovery	10.0
Secondary shredding	10.8
CFS and steel construction	4.5
Dewatering	31.8
Process water treatment	4.0

Knowing the total capital expenditure, it was possible to calculate the operational costs of the designed plant. If operated with two shifts for five days a week, these costs could be estimated to be 150 € per ton of feedstock or 37.5 € per ton of light fraction, respectively. The distribution of the operational costs per section can be seen in Table 18.

Table 18. Operational expenditure per section.

Section	Share of Operational Costs (%)
Plant depreciation	35.2
Water supply	16.0
Wastewater disposal	13.6
Maintenance	8.1
Staff	17.3
Energy	9.8

To determine if such a plant could be operated economically, the potential income and costs of the feedstock, main products, and by-products must be considered according to their expected mass shares. Table 19 shows this estimation according to income (negative sign), costs (positive sign), and the input and output streams' mass share. When these incomes and costs were summed according to the expected mass-share, an income of 54.1 € per ton of manufactured light fraction (output perspective) or of 205.5 € per ton of processed feedstock (input perspective) could be estimated.

Table 19. Income and costs of the feedstock and produced main and by-products.

Material Stream	Mass Share (%)	Usage	Income/Costs (€/t) Per Ton of		
			Material	Feedstock	LF
Homogeneous residues	23	-	−50 [1]	−11.5	−43.7
Heterogeneous residues	77	-	−75 [1]	−57.8	−219.5
Ferrous metals	3	Mechanical recycling	−105 [2]	−2.9	−11.0
Non-ferrous metals	2	Mechanical recycling	−550 [2]	−10.1	−38.3
PVC-concentrate	2	Thermal disposal	200 [3]	3.7	13.9
Light fraction (LF)	25	Chemical recycling	0 [4]	0.0	0.0
Middle fraction (MF)	51	Energy recovery	20 [5]	10.3	39.0
Heavy fraction (HF)	9	Landfilling	35 [5]	3.2	12.2
Fine fraction	5	Thermal disposal	80 [5]	3.7	13.9
Sludge	4	Thermal disposal	200 [3]	7.3	27.9
Total				−54.1	−205.5

[1] Deducted from common co-payments for secondary fuels; [2] Deducted by interviews with scrap dealers and literature; [3] Estimation for disposal in an incineration plant; [4] Price is unknown and therefore set to 0; [5] Estimation by interviews with plant operators and literature [17–20].

By subtracting the operational costs of the income derived from the feedstock, by-products, and main products, a possible profit of about 16.6 € per ton of feedstock or 55.5 € per ton of light fraction could be estimated. With these values, the amortization time of the designed plant could be assessed to be around six years using a discounted cash-flow method.

4. Discussion

Regarding the results of the test series conducted with the lab-scale plant, the recovery of a polyolefin-rich light fraction (LF) by using the combination of a jig and two centrifugal force separators could be accomplished. The polyolefin share of these LF was at least 90 wt% whether a feedstock with high (>50 wt%), middle (50–30 wt%), or low (>30 wt%) initial share was used. This indicates that the discussed process should be suitable to recover polyolefins from a broad variety of feedstocks for chemical recycling by thermo-chemical conversion. However, the initial polyolefin share will determine the resulting proportion of middle (MF) and heavy fractions (HFs). If only a single feedstock with consistent composition rather than a mixture is used, it should also be assessed if all three separations stages are necessary. For instance, feedstocks with very few heavy impurities like metals, glass, and stones would probably not require a jig.

A plant using this technology and processing 95,000 t/y of mixed waste fractions features a capital expenditure of 10 million € in a rough estimation for a green-field scenario. This capital expenditure could be reduced by 1.2 million € if the processing step for sensor-based separation of PVC is not necessary. Additionally, about 30% of the capital expenditure could be cut if the plant was not built as a green-field scenario but as an expansion of an existing site. If the produced wastewater would exceed local wastewater treatment plant capacities, an additional 2 million € would be needed for the construction of such a facility.

The operational costs of such a plant are mainly influenced by the number of operational hours changing the amount of specific material throughput. A continuous operation of the plant would, for example, cut the operational costs by 33%. Of course, the biggest influence lies within the unknown income or costs of the light fraction. Hence it can be expected that the LF would generate a least a small revenue; this would substantially improve the economic basis of such a plant and, by that, lower the payback period from the calculated six to around four years.

With these combined results, the construction of such a pilot plant can, in principle, be recommended from an ecological as well as an economic point of view. However, a comprehensive assessment of the compositions and long-term availability of the intended feedstock will have to be conducted to get a sufficient basis for decision-making.

Author Contributions: Conceptualization, D.S. and M.B.; methodology, D.S. (experiments) and M.B. (plant design); validation, D.S. (plant design) and M.B. (experiments).; investigation D.S. (experiments) and M.B. (plant design); data curation D.S. (experiments) and M.B. (plant design); writing—original draft preparation D.S.; writing—review and editing M.B. and M.L.; visualization D.S. (experiments) and M.B. (plant design).; supervision M.B. and M.L.; project administration D.S.; funding acquisition M.B. and M.L. All authors have read and agreed to the published version of the manuscript.

Funding: This research was funded by the Österreichische Forschungsförderungsgesellschaft (FFG), grant number 844713 and 866712, and by OMV (Österreichische Mineralölverwaltung) Refining and Marketing GmbH.

Institutional Review Board Statement: Not applicable.

Informed Consent Statement: Not applicable.

Data Availability Statement: The data presented in this study are available on request from the corresponding author. The data are not publicly available due to their economic utilization in the start-up company of the corresponding author.

Acknowledgments: We would like to thank the teams of the chairs for Process Technology and Industrial Environmental Protection, Mineral Processing and Waste Processing Technology and Waste Management at Montanuniversitaet Leoben for their support during ten years of research. We thank the three heads of the chairs, Markus Lehner, Helmut Flachberger, and Roland Pomberger, as well as our long-time co-researcher Lukas Kranzinger. We also thank Wolfgang Hofer from OMV GmbH for supporting us with his knowledge, funding, ideas, and sample materials throughout the years.

Conflicts of Interest: The authors declare no conflict of interest. The funders had no role in the design of the study; in the collection, analyses, or interpretation of data; in the writing of the manuscript, or in the decision to publish the results.

References

1. Plasticseurope. Available online: https://www.plasticseurope.org/ (accessed on 13 October 2020).
2. Kumar, S.; Panda, A.K.; Singh, R.K. A review on tertiary recycling of high-density polyethylene to fuel. *Resour. Conserv. Recycl.* **2011**, *55*, 893–910. [CrossRef]
3. Ragaert, K.; Delva, L.; Van Geem, K. Mechanical and chemical recycling of solid plastic waste. *Waste Manag.* **2017**, *69*, 24–58. [CrossRef]
4. Ignatyev, I.A.; Thielemans, W.; Vander Beke, B. Recycling of Polymers: A Review. *ChemSusChem* **2014**, *7*, 1579–1593. [CrossRef] [PubMed]

5. Ragaert, K.; Hubo, S.; Leite, L.; Martins, C. Evaluation of post-industrial and post-consumer polyolefin-based polymer waste streams for injection moulding. In Proceedings of the 6th Polymers & Mould Innovations International Conference, Guimaraes, Portugal, 10–12 September 2014; pp. 201–206.
6. Kranzinger, L.; Pomberger, R.; Schwabl, D.; Flachberger, H.; Bauer, M.; Lehner, M.; Hofer, W. Output-Oriented Analysis of the Wet Mechanical Processing of Polyolefin-Rich Waste for Feedstock Recycling. *Waste Manag. Res.* **2018**, *36*, 445–453. [CrossRef] [PubMed]
7. Bauer, M.; Lehner, M. Review of post-consumer plastic preparation in Austria and new approaches for feedstock recycling. In Proceedings of the CSIR-Indian Institute of Petroleum (IIP) (Hrsg.): 7th International Symposium on Feedstock Recycling of Polymeric Materials (7th ISFR 2013), New Delhi, India, 23–26 October 2013; pp. 1–2.
8. Kunter, A.; Wellacher, M. Neue Entwicklungen bei der Ersatzbrennstoffaufbereitung. In Proceedings of the 10th Recy & DepoTech-Conference, Leoben, Austria, 3–5 November 2010; Chair of Waste Processing Technology and Waste Management: Leoben, Austria, 2016; pp. 653–656.
9. EC (European Commission). *EC (European Commission): Proposal for the Directive 94/62/EG of the European Parliament and of the Council 5 December 2015 on Packaging and Packaging Waste*; EC (European Commission): Brussels, Belgium, 2015.
10. Tukker, A. *Plastics Waste: Feedstock Recycling, Chemical Recycling and Incineration*; Smithers Rapra Technology: Shrewsbury, UK, 2002.
11. Bauer, M.; Lehner, M.; Hofer, W. *Prozesskette zum Stofflichen Recycling von Kunststoffabfällen, "Berg- und Hüttenmännische Monatshefte (BHM)"*; Austrian Society for Metallurgy and Materials (ASMET) and Bergmännischer Verband Österreich (BVÖ): Leoben, Austria, 2002; pp. 246–251.
12. Kranzinger, L.; Pomberger, R.; Schwabl, D.; Bauer, M. Quo vadis Kunststoffrecycling—Bestandsaufnahme der polyolefinen Kunststoffe in der österreichischen Abfallwirtschaft. In Proceedings of the 13th Recy & DepoTech-Conference, Lezoben, Austria, 8–11 November 2016; Chair of Waste Processing Technology and Waste Management: Leoben, Austria, 2016; pp. 583–588.
13. Schwabl, D.; Flachberger, H.; Kranzinger, L.; Bauer, M.; Hofer, W. Innovatives Verfahren zur Anreicherung von Polyolefin-Konzentraten aus industriellen Reststoffströmen zum Zwecke einer stofflichen Verwertung. In Proceedings of the 10th "Recycling und Rohstoffe" Conference, Berlin, Germany, 6–7 March 2017; Thomé-Kozmiensky, K.J., Goldmann, D., Eds.; TK-Verlag: Berlin, Germany, 2017; pp. 199–218.
14. Bauer, M.; Lehner, M.; Schwabl, D.; Flachberger, H.; Kranzinger, L.; Pomberger, R.; Hofer, W. Sink-Float Density Separation of Post-Consumer Plastics for Feedstock Recycling. *Mater. Cycles Waste Manag.* **2018**. [CrossRef]
15. Strzalkowski, J. Inbetriebnahme der Schwingesetzmaschine SK 3–10 von Siebtechnik mit Primärem sowie Sekundärem Aufgabegut. Master's Thesis, Chair for Mineral Processing, Montanuniversitaet Leoben, Leoben, Austria, 2017.
16. Schwabl, D.; Flachberger, H.; Kranzinger, L.; Bauer, M.; Hofer, W. Weiterentwicklung eines nassmechanischen Aufbereitungsverfahrens für Altkunststofffraktionen. In Proceedings of the 13th Recy & DepoTech-Conference, Leoben, Austria, 8–11 November 2016; Chair of Waste Processing Technology and Waste Management: Leoben, Austria, 2016; pp. 173–178.
17. Pretz, T.; Feil, A. Ungenutzte Potentiale in der Abfallwirtschaft. In Proceedings of the 14th Recy & DepoTech-Conference, Leoben, Austria, 14–16 March 2018; Chair of Waste Processing Technology and Waste Management: Leoben, Austria, 2018.
18. Thomè-Kosmiensky, K.J. *Thermische Abfallbehandlung*; EF-Verlag für Energie- und Umwelttechnik GmbH: Berlin, Germany, 1994.
19. Kranert, M.; Cord-Landwehr, K. *Einführung in die Abfallwirtschaft. 4., Vollständig Aktualisierte und erw*; Aufl. Wiesbaden; Vieweg+Teubner Verlag/GWV Fachverlage: Wiesbaden, Germany, 2010.
20. Peter, M.S.; Timmerhaus, K.D.; West, R.E. *Plant Design and Economics for Chemical Engineers*; McGraw-Hill Chemical Engineering Series; McGraw-Hill: New York, NY, USA, 2003.
21. Bauer, M.; Lehner, M. Abtrennung von Polyolefinen aus kunststoffreichen Abfallströmen mittels NIR-sensorbasierten Sortiertechnologie und Sortierzentrifuge. In Proceedings of the 9th Minisymposium der Verfahrenstechnik, Leoben, Austria, 17–18 April 2013; Chair of Process Technology and Environmental Protection: Leoben, Austria, 2013; pp. 92–96.
22. Schwabl, D.; Bauer, M.; Peyha, M.; Lehner, M. Überführung eines nass-mechanischen Aufbereitungsverfahrens für Altkunststoffe in den Pilotmaßstab. In Proceedings of the 15th Recy & DepoTech-Conference, Leoben, Austria, 18–20 November 2020; Chair of Waste Processing Technology and Waste Management: Leoben, Austria, 2020; pp. 129–134.

processes

MDPI

Article

Lithium-Ion Batteries as Ignition Sources in Waste Treatment Processes—A Semi-Quantitate Risk Analysis and Assessment of Battery-Caused Waste Fires

Thomas Nigl [1,*], Mirjam Baldauf [1], Michael Hohenberger [2] and Roland Pomberger [1]

[1] Chair of Waste Processing Technology and Waste Management, Montanuniversitaet Leoben, 8700 Leoben, Styria, Austria; mirjam.baldauf@unileoben.ac.at (M.B.); roland.pomberger@unileoben.ac.at (R.P.)
[2] Chair of Thermal Processing Technology, Montanuniversitaet Leoben, 8700 Leoben, Styria, Austria; michael.hohenberger@unileoben.ac.at
* Correspondence: thomas.nigl@unileoben.ac.at

Abstract: Increasing occurrences of waste fires that are caused by improperly discarded lithium-based portable batteries threaten the whole waste management sector in numerous countries. Studies showed that high quantities of these batteries have been found in several municipal solid waste streams in recent years in Austria. This article reveals the main influence factors on the risk of lithium-based batteries in their end-of-life and it focuses on the quantification of damages to portable batteries during waste treatment processes. Hazards are identified and analysed and potential risks in waste management systems are comprehensively assessed. In two scenarios, the results showed that the potential risks are too high to maintain a sustainable form of waste management. According to the assessment, a small fire in a collection vehicle is located in the risk graph's yellow region (as low as reasonably practicable, ALARP), while a fully developed fire in a treatment plant has to be classified as an unacceptable risk (red region of risk graph). Finally, basic recommendations for action were made.

Keywords: risk modelling; portable batteries; lithium batteries; fire hazards; waste management

Citation: Nigl, T.; Baldauf, M.; Hohenberger, M.; Pomberger, R. Lithium-Ion Batteries as Ignition Sources in Waste Treatment Processes—A Semi-Quantitate Risk Analysis and Assessment of Battery-Caused Waste Fires. *Processes* **2021**, *9*, 49. https://doi.org/10.3390/pr9010049

Received: 30 November 2020
Accepted: 24 December 2020
Published: 29 December 2020

1. Introduction

1.1. Background

In Austria and other European countries, like Germany, France or Sweden, the occurrences of waste fires has increased enormously in the last years [1–3]. The subsequent monetary and infrastructural losses have reached a new peak and they are threatening the whole waste industry sector [4].

Recent research showed that high amounts of portable batteries are improperly discarded in different municipal solid waste streams, such as residual household waste, lightweight packaging waste, or metal packaging waste [5]. While there is a plethora of different electrochemical systems, the average distribution is shifting more and more towards metal lithium and lithium-ion. That shift is accompanied by increased fire hazards and other safety challenges all along the value chain in the batteries' end-of-life.

1.2. Battery Content and Battery Composition

In Austria, the collection rate for portable batteries sank from 55% in 2015 to 45% in 2017 and it has stayed on that low level ever since [6,7]. A lot of the batteries that are not collected properly are falsely disposed in different municipal solid waste fractions [5]. Material flow analysis revealed that: (1) about 718 tonnes end up in residual household waste, (2) about 41 tonnes in lightweight packaging waste, and (3) about 17 tonnes in metal packaging waste.

For material flow analysis (MFA), [4] investigated the battery content and battery composition of the following waste streams: (1) residual household waste, (2) lightweight packaging waste, and (3) metal packaging waste. In addition to the quantitative weight-related values for MFA, particle-related data were published in [4] and they are given in Table 1.

Table 1. Particle-based data for portable and lithium batteries in various municipal waste streams (data from [4]).

Waste Streams	Waste Generation 2016 [t]	Portable Batteries [items/t]	Portable Batteries 2016 [items]	Lithium Batteries [items/t]	Lithium Batteries [items]
Residual waste	1,436,700	20	28,734,000	1	1,436,700
Lightweight packaging waste	156,700	14	2,193,800	0.8	125,360
Metal packaging waste	30,000	33	990,000	1	30,000

1.3. Hazards of Lithium-Ion Batteries in End-of-Life

Lithium-ion batteries may undergo thermal runaway and uncontrolled heat release when they are electrically (e.g., over-charged, deep discharged), mechanically (e.g., penetration, crushing, vibration), or thermally (e.g., externally heated) damaged due to the increased energy density [8,9]. Hence, they may act as ignition sources leading to fire incidents during waste management processes, especially in the case of thermal or mechanical abuse [4].

According to [10], even severe crushing of cells that are below approximately 50% state-of-charge (SOC) will not lead to a severe reaction. Golubkov et al. [11] further investigated the influence of SOC on the thermal runaway behaviour of the two lithium-ion battery subtypes and found out that, after thermal abuse, lithium-ion cells with $Li_x(Ni_{0.80}Co_{0.15}Al_{0.05})O_2$ cathodes (NCA) displayed an unmistakable thermal runaway when SOC was $\geq$25%. In the same way, lithium-ion cells with Li_xFePO_4 cathode (LFP) showed mild exothermic reactions when the SOC was $\geq$25% and pronounced thermal runaway when SOC was $\geq$50%. Liu et al. [12] revealed that lithium-ion batteries with a lithium nickel manganese cobalt oxide cathode (NMC) show thermal runaway behaviour when the SOC is $\geq$25%.

Furthermore, due to significant differences in safety performance, waste batteries impose a higher safety hazard than new batteries, the majority of which are thoroughly tested before market input [8].

1.4. State-of-Charge

Based on that, ref. [13,14] shed light on the distribution of SOC in end-of-life portable lithium-based batteries. ref. [13] found out that 11.4% of tested lithium-based batteries showed an SOC that was higher than 25% in the end-of-life. However, it is unknown how many batteries or cells were tested in the study. According to the results of [14], where 980 batteries were analysed, approximately 24% of the tested batteries showed an SOC higher than 25% (Table 2).

Table 2. Distribution of state of charge (SOC) of end-of-life lithium-ion batteries.

Share of Batteries/Cells with SOC $\geq$ 25%	Share of Batteries/Cells with SOC $\geq$ 50%	Number of Batteries/Cells Analysed	Reference
11.4	4.9	NA	[13]
23.6	12.0	980	[14]

1.5. Risk Analysis and Risk Assessment

Risk may be defined as consequences and its associated uncertainty or the combination of frequency or probability of occurrence and the severity or extent of damage [15]. When considering waste fires as the generalised hazard, the probability of an occurring fire incident is the first term, the expected financial losses the second term of the equation (Equation (1)):

$$R = probability\ of\ fire\ incident\ \times (expected)\ loss\ in\ case\ of\ the\ incident \quad (1)$$

If not, a single event is considered, but the sum of risks (e.g., for a single waste management company or the whole waste industry), Equation (2) is applied:

$$R = \sum_{\substack{for\ all \\ incidents}} probability\ of\ fire\ incident \times (expected)\ loss\ in\ case\ of\ the\ incident \quad (2)$$

1.6. Study Objectives

The article presents a new method for risk evaluation and the assessment of portable batteries and. especially, lithium-based batteries in waste management systems strongly focused on the first term of Equations (1) and (2).

The following research questions are answered:

1. What are the primary influence factors regarding the risk of portable batteries in waste management systems?
2. What are the qualitative hazards of portable batteries in different waste management systems?
3. What are the quantitative risks of portable batteries in different waste management systems?

2. Materials and Methods

Risk analysis is the systematic use of information in order to identify hazards and estimate risks to individuals, property, and the environment. There are three main steps in the analysis of risks: (1) hazard identification, (2) frequency analysis, and (3) consequences analysis [16].

Because the exact number of fires triggered by lithium-ion batteries is not even approximately known [1], no direct probability of occurrence can be achieved [13]. Therefore, an alternative approach is taken in this study to quantify the risks of lithium-ion batteries in waste management systems.

The probability of a battery-caused fire incident (FI_{BC}) can be expressed as the product of two terms: (1) the probability of the presence of a hazardous battery and (2) the probability of critical damage done to the present battery (Equation (3)):

$$FI_{BC} = P(presence\ of\ a\ hazardous\ battery) \times P(critical\ damage\ to\ the\ present\ battery) \quad (3)$$

Figure 1 displays the overall approach and the boundaries of this study. After the basic step of literature research, two different workflows were followed. The first was covered by two previous publications and it is about the presence of a hazardous battery; the second, is about the critical damage done to a present battery (which, consequently, leads to a battery-caused fire) and it is covered by this study. Afterwards, the results are combined in the overall risk modelling and assessment.

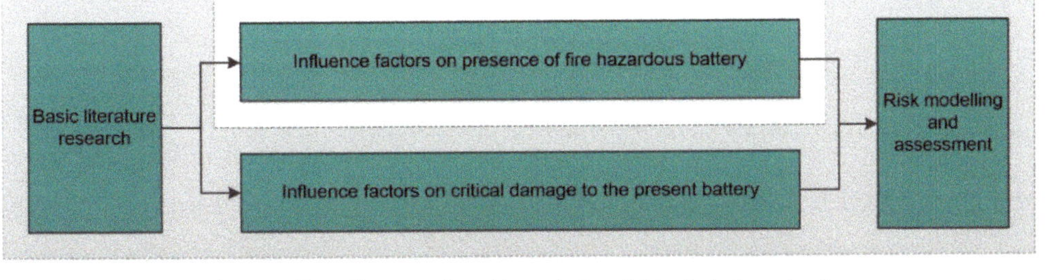

Figure 1. Overall approach and boundaries of this study (grey box).

More specifically, the first part of Equation (3) depends on the following influence factors (upper part of Table 3): (1) battery content in the respective waste stream [items/tonne],

(2) percentage of lithium batteries within these batteries (both [5]), and (3) the percentage of lithium batteries that contain a critical state of charge (SOC) [13]. The second part of the equation depends on the (1) number and intensity of potential damaging incidents, (2) the size and shape of the respective batteries, and (3) the energy content of the batteries and ambient conditions (lower part of Table 3).

Table 3. Overview of the relevant influence factors on the presence of a fire-hazardous battery (1.1–1.3) and on critical damage to the present battery (2.1–2.3).

Number	Influence Factor	Reference
(1.1)	Battery content in waste stream	[5]
(1.2)	Battery composition (share of Li-based batteries)	[5]
(1.3)	Share of Li-based batteries containing critical state-of-charge	[13,14]
(2.1)	Amount and intensity of potential damaging incidents	this study
(2.2)	Size and shape of respective portable batteries	this study
(2.3)	Energy content of the respective portable batteries and the surrounding conditions	this study

The parameters (1) the degree of damage, (2) size, and the shape (construction type) of the portable batteries were investigated for this study. Portable batteries that were collected during waste sorting and the characterisation campaign of the input fractions residual waste, lightweight packaging waste, metal packaging waste (compared to previous publication of [5]), and commercial waste and specific output waste fractions (of treatment plants for residual waste and lightweight packaging).

2.1. Degree of Damage

In Table 4, the description of the defined damage classes (DC) and respective degree of damage is given. DC 1 to 3 are considered as minorly damaged and DC 4 to 7 as majorly damaged. The classification was made visually during the manual documentation of the portable batteries in the laboratory. Sample pictures of the damage classes of portable batteries are given in Supplementary Materials.

Table 4. Overview of damage classes and description of the degree of damage.

Damage Class	Degree of Damage
DC 1	Undamaged or slightly scratched
DC 2	Superficially damaged (e.g., label torn off), slightly deformed (<10 vol.-%)
DC 3	Slightly deformed (<50 vol.-%), outer shell undamaged
DC 4	Severely deformed (>50 vol.-%), outer shell undamaged
DC 5	Deformed, outer shell damaged
DC 6	Crushed, destroyed
DC 7	Heavily bloated or post-thermal runaway

Thereafter, the distribution of the degree of damage in the investigated waste fractions was used in order to model the potential damages, the portable batteries must endure between both (1) the intention to dispose of and the arrival at the waste treatment site and (2) the arrival at the treatment site and after running through the treatment process. Because of financial and time restraints, this was done for the fractions of residual household waste and lightweight packaging waste only.

As a background reference for the degree of damage that is occurring already during the use phase, the separated collection of portable batteries was also sampled and investigated accordingly.

2.2. Size and Shape of Batteries

The size was measured as the maximal edge length (mm) of the portable batteries, and the shape (the construction type, e.g., round cell, button cell, prismatic cell, and pouch cell) of the portable batteries was documented. These parameters are considered to have an influence on the likelihood of critical damage (causing a waste fire). Hence, the distribution patterns of these parameters were investigated and documented.

2.3. Energy Content and Surrounding Conditions

When considering the various possible scenarios, the exact thermodynamic modelling of the potential ignition of waste that is caused by a portable battery must be based on many parameters. Thus, many underlying assumptions would be necessary.

For example, the ignition temperature of different plastic types ranges from 350 °C (polyethylene) to 580 °C (polytetrafluoroethylene) [17]. Furthermore, [18] presented significantly lower values for the initial combustion temperature for plastics (254–354 °C) and paper materials (240–260 °C). Golubkov et al. [11] showed that these temperatures are easily met when lithium-ion batteries (of type 18650) with an SOC >25% (for NCA) or SOC >50% (for LFP) undergo thermal runaway. However, it is known that a lower heating rate provides a lower ignition temperature, and vice versa [19]. A significant discrepancy in that regard can be seen, as [16] used a constant heating rate of 20 K/min. for defining the ignition temperature. Jhu [20] showed, under adiabatic conditions in a closed test can, that the peak temperature of the thermal runaway reaction of a fully charged 18650 lithium-ion cells is able reach its maximum temperature of 654.3 °C in 0.49 s starting from 125.2 °C. In these experiments, four cells from different worldwide battery producers have been investigated.

Regarding the released energy from the runaway reaction, [21] measured 102 to 218 kJ per Ah, depending on cell chemistry and respective SOC for LFP and NMC cells. Consequently, a typical lithium-ion cell (type 18650, 2.6 Ah) would release thermal energy in the range of 265 and 567 kJ. This energy can easily ignite various waste materials; still, the duration of the thermal runaway and ambient parameters will influence the ignition's success. The two main types of parameters can be distinguished between battery- and ambient-based.

The battery-based parameters depend on the type and SOC of the battery and they affect the following:

- the thermally convertible energy content of the battery,
- the reaction rate and temperature rise, and
- the heat release rate of the reaction.

Ambient-based parameters are:

- the ambient temperature,
- the distance between battery and other waste material,
- the moisture content of air and waste material,
- the composition of the surrounding waste material,
- the specific heat capacity and transfer coefficients, and
- the geometry of the loose or bulk material.

2.4. Risk Modelling and Assessment

Regarding risk modelling, the following assumptions were made:

- A stable content of portable batteries in the waste streams was assumed in the observed (2016) and projected period (2020) (cf. influence factor 1.1 of Table 3).
- Regarding the estimation of the risk potential in 2020, the content of lithium-based battery was assumed to raise (1) from one to two items/tonne for residual household waste and from (2) 0.8 to 1.5 items/tonne for lightweight packaging waste (cf. influence factor 1.2 of Table 3).

- Because of methodological reasons, only lithium-ion batteries were tested for their SOC in the end-of-life [13]. It is assumed that the SOC of primary lithium batteries generally follows the same pattern. Hence, the distribution pattern of the investigated batteries was applied to all lithium-based batteries (cf. influence factor 1.3 of Table 3).
- The number and intensity of potential damaging incidents, which was derived from the degree of damage of the investigated portable batteries (n = 848), is assumed to be representative for the lithium-based batteries, which were not found in high numbers (n = 66) (cf. influence factor 2.1 of Table 3).
- Portable batteries that are below a specific size are less likely damaged, due to (1) their relatively small maximal edge length and (2) their preferable mass-ratio of battery casing to active material. Furthermore, regarding the shape (construction type) of portable batteries, it is assumed that the stability of cells against mechanical abuse is as follows: button cells >> round cells/batteries, prismatic cells/batteries, battery packs >> pouch cells/batteries. The assumed correction factors for size and shape were chosen accordingly (cf. influence factor 2.2 of Table 3).
- It is assumed that a battery undergoing thermal runaway or uncontrolled heat release is causing waste ignition and a sustained spreading of fire in 10% of the cases. In the other 90%, it is assumed that either (1) the respective battery did not have enough energy to ignite flammable material, (2) there was not enough flammable material to be ignited, or (3) the flammable material could not sustain a chemical chain reaction and self-extinguished (cf. influence factor 2.3 of Table 3).
- For risk modelling, the generated amounts of waste (of residual waste and lightweight packaging waste) were taken for the year 2016, according to [22], and extrapolated accordingly from the data 2012–2018 for the year 2020 [23].

The probability of occurrence followed typical grading schemes for risk graphs/risk matrices (e.g., as low as reasonably practicable, ALARP) e.g., [13,24]:

The probability of occurrence:

- likely <1:100;
- rarely 1:100—1:1000;
- very rarely 1:1000—1:10,000;
- unlikely 1:10,000—1:100,000;
- very unlikely 1:100,000—1:1,000,000; and,
- extremely unlikely >1:1,000,000.

Property damage/losses:

- negligible no noticeable consequences of fire;
- minor no appreciable fire damage;
- severe low property damage;
- major high property damage; and,
- catastrophic very high property damage/

3. Results

Table 5 shows the identification and risk assessment of the possible hazards and threats of portable batteries for different facility areas or processes along the value chain of a residual household waste system.

3.1. Damage Degree

Table 6 presents the distribution of investigated portable batteries' maximal edge length. The main findings are:

- It is relevant to mention that many of the investigated portable batteries have just two well-known form factors: AA and AAA. That is why the high frequencies of the maximal edge length intervals are 30.0–44.9 mm and 45.0–59.9 mm.
- Small portable batteries (maximal edge length <30 mm) have a relatively low share of the DC 4 to 7.

- Larger portable batteries (maximal edge length >30 mm) show a relatively high share of the DC 4 to 7.

Table 5. Qualitative risk assessment of possible hazards and threats of portable batteries (waste stream: residual household waste).

Facility Area/Process	Possible Hazards and Threats	Risk Assessment
Collection bins	Damage due to external short-circuit	low
Loading activity	Damage due to external short-circuit	low
Collection vehicle	Mechanical damage due to compaction	medium
Unloading activity	Mechanical damage due to tip-off	low
Waste bunker/input storage	Damage due to external short-circuit Damage due to external heating (self-heating of waste)	medium–high
Waste transfer activity	Mechanical damage due to (wheel) loader or gripper	medium
Treatment facility	Mechanical damage due to pre-shredding process Mechanical damage due to post-shredding process Dangerous heat generation after damage Carry-over through the processing facility	high–very high
Output storage	Damage due to external short-circuit Damage due to external heating (self-heating of waste) Dangerous heat generation after damage	low–medium

Table 6. Distribution of maximal edge length of the investigated portable batteries.

Maximal Edge Length [mm]	Amount of Batteries [items]	Amount in damage Classes 4–7 [items]	Share of Damage Classes 4–7 [%]	Assumed Correction Factor
–14.9	41	2	4.9%	0.40
15.0–29.9	54	4	7.4%	
30.0–44.9	292	100	34.2%	0.75
45.0–59.9	430	127	29.5%	
60.0–	30	8	26.7%	0.95
Total	847	241	28.5%	

In Figure 2, the distribution of the portable batteries' damage degree according to the waste stream in which they were found.

Only two per cent of the spent batteries show a relevant degree of damage, according to the background reference sample (separated portable battery collection, SBC) (DC 4 to 7). The vast majority of over 97% of the batteries are undamaged (DC 1). No portable battery was found, which was heavily bloated or in a post-thermal runaway state (DC 7).

The input fractions results show that, on the one hand, portable batteries can already be subject to varying degrees of mechanical damage during collection, e.g., due to the force that is applied in press containers of collection vehicles. That is particularly remarkable when compared to the separate collection of portable batteries, as batteries are or will be damaged only to a minor extent in this collection system.

On the other hand, there is a significant degree of damage to portable batteries in processing plants, for example, during the conditioning of the waste stream by a pre-shredder (e.g., in residual waste treatment) or a bag opener (e.g., in lightweight packaging waste sorting), but also during the manipulation of the waste stream (while using a wheel loader or gripper). Other impacts, such as (1) shock loads, which may occur due to different drop heights in treatment processes (e.g., by falling into an output box or from one conveyor belt to another), or (2) vibration of vibrating conveyors can also damage portable batteries.

Moreover, when comparing the residual household waste before (RHW (in)) and after (RHW (out)) treatment in a respective facility, it is noticeable that the percentage of the lower degrees of damage (DC 1 to 3) are reduced, while the percentage of the higher damage classes (DC 4 to 6) are increased (from approximately 6% to over 40% in total).

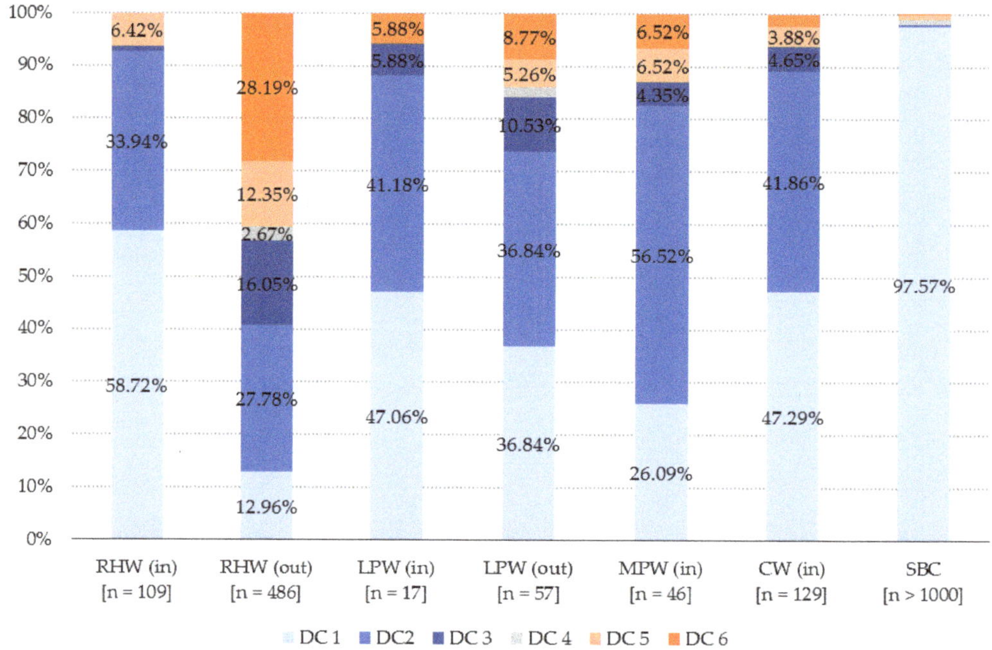

Figure 2. Distribution of damage degree according to the waste stream in which the batteries were found (Legend: RHW = residual household waste, LPW = lightweight packaging waste, MPW = metal packaging waste, CW = commercial waste, SBC = separate portable battery collection; values < 2% not displayed).

When comparing lightweight packaging waste before (LPW (in)) and after (LPW (out)) treatment (sorting plant), the effect is similar, but yet not that severe, which results in an increase of DC 4 to 6 from almost 6% to 14% in total. The share of DC 4 to 6 in commercial waste (CW (in)) is very similar to the input fractions of residual household waste and lightweight packaging waste, and it is notably much higher in metal packaging waste (MPW (in)).

3.2. Shape Distribution Pattern

Regarding the shape (construction type) of the portable batteries, the distribution pattern is shown in Table 7. The share of round cells/batteries is very high (about 85%), due to the aforementioned high quantities of portable batteries with the format factors AA and AAA. It was observed that lithium-ion batteries are often prismatic or pouch-shaped.

Table 7. Shape/construction type of the analysed portable batteries.

Battery/Cell Type	No. of Cells Analysed [items]	Share of Cells Analysed [%]	Assumed Correction Factor
Button cell	93	11.0	0.20
Round cell/battery	717	84.6	
Prismatic cell/battery	27	3.2	0.75
Battery pack	5	0.6	
Pouch cell/battery	5	0.6	
unknown	1	0.1	0.95
Total	848	100	

3.3. Risk Modelling

Table 8 presents the results of the modelling of potential risk (in potential fire incidents per year), comparing residual household waste (RHW) and lightweight packaging waste (LPW). The values are based on fictitious waste systems with 100,000 tonnes per year of the respective fractions of residual household waste and lightweight packaging waste (2016). For 2020, the assumed growth was included, which was linearly extrapolated, which resulted in a two per cent increase for residual waste and a four percent increase for lightweight packaging waste.

Table 8. Risk modelling for portable and lithium-based batteries in residual household waste (RHW) and lightweight packaging waste (LPW).

Modelling of Risk Potential	RHW Projection 2016	RHW Estimate 2020	LPW Projection 2016	LPW Estimate 2020
Waste generation [t]	100,000	102,000	100,000	104,000
in waste collection [pot. fire incidents/year]	57	116	42	82
Probability of occurrence	1:1754	1:1758	1:1904	1:1902
in waste treatment [pot. fire incidents/year]	363	741	103	202
Probability of occurrence	1:275	1:275	1:777	1:743

The risk potential is shown in potential fire incidents per year. The waste collection scenarios are very similarly related to the potential risk of fire, while the waste treatment scenarios differ more. Remarkably enough, the risk almost doubles in all of the observed scenarios.

The probability of occurrence is presented in a particle-based form and it reveals the high risks of battery-caused fire incipiencies in terms of likelihood.

3.4. Risk Assessment

The modified risk graph presented Figure 3 visualises the assessment of the risk of portable lithium-based batteries. Two specific scenarios were chosen representing the observed areas of (1) collection and (2) treatment of waste from Table 8.

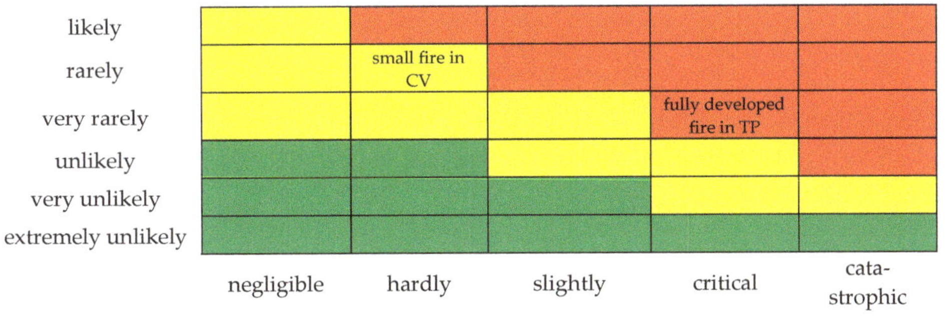

Figure 3. Modified risk graph for the two scenarios of Table 8: Waste collection (a small fire in a collection vehicle, CV) and waste treatment (a fully developed fire in a treatment plant, TP).

Keeping the grading of the probability of occurrence in mind, the various probabilities do not lead to different results in the risk graph, whether residual waste or lightweight packaging waste; however, they vary considerably in Table 8.

4. Discussion

A broad spectrum of potential hazards of batteries was considered within this study's investigations and it is presented in Table 5, being exemplary for residual household waste. However, from a holistic viewpoint, in complex systems, such as waste management systems, it is unlikely to cover all of the possible hazards that are posed by lithium-based batteries.

As [25] stated, risk matrixes may be subject to biases or errors. A risk matrix can compare only a small fraction of more or less randomly selected hazards due to their low resolution. In general, that fact has to be kept in mind when risk graphs are created or the basis of decision-making.

The scenarios that were observed in detail reveal that the risks are either in the ALARP region (yellow area) or in the region of inacceptable risks (red area of the risk graph). For the latter, measures of risk reduction are inevitable.

Fires in waste collection vehicles have become a well-known phenomenon in recent years. The smoke of the fires is usually recognised by the vehicle drivers in an early stage. An unspoken best-practice measure of the respective vehicle driver is to choose a known asphalt or concrete surface that is big enough to unload the burning waste (e.g., parking area of a nearby supermarket) and call the local fire brigade. The waste is usually pulled apart during extinguishing in order to speed up the subsequent fire-fighting operations and to locate the origin of the fire.

The different results for RHW and LPW indicated that waste shredders are the main hot-spot for battery-caused waste fires. In contrast to shredders, bag openers in LPW sorting plants are not crushing the waste. That is also confirmed by a recently published study [26], revealing high numbers of battery-caused ignitions after pre- and post-shredders in treatment facilities for RHW and commercial waste.

The comparison of the data of potential fire incidents per year with values from literature allows for an overall evaluation of these results. In contrast to the statistical data of waste fire incidents [1,27], where relatively low numbers are determined battery-caused fires, the results of this study seem to be very high. The data of [1,27] are characterised by high shares of unknown fire causes, which probably underestimates the relevance of battery-caused fires. In contrast to the high numbers of heat-related threshold exceedances in waste shredding machines found by [26], which are caused by batteries to a large extent, the results of this study seem appropriate. Autischer et al. [26] detected 0.1 to 1.5 threshold exceedances per eight-hour-work shift, depending on the respective measuring point in one of the two observed treatment plants. That means 260 to 1.580 threshold exceedances per year and treatment facility. In comparison to the results, it should be noted that (1) both treatment plants process significantly less than 100,000 tonnes per year and (2) temperature-related threshold exceedances are detected—not fire incidents.

5. Conclusions

In this article, the main influence factors on lithium-based portable batteries' risk were investigated in their end-of-life phase. The degree of damage that was happening to portable batteries was investigated and compared for different waste streams. After that, the hazards and potential risks were modelled and assessed, including the present and previous studies' findings.

The results included a detailed analysis of the damage degree, the distribution pattern of influencing characteristics of portable batteries, and profound risk analysis and assessment.

The assessment reveals that the risk of lithium-based portable batteries is significantly too high, which makes it difficult to maintain modern waste management in a sustainable way. Primarily, municipal solid waste treatment plants are at increased risk. The probability that treatment plants burn to the ground is far too high, according to the available assessment. The increased number of major fires in waste management in recent years is clear and undeniable evidence.

No other substance or material has ever comparably endangered the whole waste industry. Hence, besides research and development activities for investigating and understanding the hazards and risks of lithium-based portable batteries, increased technological development and innovation efforts are indispensable for reducing the risk potential of end-of-life portable batteries.

In order to reduce risk, the waste sector has to aim to collect as many batteries as possible in the separate collection systems and take-back schemes, as only this collection system guarantees a damage-free return system. That requires increased effort in public relations and consumer awareness-raising. However, a 100% separate collection rate for portable batteries is highly unrealistic without a comprehensive deposit system. Hence, operators of treatment facilities have to find ways to (1) protect critical infrastructure and treatment processes (e.g., including new detection and extinguishing methods) or (2) preferably detecting and separating portable batteries in the course of their treatment processes.

Further research is necessary in order to gain more specific knowledge on the influence factors, where assumptions had to be made in the present study. Especially, the probability of waste ignition due to the thermal runaway of portable batteries has to be investigated.

Supplementary Materials: The following are available online at https://www.mdpi.com/2227-9717/9/1/49/s1.

Author Contributions: Conceptualization, T.N. and R.P.; Data curation, T.N. and M.B.; Formal analysis, T.N. and M.B.; Funding acquisition, R.P.; Investigation, T.N. and M.B.; Methodology, T.N. and M.H.; Project administration, T.N.; Resources, T.N. and R.P.; Software, T.N.; Supervision, R.P.; Validation, T.N., M.H. and R.P.; Visualization, T.N.; Writing—original draft, T.N. and M.H.; Writing—review & editing, T.N. All authors have read and agreed to the published version of the manuscript.

Funding: This research was funded by the Austrian Research Promotion Agency, grant number 850753.

Data Availability Statement: The data presented in this study are available on request from the corresponding author. The data are not publicly available due to privacy restrictions.

Acknowledgments: The authors thank T. Bäck, L. Kranzinger, P. Puchbauer, W. Rübenbauer, S. Schlögl, B. Steiner, and the members of the Chairs working group 'Environmental Analytics' for their support in the sampling and sorting campaign. Furthermore, the authors thank the Austrian Research Promotion Agency for funding and the project partners for support.

Conflicts of Interest: The authors declare no conflict of interest. The funders had no role in the design of the study; in the collection, analyses, or interpretation of data; in the writing of the manuscript, or in the decision to publish the results.

References

1. Nigl, T.; Rübenbauer, W.; Pomberger, R. Cause-Oriented Investigation of the Fire Incidents in Austrian Waste Management Systems. *Detritus* **2020**, *9*, 213–220. [CrossRef]
2. LANUV (Landesamt für Natur, Umwelt und Verbraucherschutz Nordrhein-Westfalen). *Brandereignisse in Abfallbehandlungsanlagen—Abschlussbericht und Schlussfolgerungen der Landesregierung*; LANUV-Fachbericht 68: Recklinghausen, Germany, 2016. (In German)
3. Ibrahim, M.A.; Alriksson, S.; Kaczala, F.; Hogland, W. Fires at storage sites of organic materials, waste fuels and recyclables. *Waste Manag. Res.* **2013**, *31*, 937–945. [CrossRef] [PubMed]
4. Nigl, T.; Pomberger, R. Brandrisiko durch Lithium-Ionen Batterien: Sind unsere Anlagen noch versicherbar? In *Berliner Recycling—Und Sekundärrohstoffkonferenz, Berlin, Germany, 2–3 March 2020*; Holm, O., Thomé-Kozmiensky, E., Goldmann, D., Friedrich, B., Eds.; Thomé-Kozmiensky Verlag GmbH: Neuruppin, Germany, 2020; pp. 482–494. ISBN 978-3-944310-51-0. (In German)
5. Nigl, T.; Schwarz, T.E.; Walch, C.; Baldauf, M.; Rutrecht, B.; Pomberger, R. Characterisation and Material Flow Analysis of End-of-Life Portable Batteries and Lithium-based Batteries in Different Waste Streams in Austria. *Waste Manag. Res.* **2020**, *38*, 649–659. [CrossRef] [PubMed]
6. EAK (Elektroaltgeräte Koordinierungsstelle). Tätigkeitsbericht 2015. Available online: https://www.eak-austria.at/presse/ (accessed on 1 November 2020). (In German).
7. EAK (Elektroaltgeräte Koordinierungsstelle). Tätigkeitsbericht 2017. Available online: https://www.eak-austria.at/presse/ (accessed on 1 November 2020). (In German).
8. Lisbona, D.; Snee, T. A review of hazards associated with primary lithium and lithium-ion batteries. *Process. Saf. Environ. Prot.* **2011**, *89*, 434–442. [CrossRef]

9. Winslow, K.M.; Laux, S.J.; Townsend, T.G. A review on the growing concern and potential management strategies of waste lithium-ion batteries. *Res. Conserv. Recycl.* **2018**, *129*, 263–277. [CrossRef]
10. Mikolajczak, C.; Kahn, M.; White, K.; Long, R.T. *Lithium-Ion Batteries Hazard and Use Assessment*, 1st ed.; Springer: New York, NY, USA, 2011; pp. 43–70. [CrossRef]
11. Golubkov, A.W.; Scheikl, S.; Planteu, R.; Voitic, G.; Wiltsche, H.; Stangl, C.; Fauler, G.; Thaler, A.; Hacker, V. Thermal runaway of commercial 18650 Li-ion batteries with LFP and NCA cathodes—impact of state of charge and overcharge. *RSC Adv.* **2015**, *5*, 57171–57186. [CrossRef]
12. Liu, X.; Stoliarov, S.I.; Denlinger, M.; Masias, A.; Snyder, K. Comprehensive calorimetry of the thermally-induced failure of a lithium ion battery. *J. Power Sources* **2015**, *280*, 516–525. [CrossRef]
13. Nordsieck, H.; Förster, A.; Martin, A.; Zepf, V.; Hertel, M. *Sicheres Sammeln von Elektroaltgeräten in Depotcontainersystemen*; University of Augsburg and Bifa: Augsburg, Germany, 2016. (In German)
14. Nigl, T.; Bäck, T.; Stuhlpfarrer, S.; Pomberger, R. The Fire Risk of Portable Batteries in Their End-of-Life—Investigation of the State of Charge of Waste Lithium-Ion Batteries in Austria. *Waste Manag. Res.* **2020**. submitted for publication. [CrossRef]
15. ISO. *Safety of Machinery—General Principles for Design—Risk Assessment and Risk Reduction*; EN ISO 12100:2010; ISO: Geneve, Switzerland, 2010.
16. Rausand, M. *Risk Assessment: Theory, Methods, and Applications*; John Wiley & Sons: Hoboken, NJ, USA, 2013; pp. 1–28. ISBN 978-1-118-28110-9.
17. VdS. Kunststoffe—Eigenschaften, Brandverhalten, Brandgefahren. *VdS* **2000**, *2516*, 2000–2012. (In German)
18. Grammelis, P.; Basinas, P.; Malliopoulou, A.; Sakellaropoulos, G. Pyrolysis kinetics and combustion characteristics of waste recovered fuels. *Fuel* **2009**, *88*, 195–205. [CrossRef]
19. Graf, S.H. *Ignition Temperatures of Various Papers, Woods and Fabrics. Engineering Experiment Station Oregon State System of Higher Education, Bulletin 26*; Oregon State College: Corvallis, OR, USA, 1949.
20. Jhu, C.Y.; Wang, Y.W.; Shu, C.M.; Chang, J.C.; Wu, H.C. Thermal explosion hazards on 18650 lithium ion batteries with a VSP2 adiabatic calorimeter. *J. Hazard. Mater.* **2011**, *192*, 99–107. [CrossRef]
21. Sturk, D.; Hoffmann, L.; Tidbald, A.A. Fire Tests on E-vehicle Battery Cells and Packs. *Traffic Inj. Prev.* **2015**, *16*, 159–164. [CrossRef]
22. BMNT. *Die Bestandsaufnahme der Abfallwirtschaft in Österreich—Statusbericht 2018*; Bundesministerium für Nachhaltigkeit und Tourismus: Vienna, Austria, 2018. (In German)
23. BMK. *Die Bestandsaufnahme der Abfallwirtschaft in Österreich—Statusbericht 2020 (Referenzjahr 2018)*; Bundesministerium für Klimaschutz, Umwelt, Energie, Mobilität, Innovation und Technologie: Vienna, Austria, 2020. (In German)
24. Franks, A.; Whitehead, R.; Crossthwaite, P.; Smail, L. *Application of QRA in Operational Safety Issues*; Research Report 25; HSE Books: Buxton, UK, 2002; ISBN 0-7176-2570-2.
25. Cox, L.A. What's Wrong with Risk Matrices? *Risk Anal.* **2008**, *28*, 497–512. [CrossRef] [PubMed]
26. Autischer, M.; Holzschuster, S.; Nigl, T. Statistische Betrachtung von Infrarot-Sensordaten in der Aufbereitung mit Relevanz zur Brandfrüherkennung. In Proceedings of the Recy & DepoTech 2020, Leoben, Austria, 18–20 November 2020; pp. 287–294, ISBN 979-3-200-07190-2. (In German).
27. Nigl, T.; Bäck, T.; Pomberger, R. Vertiefende Ursachenforschung zu Brandereignissen in der österreichischen Abfall-, Entsorgungs- und Recyclingwirtschaft. In Proceedings of the Recy & DepoTech 2020, Leoben, Austria, 18–20 November 2020; pp. 279–286, ISBN 979-3-200-07190-2. (In German).

processes

MDPI

Article

Recovery of Raw Materials from Ceramic Waste Materials for the Refractory Industry

Severin Seifert [1,*], Sebastian Dittrich [1] and Jürgen Bach [2]

[1] Fraunhofer Institute for Building Physics IBP, 83626 Valley, Germany; sebastian.dittrich@ibp.fraunhofer.de
[2] Independent Researcher, 65549 Limburg, Germany; jottbach@t-online.de
* Correspondence: severin.seifert@ibp.fraunhofer.de

Abstract: Products of the refractory industry are key for the production of heavy industry goods such as steel and iron, cement, aluminum and glass. Corresponding industries are dependent on thermal processes to manufacture their products, which in turn would not be possible if there were no refractory materials, such as refractory bricks or refractory mixes. For the production of refractory materials, primary raw materials or semi-finished products such as corundum, bauxite or zircon are used. Industrial recycling of refractory raw materials would reduce dependencies, conserve resources and reduce global CO_2 emissions. Today, only a small quantity of the refractory materials used can be recycled because raw materials (regenerates) obtained from end-of-life materials are of insufficient quality. In this study, regenerates from different refractory waste products could be obtained using the innovative processing method of electrodynamic fragmentation. It was shown that these regenerates have a high chemical purity and are therefore of high quality. It could be confirmed that the use of these regenerates in refractory materials does not affect the characteristic properties, such as refractoriness and mechanical strength. Thus, electrodynamic fragmentation is a process, which is able to provide high-quality raw materials for the refractory industry from used materials.

Keywords: recycling; refractory; regenerate; electrodynamic fragmentation; innovative process

Citation: Seifert, S.; Dittrich, S.; Bach, J. Recovery of Raw Materials from Ceramic Waste Materials for the Refractory Industry. *Processes* **2021**, *9*, 228. https://doi.org/10.3390/pr9020228

Academic Editor: Daniel Vollprecht

Received: 23 November 2020
Accepted: 21 January 2021
Published: 26 January 2021

1. Introduction

Today, refractory materials are used in a wide variety of industrial sectors. These ceramic materials are products that are generally used under a high-temperature load of >1500 °C [1] for lining and delivery of thermotechnical facilities, such as blast furnaces or converters, as well as for transport devices in iron and steel, aluminum, cement and ceramic industries, incineration plants and refineries. The material separates the reaction zone from the outer parts of the process devices and are in contact with solid and liquid but also gaseous, partly very aggressive reaction components and reaction products [2]. Without these refractory materials there would be no technical thermal processes, which are fundamental to the production of steel, iron, aluminum, cement and glass. For example, in 2016 the German steel industry produced a total of 40.2 million t of steel products, requiring 0.5 million t of refractory materials [3]. This means approximately 10 kg of refractory materials were needed to produce 1 t of steel.

The necessary refractability as well as other important properties such as zero shrinkage, high thermal shock resistance, chemical resistance and mechanical or temperature-dependent strength are given to refractory materials by their non-metallic, inorganic composition. The main components consist of the six basic oxides SiO_2, Al_2O_3, MgO, CaO, Cr_2O_3 and ZrO_2, often in combination with carbon (e.g., SiC) [4]. These oxides form refractory compounds such as bauxite, corundum or white corundum, tabular alumina, zirconium (zirconium silicate), fireclay or silicon carbide through complex and high-emission thermal processes (dehydrogenation, sintering reactions or melt flow processes). These raw materials or semi-finished products are then used to produce the refractory materials. The largest raw material resources are located mainly in China, Russia, South Africa, South America and Australia. Accordingly, the German and European refractory industry is highly dependent on these imports. The refractory materials are exposed to high thermal, physical and chemical loads and serve as wear material during the use in industrial processes.

The average lifetime of refractory materials depends on the application and ranges from few days (e.g., purging plugs) to a few weeks (e.g., steel-casting ladles and converters) to long-term applications from one to several years (e.g., converters and heat-exchangers/preheaters). After the utilization phase, the refractory lining must be repaired or at least relined. The majority of the lining wears out completely and is no longer usable or is chemically contaminated such that it requires an extensive replacement. A smaller amount of the material (approximately 5–10% of the waste material) is currently recycled after stripping and used as so-called regenerates for the production of new refractory materials. For this purpose, the stripped material is pre-sorted and afterwards crushed in a jaw crusher, for example. The subsequently sieved material can partly be used as aggregate in new refractory products. However, a higher chemical purity of these regenerates required for the refractory industry cannot be achieved. For these reasons, little recycled material is used in the refractory industry today. Predominantly waste material and thus valuable secondary raw material for the refractory industry is deposited in special landfills for final disposal. A comprehensive industrial recycling process for refractory materials is a major challenge that offers the chance to reduce the dependency on primary raw material imports and the global CO_2 emissions.

A suitable method for processing refractory waste material in order to enable a high rate of recycling could be electrodynamic fragmentation. This innovative method uses pulsed high voltage discharges to separate bulky multi-phase material selectively along grain boundaries. The discharge has to take place underwater to enable the solid material to be penetrated. Thus, the whole approach is a "wet process". The potential of this pulsed power processing approach was demonstrated for the recycling of waste concrete [5] or the separation of municipal waste incineration ash [6,7]. Like refractory material, the investigated ash is a mineral-based compound material. Reusable components like iron metal, glass or ceramics are enclosed by a siliceous matrix. It was proved that after processing the ash with electrodynamic fragmentation the individual components of the ash could be recovered selectively.

The aim of the study presented here was to use the innovative process of electrodynamic fragmentation as a recycling strategy for refractory waste products in order to obtain regenerates with a high chemical purity. With these high-quality regenerates, new sources of raw materials for the refractory industry could be made accessible, and thus primary resources are conserved. The increased use of such high-purity refractory regenerates would lower the cost of refractory materials and could reduce dependencies on world markets.

2. Materials and Methods

2.1. Electrodynamic Fragmentation

Electrodynamic fragmentation is a technique using pulsed power discharges to separate compound materials selectively. The method itself was first investigated and described at the University of Tomsk in the late 1940s [8]. The efficiency of the method is the higher the more electrical discharges are generated over time. Usually, a so-called Marx generator is used to enable a high rate of high voltage discharges [9]. Besides a high voltage, a high slew rate is of special interest. The slew rate describes how fast the maximum voltage of a discharge can be achieved and thus determines the possible pulse length. With a Marx generator a slew rate that enables pulse lengths of a few nanoseconds can be realized. This is very important in order to force the electrical discharge into a solid material. Whether an electrical discharge can penetrate a material or not depends on the dielectric strength of the given material. The dielectric strength of a material is not a constant but varies in dependency with the slew rate and therefore the length of a discharge [10]. At a pulse rise time lower than 500 nanoseconds, the dielectric strength of water is higher than that of solid material (see Figure 1). Electrodynamic fragmentation utilizes this physical principle.

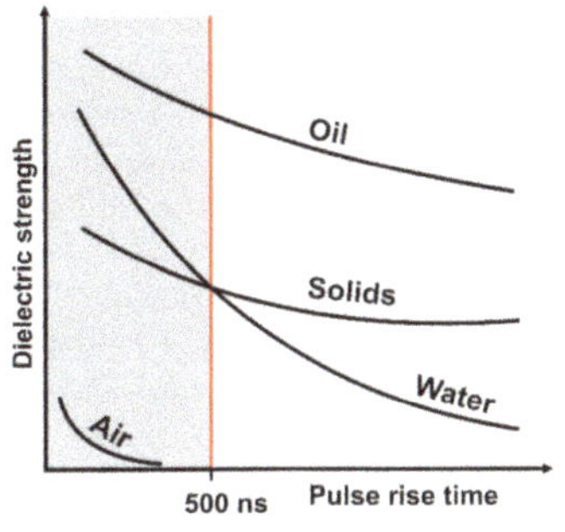

Figure 1. Dielectric strength of different materials as a function of the pulse rise time (modified from [10]).

In the process vessel of the fragmentation unit, a solid material sample is placed underwater. The bottom of the process vessel serves as an electrode. A second electrode is placed on top of the sample. The distance between sample and upper electrode can be varied. After voltage application, a polarization of the sample takes place. This leads to electrostriction, meaning micro stress within the sample as the charge carrier (e.g., electrons) of the solid material are not freely movable [11]. As different components in a compound material, such as refractory waste material, differ in their dielectric strength, the grain boundaries represent the regions with the biggest contrast in polarization. The grain boundaries therefore are the path of least resistance for the discharge. Along the grain boundaries, so-called streamers infiltrate the material during a first stage of the pulse discharging. When the first streamer reaches the electrode at the bottom of the process vessel, the complete energy of the discharge runs along the corresponding path. A short-lived plasma channel is generated reaching temperatures of up to 10^4 K. During the collapse of the plasma channel, a shockwave of up to 10 GPa shatters the material having the strongest influence for the separation. The shockwave is reflected at the process vessel and penetrates the material a second time. As the described procedure occurs with each discharge, a fast and efficient separation of a solid material is achieved (Figure 2).

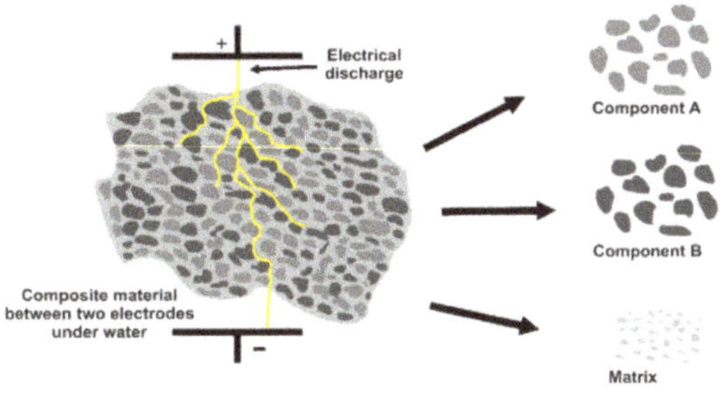

Figure 2. Principle of electrodynamic fragmentation of a composite material.

In this study a laboratory plant was used, which works in batch mode with a five liter process vessel [12]. The general setup of the lab plant allows a processing of samples with a diameter of around 40 mm. The maximum volume of sample material for a single fragmentation step depends on the density of the material and is about several hundred grams of material generally. The default operating parameters for the lab plant are a pulse rate of 5 Hz and a voltage of 180 kV per pulse, though these parameters are adjusted to achieve an optimum in separation for each sample processed.

2.2. Sample Material

In this study, different refractory waste or break-out material was investigated and chosen from different industrial fields of application that were as varied as possible. It was also important that the selected materials were exposed to different temperatures during the manufacturing process or during use. In refractory materials, higher temperatures lead to a stronger sintering of the individual components, which makes it considerably more difficult to process or cleanly separate the individual components with conventional processing methods.

For this study, three different refractory waste materials were collected (Figure 3), from different fields of application and containing potential regenerates, such as bauxite or zircon. A brief description of the sample material with details of the potential regenerates is given in Table 1. All materials used in this study originate from shaped bricks.

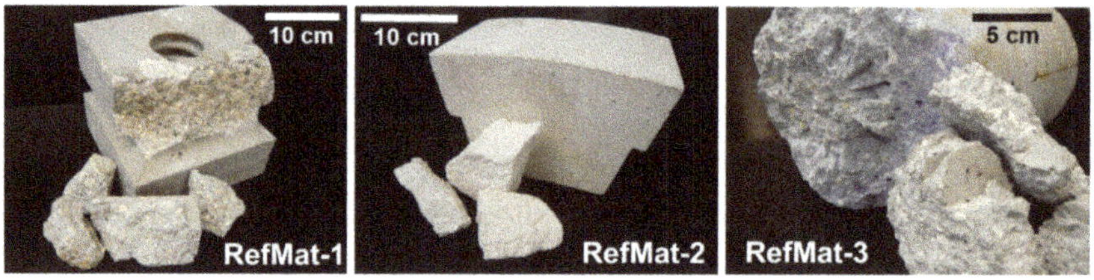

(a) (b) (c)

Figure 3. Refractory waste materials from various applications. (**a**) Highly sintered brick for sintering or rotary kilns; (**b**) corundum stone for melting tanks; (**c**) functional refractory ceramic from the steel industry.

Table 1. Brief description of the sample material used.

Sample	Material	Potential Regenerates
RefMat-1	Sintered brick for inlet chambers for rotary kilns	Bauxite, zirconia-alumina fused grain (ZAC)
RefMat-2	Corundum stone for aluminum melting furnace	White corundum
RefMat-3	Functional refractory ceramic from the steel industry	Tabular alumina, white corundum

The sample material RefMat-1 (Figure 3a) consists of preformed bricks, which are fired at high temperatures (1250 °C) before being used in a sintering furnace. Only after the firing is the stone used for the kiln lining. These stones are mainly composed of bauxite, zirconia-alumina fused grain (ZAC), SiC and microsilica as well as a matrix of alumina-cement-based binder. Of this refractory material, there is interest above all in recovering bauxite and ZAC as regenerates.

The sample material RefMat-2 (Figure 3b) is a pure corundum stone, which is used at very high temperatures and has a high chemical load. The main components are white corundum and alumina-cement-based binder. These highly sintered bricks are used as lining for melting furnaces, e.g., for aluminum melts. The recovery of the white corundum as regenerate would considerably save primary resources.

The sample material RefMat-3 (Figure 3c) consists of functional refractory ceramics from the steel industry. These functional refractory ceramics have fine channels and are used in the furnace linings or at the bottom of the blast furnaces and ladle linings of steel-works to inject various gases into the molten steel via the channels. The main components of these functional refractory ceramics are tabular alumina and white corundum in a binder matrix of alumina cement. Potential regenerates with this sample material would be tabular alumina and the white corundum.

2.3. Fragmentation of the Sample Material

In order to obtain coarse sample fragments, prior to the fragmentation experiments, the three different blocks were crushed down using a sledgehammer. The resulting size of sample material was about 4 to 5 cm each. At the beginning of the fragmentation experiments (Figure 4), several small samples of each material were processed, whereby the parameters voltage, electrode-sample gap, number of pulses and pulse frequency were varied. The goal was to identify the optimal fragmentation parameters for each sample material in order to subsequently process a larger amount of sample material with the optimized parameters.

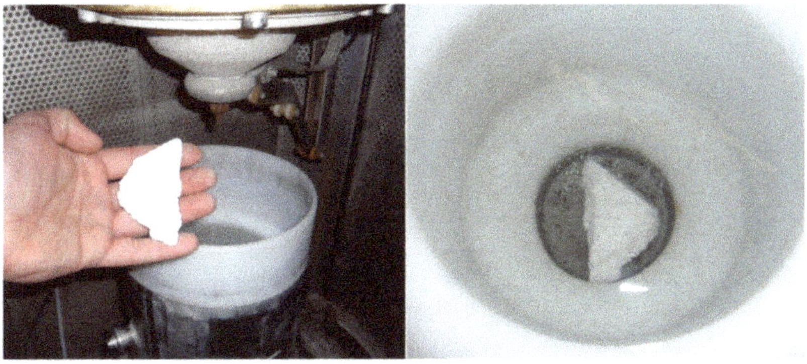

(a) (b)

Figure 4. Fragmentation of the RefMat-2 sample. Placing the sample in the process vessel (**a**) and underwater sample in the process vessel (**b**).

After the optimal fragmentation parameters for the individual sample materials had been determined in the preliminary tests, larger quantities of the individual materials could be processed in the next step in order to obtain sufficient material or regenerates for the subsequent recycling tests. Finally, between 20 and 32 kg of each of the three samples was fragmented in batches. After fragmentation, the separated material was dried and classified into different grain size fractions by means of sieve classification. In addition, the process water from the individual tests was filtered and the filter residue dried so that a fine fraction could be obtained.

In the next step, the fragmented and sieve-classified material was sorted in order to keep the separated aggregates, or regenerates, sorted by type. Therefore, the fragmented material was sorted by optical sorting, meaning by differences in color and translucency. The sorting was carried out with a laboratory system for optical bulk material sorting (TableSort), which was developed at the Fraunhofer IOSB [13,14]. Equipped with a RGB filter camera and a filigree blow-out device, this system is suitable for small amounts of material. The electro-optical sorting was carried out in several steps or passes for each individual grain size fraction. In each pass, the optical filter was adjusted in such a way that the desired material, meaning the regenerate in question, was blown out from the bulk mass flow.

2.4. Methods of Investigation

To characterize the collected material, all samples were analyzed before the fragmentation process by X-ray phase analysis (XRD) to determine the mineralogical phase composition and by X-ray fluorescence analysis (XRF) to investigate the chemical composition. For the XRD analysis, a D2 Phaser (Bruker) was employed, and for the XRF analysis, an Epsilon 3 XL (Panalytical) was used. The XRD analysis was performed using powder samples. For the XRF analysis, powder compacts were used. Before sample preparation, all materials were crushed and milled down. For the analysis before the fragmentation experiments, the fragments obtained from crushing the refractory blocks with a sledgehammer were milled in a two-step process. Firstly, a vibratory disc mill was used to achieve a grain size below 1 mm. Secondly, the material was milled down using a McCrone micronizing mill. With this wet-milling process a powder fineness was achieved suitable for XRD analysis. After the material was fragmented and sorted, the obtained regenerates were examined by optical microscopy to evaluate the degree of the achieved detachment. To determine the chemical purity of the recovered material, the regenerates were analyzed by XRD and XRF. Subsequently, the analytical results were compared with the analysis of primary aggregates (e.g., ZAC or tabular alumina).

The sample material obtained from the fragmentation experiments was milled down in a two-step process as well. The aforementioned McCrone mill was used after the material was pretreated in a ball mill to obtain the particle size required in the wet-milling process.

Furthermore, the recovered material was used to produce the refractory test specimen (prisms with a dimension $4 \times 4 \times 16$ cm) in accordance with industrial requirements. The cold compressive strength as well as the cold bending tensile strength of all prisms was determined using a Z100 testing machine from ZwickRoell and an Alpha 3-3000 S testing machine from Form + Test, respectively.

3. Results

3.1. Characterization of the Refractory Waste Material

As expected, all samples show a very high Al_2O_3 content, which is mostly distributed between the mineralogical phases α-alumina (corundum) and β-alumina. Therefore, all samples are high alumina refractories.

In addition to the high Al_2O_3 content, the sample material RefMat-1 shows high contents of SiO_2 and ZrO_2 (Table 2). This chemism is reflected in the mineralogical phases corundum (Al_2O_3), baddeleyite (ZrO_2) and mullite ($3Al_2O_3$-$2SiO_2$). Minor phases are potassic and alkali feldspars, SiC and grossite ($CaAl_4O_7$) and some zirconium ($ZrSiO_4$). The two mineral phases corundum and baddeleyite are components of the zirconia-alumina fused grain (ZAC), which is one of the main components of these refractory materials. Furthermore, the two highly refractory mineral phases corundum and mullite are components of calcined bauxite, which is also used as an important raw material in many shaped and unshaped refractory products.

Table 2. Chemical and mineralogical composition of the sample material RefMat-1.

Chemical Composition		Mineralogical Composition	
Oxide	**wt.%**	**Phase**	**wt.%**
Al_2O_3	52.8	α-Alumina (Corundum)	67.7
SiO_2	26.6	Baddeleyite	11.2
ZrO_2	12.0	Mullite	6.4
CaO	2.4	Feldspar	3.4
K_2O	0.1	SiC	7.4
Fe_2O_3	1.3	Grossite	1.3
TiO_2	1.4	Zirconium	1.1

The sample material RefMat-2 is an almost pure Al_2O_3 product (Table 3). Besides the very high Al_2O_3 content, only very low contents of Fe_2O_3 and CaO could be detected. The Al_2O_3 is mainly found in the mineral phase corundum or in white corundum. White corundum, a chemically pure alumina, is extracted from the melt and is used both in unshaped and shaped refractory products. Figure 5 shows an example of the X-ray diffractogram of sample RefMat-2.

Table 3. Chemical and mineralogical composition of the sample material RefMat-2.

Chemical Composition		Mineralogical Composition	
Oxide	**wt.%**	**Phase**	**wt.%**
Al_2O_3	93.6	α-Alumina (Corundum)	88.2
Fe_2O_3	1.0	β-Alumina	3.1
CaO	3.4	Grossite	8.7

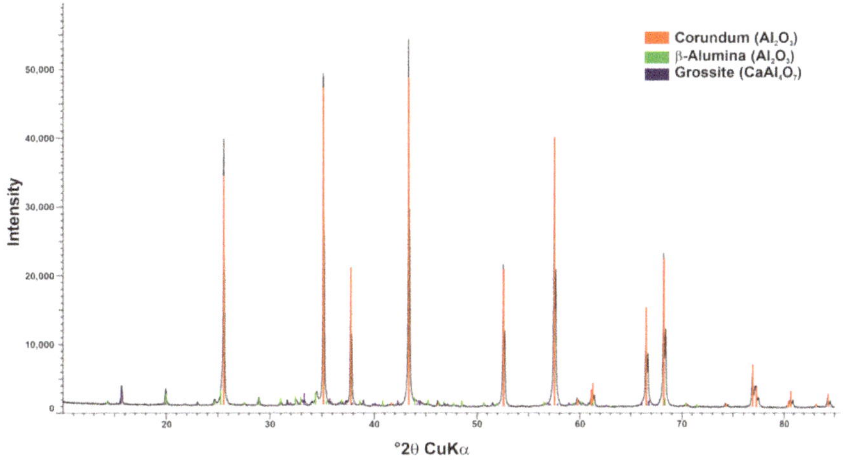

Figure 5. X-ray diffractogram of sample RefMat-2.

Sample material RefMat-3 has a slightly lower Al_2O_3 content than sample RefMat-2. In addition, contents of MgO and traces of SiO_2, Fe_2O_3, CaO and Cr_2O_3 were analyzed (Table 4). The two mineral phases α-alumina (corundum) and spinel ($MgAl_2O_4$) could be detected by phase analysis. The α-alumina can be attributed to the tabular alumina contained in the sample, but also to the white corundum. The two phases can hardly be distinguished by X-ray diffraction. Tabular alumina is chemically equivalent to the white corundum described above. The only difference is that white corundum is obtained from the melt and tabular alumina is recrystallized or sintered α-alumina with a high density, the morphology of which consists of large (50–200 μm), tabular corundum crystals. The two mineral phases catoite ($Ca_3Al_2(OH)_{12}$) and grossite ($CaAl_4O_7$) are residual components of the alumina-cement-based binder.

Table 4. Chemical and mineralogical composition of the sample material Ref-Mat-3.

Chemical Composition		Mineralogical Composition	
Oxide	**wt.%**	**Phase**	**wt.%**
Al_2O_3	87.3	α-Alumina (Corundum)	69.9
MgO	6.1	Spinel	23.4
SiO_2	0.5	β-Alumina	2.6
Fe_2O_3	1.2	Catoite	2.5
CaO	4.0	Grossite	1.0
Cr_2O_3	0.2	Quartz	0.5

3.2. *Fragmentation of the Refractory Waste Material*

Based on the fragmentation pre-tests, the individual parameters could be determined for each sample material, allowing a larger amount of material to be processed in the next step. The resulting material was then dried and subsequently screened into four grain size fractions (<3, 2–3, 1–2 and >1 mm; see Figure 6). In addition, a fine fraction was obtained from the process water by filtration.

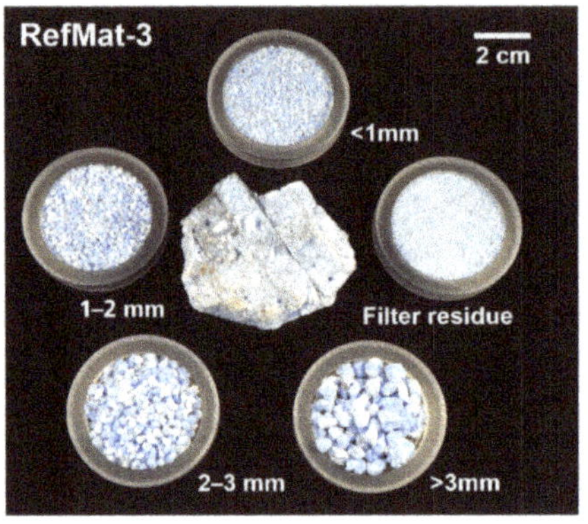

Figure 6. Sample material RefMat-3 before (central) and after fragmentation and subsequent classification by grain size.

Table 5 shows the mass balance of the individual grain size fractions. It can be seen from this table that the proportion of the fine fraction (filter residue) from the process water for sample MatRef-1 and sample MatRef-3 is almost 45 wt.% and for sample MatRef-2 even 60 wt.%. This high proportion is due to the relatively high number of pulses per material input (Table 6). In order to separate the individual components from the binder matrix, a high number of pulses with a high voltage had to be applied. This resulted in a lot of fine material. After drying, it was also observed that the fine material from the process water partially solidified. One reason for this is that a large amount of binder accumulates in the filter residue, and the binder also has a residual hydraulic activity.

Table 5. Weight percentages of the individual fractions after fragmentation.

Sample	RefMat-1	RefMat-2	RefMat-3
	Weight Percentages (wt.%)		
Filter residue	44.8	60.0	44.2
<1 mm	1.5	0.7	1.1
1–2 mm	21.3	21.4	20.6
2–3 mm	15.4	10.4	17.0
>3 mm	16.3	7.0	16.7
Loss of material	0.7	0.5	0.4

Table 6. Process data.

Sample	RefMat-1	RefMat-2	RefMat-3
Material input (kg)	20.2	23.7	31.5
Number of pulses	4484	5955	6017
Pulses per kg	222	251	191

The various aggregates of the fragmented refractory ceramics, which are to be recycled as regenerates, have mostly accumulated in the coarser fractions (1 to >3 mm). In addition to the very cleanly exposed aggregates, binder residues can also be detected in the coarser fractions.

3.3. Sorting of Fragmented Material

In the next step, the fragmented and screened materials were sorted according to optical criteria in order to obtain the exposed regenerates in the mono-fraction. Due to the significant color differences, the desired regenerates of the sample material RefMat-1 (Figure 7), the zirconia-alumina fused aggregate (ZAC) and the bauxite could be separated from the remaining material of the sample with high accuracy for all grain size fractions. The optical sorting was performed in several runs. In the first run, the yellow or flesh colored ZAC was sorted out with as little residual material as possible. Afterwards, the greyish brown to black colored bauxite was separated from the residual material. The remaining residual material consists mainly of coarse grains of light grey colored binder matrix and aggregates with residual material on its surface (ZAC and bauxite). The largest proportion of regenerates could be obtained from the fraction >3 mm. A total of approximately 5.3 kg (26 wt.%) of ZAC and approximately 2.7 kg (13 wt.%) of bauxite were exposed from 20 kg of the refractory material (RefMat-1). However, the results also showed that there was still a high proportion of residual material in the 2–3 mm fraction.

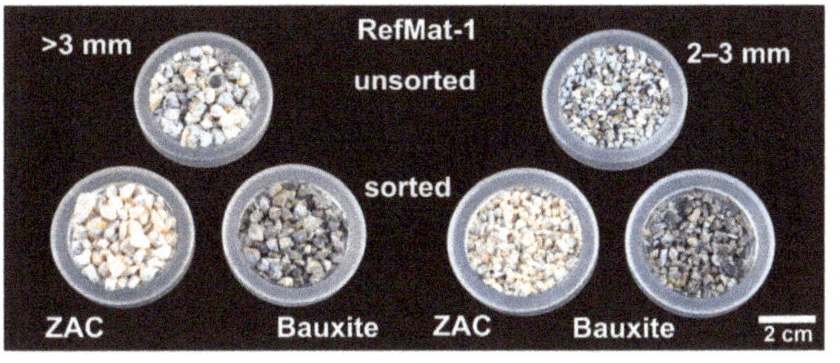

Figure 7. Sample RefMat-1 before and after optical sorting.

In case of the sample material RefMat-2, it was possible to separate the desired regenerate white corundum (glassy transparent grains) for each grain size fraction in just one pass (Figure 8). However, the output in all three fractions was very low, and the proportion of residual material, i.e., white colored binder matrix and not cleanly exposed white corundum, was relatively high. In all three fractions, only approximately 3.3 kg (13.8 wt.%) could be separated cleanly from the binder matrix, whereby approximately 23.6 kg of sample material (RefMat-2) was fragmented. The largest proportion of white corundum could be found in the 1–2 mm fraction.

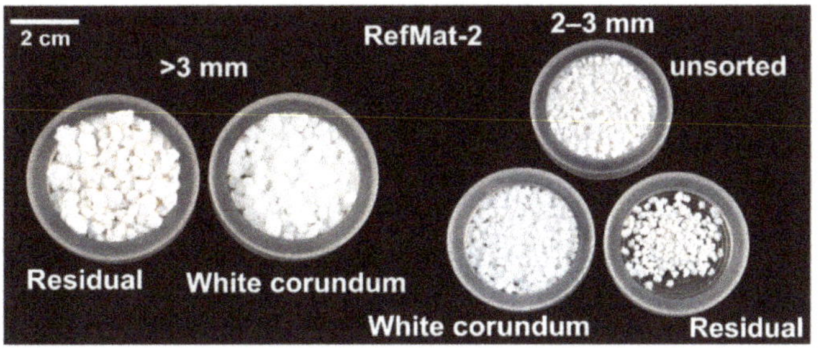

Figure 8. Sample RefMat-2 before and after optical sorting.

From the RefMat-3 sample, two different regenerates (tabular alumina and white corundum) needed to be sorted out, as for the RefMat-1 sample. However, it was discovered that due to the optical properties, the glassy transparent white corundum could hardly be distinguished from the grayish colored binder grains by the laboratory sorting system and could therefore not be sorted out. The optical filters could not be adjusted accordingly on the laboratory equipment, so a small amount of white corundum from the 2–3 mm fraction was sorted out by hand. Nevertheless, it was possible to sort out the white colored tabular alumina from all fractions from the sample material RefMat-3. This regenerate could be easily distinguished from the remaining material by optical sorting and could be sorted out for all three fractions (Figure 9). In addition to the non-sortable white corundum, the remaining material consists mainly of binder matrix and not cleanly exposed tabular alumina. With this sample material, the yield of regenerates is not very high. From the total fragmented sample material (RefMat-3) of approximately 31.5 kg, only approximately 4.8 kg (15.3 wt.%) of regenerates distributed over three grain size fractions could be obtained.

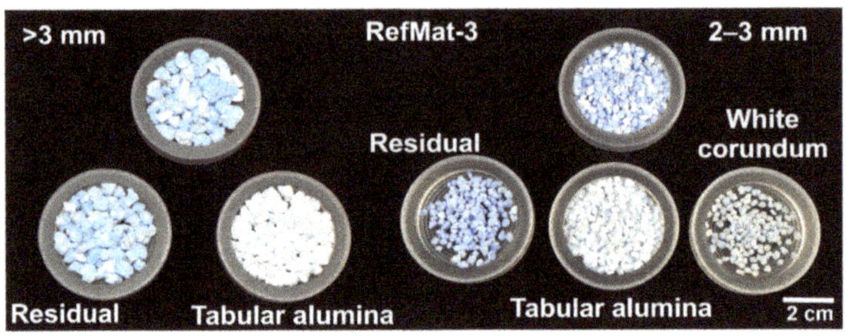

Figure 9. Sample RefMat-3 before and after optical sorting.

It is obvious that the desired regenerates from the three different samples could be cleanly separated out of each sample material. However, it can also be seen that the yield of regenerates is not very high. Regenerates can still be found in the remaining materials that have been sorted out, but these have not been sufficiently liberated from the binder matrix by the fragmentation.

3.4. Characterization of the Fragmented Material

On the basis of the microscopic examinations, the extent of binder matrix adhesions to the regenerates obtained was investigated. In the recovered material from the sample material RefMat-1, namely the sorted out ZAC (Figure 10a) and bauxite (Figure 10b), binder adhesion could be detected on rare occasions. Generally, the individual grains were very effectively separated from the binder matrix using electrodynamic fragmentation.

(a) (b)

Figure 10. Micrographs of the fragmented sample RefMat-1, sorted out ZAC (**a**) and sorted out bauxite (**b**).

The sorted out white corundum regenerates, which were obtained from the sample material RefMat-2, also show hardly any binder residues on the grain surfaces (Figure 11a).

Only sporadic white binder residues are still visible (Figure 11b). The degree of detachment for these regenerates is thus evaluated as very high in purely optical terms.

(a) (b)

Figure 11. Micrographs of the fragmented sample RefMat-2. Sorted white corundum (**a**), single grain of white corundum (**b**).

The microscopic images of the fragmented and unsorted RefMat-3 sample (Figure 12a) clearly show why the white corundum could not be clearly detected by optical sorting. Due to the different blue coloration of the binder, the glassy transparent white corundum cannot be clearly distinguished optically. On the other hand, the exposed white colored tabular alumina can be recognized and can therefore be sorted out without difficulty by optical sorting. The sorted out tabular alumina (Figure 12b) shows only slight binder adhesion.

(a) (b)

Figure 12. Micrographs of the fragmented sample RefMat-3. Unsorted material (**a**) and sorted out tabular alumina (**b**).

For a better evaluation of the chemical purity of the recovered regenerates, they were investigated in terms of their chemistry and mineralogical phase composition. Table 7 shows the results of the XRF measurement of the ZAC regenerates and bauxite regenerates from the sample material RefMat-1. For comparison, primary materials (ZAC and bauxite) were also analyzed and then cross-checked with the results of the recovered materials.

Table 7. Comparison of the chemical composition (XRF) of ZAC regenerates and bauxite regenerates with primary material (b.d.l.—below determination limit).

Oxide	ZAC Regenerate	Primary ZAC	Bauxite Regenerate	Primary Bauxite
	wt.%	wt.%	wt.%	wt.%
Al_2O_3	52.5	47.7	89.1	87.8
ZrO_2	27.6	34.7	0.4	0.2
SiO_2	16.1	14.0	5.6	5.5
Fe_2O_3	0.2	0.05	0.8	1.6
CaO	0.2	0.3	0.3	0.4
TiO_2	0.4	0.1	3.3	3.5
K_2O	0.1	0.1	0.1	0.3
Na_2O	2.4	2.2	b.d.l.	b.d.l.

It becomes clear that the chemical composition of the regenerates does not differ greatly from the primary material. There are no significant differences between the regenerate and the primary bauxite, so the chemical purity of the obtained bauxite regenerate can be confirmed. The recovered ZAC has a slightly higher Al_2O_3 content and slightly lower ZrO_2 and SiO_2 contents compared to the primary material. As these are the three main elements in ZAC, and no unusually high contents of CaO, Na_2O or Fe_2O_3 or other impurities were measured, the deviating element contents are due to the inhomogeneity of the material. The material can therefore be regarded as a regenerate with a high chemical purity.

Additionally, the mineralogical phase composition does not show high impurities (Table 8). Only the grossite ($CaAl_4O_7$) detected in both regenerates, which is a relic of the hydrated alumina cement, indicates that there is a small amount of binder residue.

Table 8. Mineralogical phase composition of the regenerates ZAC and bauxite without X-ray amorphous content.

Phase	ZAC Regenerate (wt.%)	Bauxite Regenerate (wt.%)
α-Alumina (Corundum)	65.1	88.3
Mullite	1.7	6.4
Anorthite	1.0	2.2
Baddeleyite	28.0	0.3
Grossite	0.9	0.5
Zircon	1.1	-
Quartz	-	0.3
β-Alumina	-	0.6
SiC	2.1	1.4

The determined chemical composition of both regenerates, white corundum from sample RefMat-2 and tabular alumina from sample RefMat-3, show no impurities (Table 9). The chemical analysis of the tabular alumina obtained could be compared to an analysis of primary material. This confirms the very high chemical purity of the tabular alumina regenerates. The phase composition (Table 10) also shows no major impurities. As with the recovered material from the RefMat-1 sample, minimal contents of grossite and spinel are detectable, which indicate binder residues.

Table 9. Chemical composition of regenerates white corundum from sample RefMat-2 and tabular alumina from sample RefMat-3 compared to a primary sample of tabular alumina (AlfaTab 30) (b.d.l.—below determination limit).

Oxide	White Corundum Regenerate from RefMat-2 (wt.%)	Tabular Alumina Regenerate from RefMat-3 (wt.%)	Primary Tabular Alumina (AlfaTab 30) (wt.%)
Al_2O_3	99.7	99.4	99.5
CaO	0.1	0.1	0.2
SiO_2	-	0.2	0.04
Fe_2O_3	-	b.d.l.	0.08
Na_2O	-	b.d.l.	0.3

Table 10. Mineralogical phase composition of the regenerates white corundum (RefMat-2) and tabular alumina (RefMat-3).

Phase	White Corundum Regenerate (wt.%)	Tabular Alumina Regenerate (wt.%)
α-Alumina (Corundum)	97.2	93.6
β-Alumina	2.1	5.1
Grossite	0.7	-
Spinel	-	1.3

In summary, based on these results, it can be seen that the regenerates obtained by fragmentation have a high chemical purity, which is a basic prerequisite for a high-quality recycling.

3.5. Reuse of the Regenerates

Only after it was proved by analysis that the regenerates obtained possess a high chemical purity were the regenerates used in new refractory ceramics. The aim was to evaluate whether the recovered materials can substitute for the primary raw materials without disadvantages in terms of rheological, mechanical and refractory properties. A refractory concrete (mixture RC) and a refractory tamped concrete (mixture TC) were selected for the recycling tests, in which the primary raw materials were substituted by the regenerates obtained in this study. In the refractory concrete, the regenerates ZAC and bauxite were used, while in the refractory tamped concrete, the tabular alumina was used. In parallel, both mixtures were produced using primary raw materials, acting as reference samples for later performance comparisons. During the production of the different mixtures (RC and TC), it was found that the rheology, and thus the workability of the fresh mixtures, did not change and was absolutely comparable to the reference mixtures. The important factor here was to reach a comparable flow behavior and thus a comparable workability without changing the water demand. Regenerates of lower quality might require more water in order to achieve satisfactory workability. A higher water demand is known to have negative effects on the strength development of refractory concretes as well as tamped concretes. This is also known from concrete technology [15].

Several test specimen prisms were produced with the mixed concretes. All prisms were stripped after 24 h, stored for another 24 h at room temperature and then dried for another 24 h in a drying oven at 120 °C. It was found that the setting and drying behavior of the regenerate masses and reference masses is close to identical. After drying all prisms, some of the samples were burned at different temperatures in a high temperature furnace. This high-temperature treatment (sintering) can be used to check the refractoriness of the samples produced. The strength development by sintering at different temperatures is also an important characteristic value for refractory materials. The specimens of mixture RC were sintered at 1000 and 1450 °C, and the specimens of mixture TC were sintered at 1000 and 1500 °C.

As a result of this sintering, no negative changes were found in the regenerate samples. Thus, it could be proved that using the regenerates provided by the fragmentation results in a comparable refractoriness compared to the refractory material made of primary raw materials.

For a final evaluation of the quality and usability of the regenerates obtained, a test of cold compressive strength and cold bending tensile strength was carried out. By previously treating the samples at different temperatures, three different strength characteristics could be determined for each mixture: the test specimens that were dried in the drying oven at 120 °C only (mixture RC and TC) and the test specimens that were sintered at 1000 °C (mixture RC and TC) as well as at 1450 °C (mixture RC) and 1500 °C (mixture TC).

The test results for the samples of mixture RC are shown in Figure 13. The direct comparison of the achieved strengths of the regenerate samples and the reference samples shows that there are no significant differences in strength. The achieved strengths of the regenerate samples are at least as high as the strengths of the reference samples. Some of the regenerate samples even show slightly higher strengths.

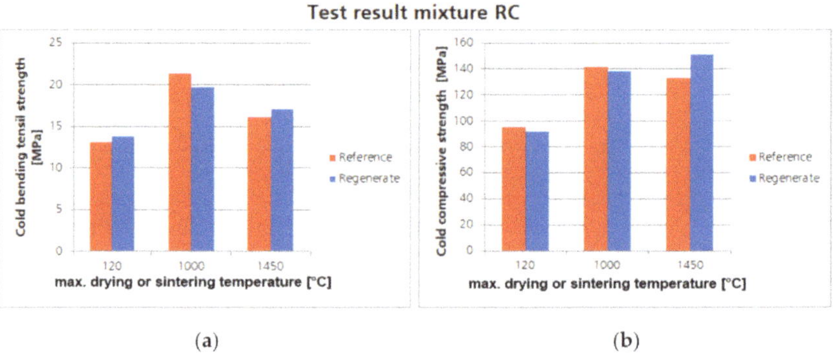

Figure 13. Comparison of the achieved cold bending tensile strength (**a**) and cold compressive strengths (**b**) of the refractory concrete (mixture RC) with regenerates (ZAC and bauxite) and the corresponding reference samples.

After testing the cold bending tensile strength, the fracture surfaces of the individual test specimens could also be checked (Figure 14). It is clear that the microstructure of the refractory concrete with regenerates is completely comparable to the microstructure of the refractory concrete with primary raw materials. Thus it is shown that the regenerates used have no negative influence on the quality of the refractory concrete and can therefore substitute for primary raw materials equally.

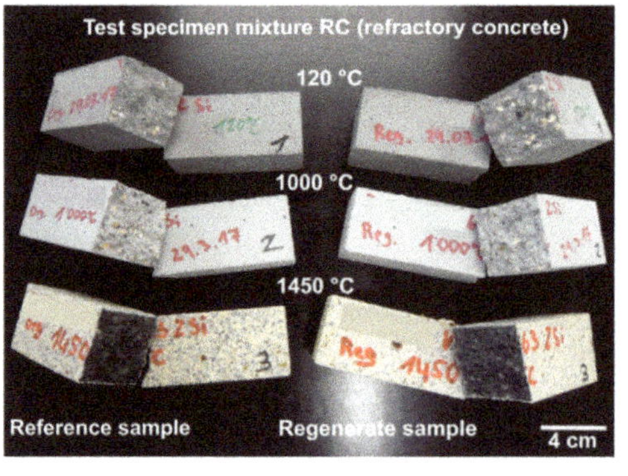

Figure 14. Test specimen of mixture RC after the cold bending tensile strength test.

A similar result was found for the samples of mixture TC with the tabular alumina. The determined strengths of the regenerate samples and the reference samples are absolutely comparable (Figure 15). The strengths achieved in the regenerated sample are even slightly higher than in the reference samples. When examining the fracture surfaces of the test specimens (Figure 16), no differences can be found. Thus, the suitability of the recovered tabular alumina used could also be proven on the basis of these results.

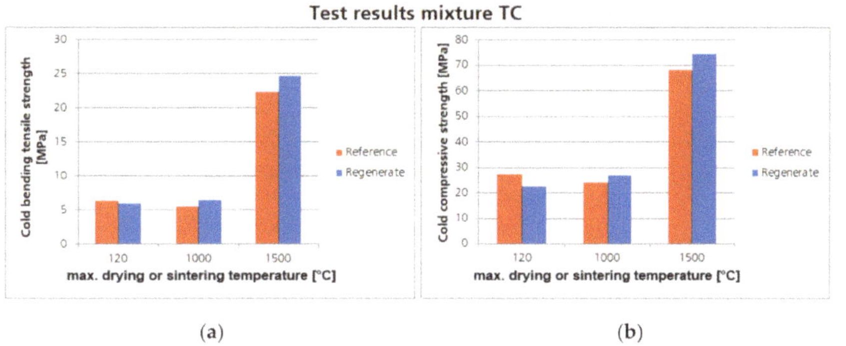

Figure 15. Comparison of the achieved cold bending tensile strength (**a**) and cold compressive strength (**b**) of the refractory concrete (mixture TC) with regenerates (ZAC and bauxite) and the corresponding reference samples.

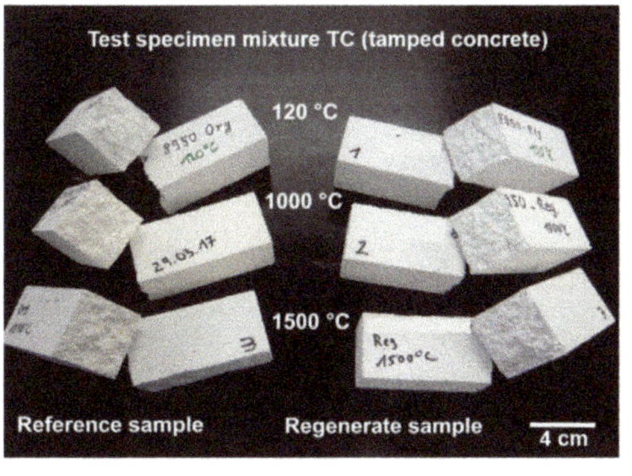

Figure 16. Test specimen of mixture TC after the cold bending tensile strength test.

4. Discussion

The described results of this study prove that the technology of electrodynamic fragmentation is suitable for breaking down the composite material refractory into its individual components. While a conventional jaw crusher can only crush composite materials to a small size but still maintain the bond, electrodynamic fragmentation allows the valuable aggregates or regenerates respectively to be separated from the composite or binder matrix selectively. The advantage of the technology used is primarily that the electrical impulses run along the grain boundaries of the material, and the composite is torn from the inside to the outside by an expanding plasma channel. Another advantage of the fragmentation technology is that it can be considered as a dust-free and, above all, contamination-free process. Metallic abrasion, as is the case with jaw crushers, cannot take place.

After fragmentation, the various regenerates could be sorted by optical sorting, which is a necessary requirement for a high-quality reuse. Only the white corundum from sample material RefMat-3 could not be sorted out because the color differences compared to the matrix material are too small. In order to optically separate these materials with slight color differences, more powerful sorting approaches must be developed and used.

It could also be shown that the separated and sorted regenerates, namely tabular alumina, bauxite, ZAC and white corundum, have a very clean surface and are almost completely free of binder residues and other adhesions. By means of chemical–mineralogical analyses, a very high chemical purity of the recovered regenerates could be achieved. A comparison of the chemical composition with original raw materials clearly showed that there are almost no differences observable.

Finally, the recovered regenerates could be reused in refractory concrete or refractory tamped concrete without sophisticated post-treatment, thus replacing primary raw materials without affecting the properties of the refractory products. The processing properties of the fresh masses as well as the mechanical test results of the sintered samples using regenerates do not show any adverse effects, as is usually the case when using conventionally crushed material as aggregate material. However, no testing of the newly developed refractory material under working conditions (e.g., lining in an aluminum melting furnace) has been taken place so far. Thus, the material must be checked for further aspects such as corrosion resistance, thermal shock resistance and abrasion resistance.

The technology of electrodynamic fragmentation can be a promising alternative to existing recycling technologies. It is possible to recover high-quality secondary raw materials for the refractory industry and thus realize a high recycling rate for the material itself. However, further research and development work is needed to make a recycling process for refractory materials economically realizable on an industrial scale. This is especially true for the fragmentation process. For this purpose, the plant must be designed in such a way that a continuous throughput of material for an industrial viable throughput is possible.

5. Patents

Based on this study, a patent was applied for with the title "Method for recycling ceramics, regenerated materials obtained thereby, and use of the regenerated materials for manufacturing ceramics" (DE 10 2017 217 611).

Author Contributions: Conceptualization, S.S., S.D. and J.B.; methodology, S.S. and S.D.; validation, S.S., S.D. and J.B.; investigation, S.S. and S.D.; resources, J.B.; data curation, S.S. and S.D.; writing—original draft preparation, S.S.; writing—review and editing, S.D. and J.B.; visualization, S.S.; supervision, S.S.; project administration, S.S. All authors have read and agreed to the published version of the manuscript.

Funding: This research received no external funding.

Institutional Review Board Statement: Not applicable.

Informed Consent Statement: Not applicable.

Data Availability Statement: Data available on request due to restrictions e.g., privacy or ethical. The data presented in this study are available on request from the corresponding author. The data are not publicly available due to ongoing patent process.

Conflicts of Interest: The authors declare no conflict of interest.

References

1. *DIN 51060:2000-06, Refractory Ceramic Raw Materials and Refractories—Definitions of the Terms Refractory, High Refractory*; Beuth Verlag GmbH: Berlin, Germany, 2000.
2. Schulle, W. *Feuerfeste Werkstoffe: Feuerfestkeramik; Eigenschaften, Prüftechnische Beurteilung, Werkstofftypen*, 1st ed.; Deutscher Verlag für Grundstoffindustrie: Leipzig, Germany, 1990; ISBN 3342003065.
3. Wirtschaftsvereinigung Stahl. *Fakten zur Stahlindustrie in Deutschland 2017*; Wirtschaftsvereinigung Stahl: Düsseldorf, Germany, 2017.
4. *Taschenbuch Feuerfeste Werkstoffe*, 3rd ed.; Routschka, G., Ed.; Vulkan-Verl.: Essen, Germany, 2001; ISBN 9783802731501.
5. Thome, V. Cementing Emissions: Recycling waste concrete with lightning bolts. *AWE Int.* **2013**, *2013*, 18–25.

6. Dittrich, S.; Thome, V.; Seifert, S.; Maier, M. Effektive Aufbereitung von Müllverbrennungsschlacken mittels Hochspannungsimpulsen. *Chem. Ing. Tech.* **2016**, *88*, 461–468. [CrossRef]
7. *Elektrodynamische Fragmentierung von MVA-Schlacken—Zerlegung der Schlacken und Abscheidung von Chloriden und Sulfaten//Aschen, Schlacken, Stäube. Aus Abfallverbrennung und Metallurgie*; Thomé-Kozmiensky, K.J. (Ed.) Berliner Schlackenkonferenz: Berlin, Germany; TK Verlag: Neuruppin, Germany, 2013.
8. Semkin, B.V.; Usov, A.F.; Kurets, V.I. The principles of electric impulse destruction of materials. *Kola Sci. Cent. Russ. Acad. Sci.* **1995**, *8*.
9. Marx, E. Versuche über die Prüfung von Isolatoren mit Spannungsstößen. *Elektrotechnische Z.* **1924**, *1924*, 652–654.
10. Bluhm, H.; Frey, W.; Giese, H.; Hoppe, P.; Schultheiss, C.; Strassner, R. Application of pulsed HV discharges to material fragmentation and recycling. *IEEE Trans. Dielect. Electr. Insul.* **2000**, *7*, 625–636. [CrossRef]
11. Sundar, V.; Newnham, R.E. Electrostriction and polarization. *Ferroelectrics* **1992**, *135*, 431–446. [CrossRef]
12. Hoppe, P.; Singer, J.; Giese, H.; Stemmermann, P.; Schweike, U.; Edinger, W. Prozessreaktor und Betriebsverfahren für die Elektrodynamische Fragmentierung. European Patent 1673172, 6 August 2004.
13. Robben, C.; Wotruba, H. Sensors-Based Ore Sorting Technology in Mining—Past, Present and Future. *Minerals* **2019**, *9*, 523. [CrossRef]
14. Maier, G.; Pfaff, F.; Pieper, C.; Gruna, R.; Noack, B.; Kruggel-Emden, H.; Langle, T.; Hanebeck, U.D.; Wirtz, S.; Scherer, V.; et al. Experimental Evaluation of a Novel Sensor-Based Sorting Approach Featuring Predictive Real-Time Multiobject Tracking. *IEEE Trans. Ind. Electron.* **2021**, *68*, 1548–1559. [CrossRef]
15. Nobis, C.; Vollpracht, A. Verwendung von Recyclingmaterial in der Betonproduktion—Sachstand. 2015. Available online: https://www.transportbeton.org (accessed on 13 November 2020).

processes MDPI

Article

Lumped Kinetic Modeling of Polypropylene and Polyethylene Co-Pyrolysis in Tubular Reactors

Andreas E. Lechleitner [1,*], Teresa Schubert [2], Wolfgang Hofer [3] and Markus Lehner [1]

1 Chair of Process Technology and Industrial Environmental Protection, Montanuniversitaet Leoben, Franz-Josef-Straße 18, 8700 Leoben, Austria; markus.lehner@unileoben.ac.at
2 Wien Energie GmbH, Thomas-Klestil-Platz 14, 1030 Wien, Austria; teresa.maria.schubert@gmail.com
3 OMV Downstream GmbH, Mannswoerther Straße 28, 2320 Schwechat, Austria; wolfgang.hofer@omv.com
* Correspondence: andreas.lechleitner@unileoben.ac.at; Tel.: +43-664-612-1025

Abstract: The recycling rates, especially those from plastic packaging waste, have to be increased according to the European Union directive in the next years. Besides many other technologies, the pyrolysis of plastic wastes seems to be an efficient supplementary opportunity to treat mixed and unpurified plastic streams. For this reason, a pyrolysis process was developed for the chemical recycling of hydrocarbons from waste polyolefins. The obtained products can be further processed and upgraded in crude oil refineries, so that also monomers can be recovered, which are used for the plastic polymerization again. However, to achieve a scale up to a demo plant, a kinetic model for predicting the yields of the plastic pyrolysis in a tubular reactor is needed. For this reason, a pilot plant was built, in which different plastics and carrier fluids can be tested. Based on the data generated at the pilot plant, a very practical and suitable model was found to describe the plastic co-pyrolysis of the carrier fluid with polypropylene (PP) and low density and high density polyethylene (HDPE and LDPE), respectively. The physical and chemical mechanisms of the co-pyrolysis in the tubular reactor are successfully investigated.

Keywords: feedstock recycling; plastic pyrolysis; lumped modeling; kinetic modeling; ReOil

Citation: Lechleitner, A.E.; Schubert, T.; Hofer, W.; Lehner, M. Lumped Kinetic Modeling of Polypropylene and Polyethylene Co-Pyrolysis in Tubular Reactors. *Processes* **2021**, *9*, 34. https://doi.org/10.3390/pr9010034

Received: 30 November 2020
Accepted: 21 December 2020
Published: 25 December 2020

Publisher's Note: MDPI stays neutral with regard to jurisdictional claims in published maps and institutional affiliations.

1. Introduction

25.8 Mio. tons of plastic waste are produced in Europe annually [1] and less than 30% of this are collected for recycling. Therefore, high amounts of plastic wastes are sent to incineration, landfilling or are sent to non-EU regions [2].

To overcome this situation, the European Union compiled a mandatory goal to reach a recycling rate of plastic packaging waste of 55% in 2030 [3]. The strategy in accomplishing this goal is to enhance a circular economy. The ways of increasing recycling rates include introducing deposit return systems, mechanical recycling, and also chemical recycling via depolymerization and anew polymerization.

However, a common challenge of plastic waste is the sorting of different types of plastics and the contamination of waste streams. Therefore, beside other technologies, pyrolysis is an attractive way to treat mixed plastic or even contaminated streams. In pyrolytic processes, valuable resources are generated out of organic matter at elevated temperatures in absence of oxygen, which can be used as feedstock for refineries and petrochemical industries again [4].

Pyrolysis is a cracking process by breaking long-chain hydrocarbons in molecules with lower molecular weight. These products are liquid and gaseous, respectively, and can be further processed in existing infrastructure, like a conventional crude oil refinery. Thermal cracking (pyrolysis) and catalytic cracking are compared in literature, for example the works [5–9] describe the advantages, disadvantages, technologies and properties of various feedstocks, products and processes. A significant advantage of pyrolysis is the resistance against impurities and the ability to process mixtures of plastic types. All kinds of polyolefins are suitable for pyrolysis because they consist of carbon and hydrogen only without heteroatoms, which then produces valuable yields. Polyolefins, such as polypropylene (PP), low density and high density polyethylene (LDPE and HDPE), are the most used plastics in packaging. Polystyrene (PS) has no heteroatoms and can, thus, be used as feedstock as well. The main challenge in pyrolytic processes is the poor thermal conductivity of plastic. In most waste streams the plastics occur as thin foils or in very voluminous shapes, what makes the handling additionally more complicated. These properties make the necessary heat transfer for thermal cracking difficult. Furthermore, when the plastic is molten, the viscosity is high, and the melt often cannot be pumped with conventional pumps used in refineries.

For operating a commercial plant, high processing capacities and a stable, continuous process is needed to guarantee the feedstock for a refinery. For this reason, a continuous co-pyrolysis process in a tubular reactor was developed [10]. The idea is to use a refinery residue as carrier to fluidize the plastic, consequently it can be treated as a liquid. However, the carrier fluid, also a hydrocarbon, has its own crackability, so the carrier fluid also has to be investigated. Valuable products, which are gaseous at process conditions, as well as solid impurities are separated at the end. The cracking conditions in the tubular reactor are at moderate temperatures below 500 °C and elevated pressures below 25 barg. The process is already verified in lab and pilot scale. A demonstration plant will follow in 2022+. Hence, a simulation and optimization tool must be provided for the upscale. To do so, lumped kinetic modeling was chosen to simplify the complex reaction mechanism [11]. The reactions that take place in the thermal degradation of plastics are radical chain reactions, which results in molecules with lower molecular weight and less saturation. A large amount of species and reactions are involved. These degradation of plastics is often described by thermogravimetric data, and hence, models are received, in which the time depending loss of mass can be measured [8,12–17]. Consequently, this kind of models just describe the overall degradation kinetics of polymers. Other models go deeper into the fundamental reactions, such as chain fission, radical recombination, end-chain β-scission and hydrogen transfers, just to mention a few. For this kind of modelling, a high experimental and analytic effort as well as high computational capacity is needed [13,14,18,19].

For upscaling and process optimization, a kinetic model which considers the main parameters, temperature, pressure and residence time, is needed. Furthermore, the model should be simple and should be capable of being integrated in a process simulation tool. Therefore, lumped modeling is used to describe the emerging and vanishing species in a thermal cracking of post-consumer plastics. This approach simplifies the reactions scheme drastically, by grouping (=lumping) species with similar properties together. In this study, a four lump kinetic model, based on boiling points of the species, will be introduced whereby the four lumps are connected with six monomolecular, irreversible, first order reactions. Even with just four reacting pseudo components, the product yields of plastic pyrolysis and the degradation of the carrier fluid can be predicted. However, a model needs reasonable data for verification, so a pilot plant was built in previous works, which can be varied in the main process conditions and feeds [20].

2. Materials and Methods

To observe the effects of temperature, pressure, residence time and plastic type on the product yield of the developed process, a pilot plant was build [20], which is also used for a kinetic study. The formed vaporous products with lower molecular weight are discharged of an unpressurized system as soon as the reaction condition exceed the boiling point. If the cracking process is performed continuously in a tubular reactor, as it is the case in the described process here, the produced gaseous species cannot leave the system, thus parameters like flow rate and pressure have a significant impact on the product yields. Since the pilot plant is diminutive (inner diameter of pipes in the tube reactor is 4.3 mm), the risk of blockages is high. Consequently, the pilot plant is used to investigate pure, virgin plastics and not real, waste plastics. Furthermore, the occurrent plastics in various waste streams are highly diverse, thus it will not be possible to examine all types and all potential mixture configurations of plastic. Hence, reference plastics from all main types are used to get an overview, how the different types of plastic behave in the process. The main feedstock for the thermal degradation process is all kind of polypropylene (PP), low density and high density polyethylene (LDPE and HDPE). Common parameters of the examined plastics types are shown in the next table (Table 1).

Table 1. Examined plastic types with some measured, descriptive parameters (own measurements).

Plastic Type	Molecular Mass (Mw)	Calorific Value	Upper Heating Value	H_2 Content	TGA Inflection Point	TGA Residue
(-)	(kg/mol)	(kJ/kg)	(kJ/kg)	(%)	(°C)	(%)
PP	364	44,510	47,343	13.8	484	0.03
LDPE	235	43,409	46,159	13.4	500	0.81
HDPE	197	43,525	46,409	14.0	509	0.22

However, the plastic is pretreated for the usage in the plant. Since no extruder is available for feeding the plastic melt into the pilot plant, a fine plastic powder (<500 μm), obtained by cryogenic grinding, is mixed with the carrier fluid in a certain mixing ratio in a stirring tank.

A test campaign combines different test runs with the specific, investigated plastic type commingled with the carrier fluid. The carrier fluid is a high boiling hydrocarbon and is available on site in the refinery. Therefore, an evaluation of the behavior of the pure carrier fluid has to be done in advance. Tests with pure carrier fluid are run at diverse temperature and residence time couples to measure the conversion rates of the carrier fluid itself under different process conditions. Subsequently, the test series with increasing content of plastic mixed with the carrier fluid are performed again at different process conditions. In this way, the share of the carrier fluid on the observed conversion can be identified. The highest ratio of plastic in the mixture reachable is 30 wt.%, at this plastics content pumping becomes impossible due to the mixture's increasing viscosity.

2.1. Experimental Setup

The scheme of the pilot plant used for the collection of the process data is shown in Figure 1.

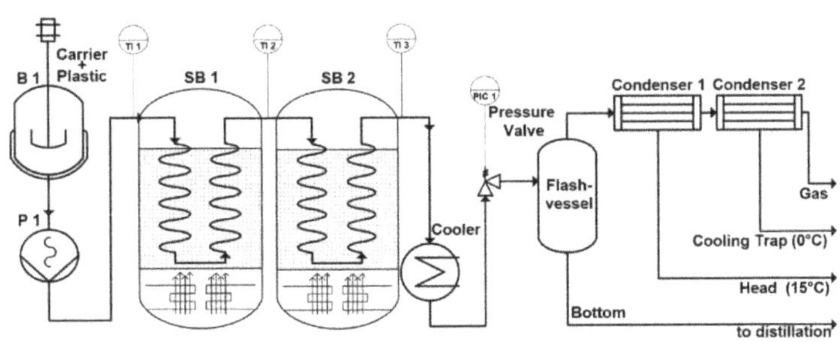

Figure 1. Scheme of the pilot plant.

In a stirring tank, B1, the plastic powder is mixed with the carrier medium in the specified mixing ratio. Then the mixture is pressed into the reactors by the eccentric screw pump P1. The reactor tubes are immersed in the sand baths SB1 and SB2. Downstream the reactor section, the medium is cooled down in order to instantaneously stop the reaction. The following pressure valve regulates the pressure in the plant. After flashing the products to atmospheric pressure, vaporized products are stripped out and condensed. All products are collected, weighted and analyzed. Density, heating value, paraffinic, aromatic, naphthenic and olefinic content, true boiling point curve are analyzed and a laboratory distillation is performed to calculate the product fraction yields.

The pressure valve can be adjusted from 0–20 barg to adjust the testing condition. This valve is also the bottleneck of the plant because it has the smallest cross section. Therefore, the operating parameters have to be adjusted for achieving a certain range of conversion. A conversion that is too high causes coking and a conversion that is too low results in unconverted plastic or very high molecular waxes. Blocking of the valve occurs in both cases. These limiting factors in the experimental investigations have a big impact on the model, which will be discussed later in the chapter "Lumped model".

Two sand baths are used for adjustment of the testing temperature. These sand baths have a very homogenous temperature distribution which is achieved by fluidization with air. The coiled reactors are immersed in this sand, whereas reactor tubes wall temperatures are approximately the same as the sand itself. The sand can be heated up to 600 °C to test a wide range of process parameters and resulting achieved conversion rates.

Finally, the residence time can be varied by two different modulations. One is the simple adjustment of the working load of the pump, the second method is to alter the reactor length. Each sand bath can be mounted with reactor coils ranging from 3 to 25 m length, which results in a maximum reactor length of 50 m, if two sand baths are used. The mass flow of the pilot plant can be regulated between 300 and 3000 g/h. Consequently, the maximal production rate of the plant is nearly 1 kg plastic per hour.

2.2. Modelling

The finding and conclusions of the test results of the pilot plant should be used for upscaling a demo plant. To do so, the obtained data should be implemented in a model which can describe engineering relevant phenomena, such as vaporizing, and predict the yield of products. As with other pyrolysis processes, the main influencing parameters of the conversion are temperature, pressure and residence time. The temperature has the major effect, which can be easily explained with the commonly used temperature dependency of reaction rates via the Arrhenius equation. A very minor effect has the pressure on the reaction, often described by Le Chatelier's principle. Due to the formation of gaseous products, a low pressure is beneficial during pyrolysis. However, since this process takes place in a tubular reactor, the pressure has a greater impact on the residence time due to reduced vaporization at elevated pressures [21]. Finally, the residence time is strongly affected by temperature and pressure. The high vaporization at high temperatures and low-pressure results in high gas phase amounts in the reactor pipes. Consequently, the residence time is reduced, but also the flow regime can change. Hence, the gas to liquid ratio in the system is also an additional physical parameter affecting the reaction by changing the residence time [22].

Since the residence time is the most uncertain parameter, the model is built on a calculation based on the reactor length instead of time. Additionally, the properties of the occurrent media have to be considered to calculate the gas phase share and the resulting volume stream. The analytics of the products and educts are used here to generate pseudo-components, which reflect the educts and products very well. For refinery simulation tools like PetroSim the density and true boiling point curve of a fluid is sufficient to describe this stream and to evaluate its physical properties.

As the temperature, pressure, reactor length and the properties of the mediums are known, the chemical reaction itself has to be described. Since the pilot plant has no continuous measurement, and the plastic pyrolysis produces a multitude of different species because of its radical complexion [22], and additionally, the plant runs at non-idealistic conditions (isothermal reactors cannot be assumed), it is obvious that fundamental reaction mechanisms cannot be determined. However, the object of this study is to find a model which describes this process sufficiently and practicably for the integration in a process simulation. Hence, Lumped Kinetic Modelling is introduced, in which the complexity of the reaction system can be reduced significantly without losing sufficient predictability.

2.2.1. Lumped Model

Lumping is a commonly used method in refinery calculations because in this applications it is impossible to measure each single molecule [23–26]. The concrete lumping approach just combines a pool of species with similar properties together, defining a single pseudo component, called lump, which has the properties of this mixed pool of species. Often just one main parameter is used to classify the lumps, commonly used properties are for example the molecular weight, densities, or true boiling points. In this work the true boiling point of the oily mixture is deciding the affiliation of species into lumps.

In a first approach a six-lump system of standard boiling cuts in refineries has been created: "Gas"(boiling range: incondensable at 0 °C), "Naphtha" (boiling range: initial boiling point to 175 °C), "Kerosene" (boiling range: 175–225 °C), "Gasoil" (boiling range: 225–350 °C), "Spindle oil" (boiling range: 350–410 °C) and "Residue" (boiling range: 410+ °C). First results show that the three light lumps "Naphtha", "Kerosene" and "Gasoil" nearly react the same way, so they are merged further together to the lump "Light Liquids" [27]. The classification of the received four-lump system is shown in Figure 2.

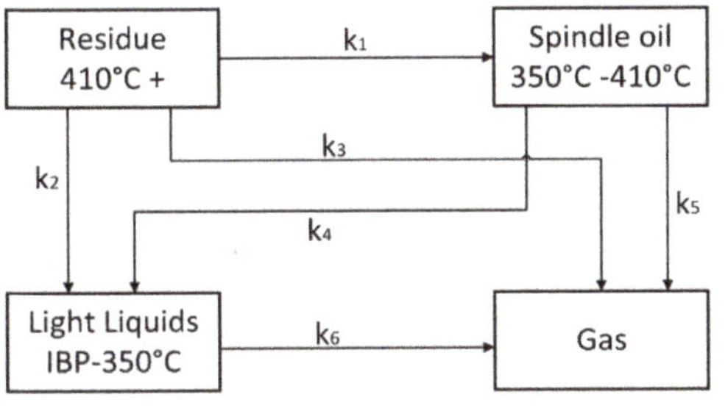

Figure 2. Four-lump system with six monomolecular, irreversible reactions. IBP means initial boiling point higher than 0 °C and Gas is specified by incondensable at 0 °C.

For a reaction scheme of pyrolysis containing four lumps, four mass balances have to be computed. Additionally, the assumption of irreversible reactions is made to reduce the complexity further. This decreases the reaction pathways to six possibilities as shown in Figure 2. The radical character of the reaction system is neglected, and no recombination reactions are considered in the model. The six reactions (k_1–k_6) are described by the Arrhenius law to consider the temperature dependency,

$$k_i = A_i * e^{\frac{-E_{A,i}}{R * T}} \tag{1}$$

in Equation (1) k is the reaction rate, A the Arrhenius constant, E_A the activation energy, R the ideal gas constant, T the temperature and i the indices of the reactions 1 to 6. As mentioned before, the residence time cannot be determined directly, so there is no possibility to draw an Arrhenius plot. Hence the two parameters activation energy $E_{A,i}$ and the Arrhenuis constant A_i are unknown for each specified reaction i. Consequently, in this four-lump system $2 \times 6 = 12$ unknown parameters have to be obtained. For a further simplification of the model, a reaction order of 1 is assumed. Otherwise the unknown parameters would increase computational effort drastically. Finally, the model consists of a four-lump system with six monomolecular, irreversible, first order reactions with twelve unknown parameters.

Solving a problem like this requires a nonlinear regression solver. Such solvers are implemented for instance in Matlab and enable to find local and global minima for fitting the model to the measured data of the pilot plant.

As a precondition for the modeling of the plastic pyrolysis, the cracking behavior of the carrier fluid has to be investigated more in detail. For that, the composition of the carrier medium is required, which is obtained with the same analytic and distillation method as for the products. Based on the obtained boiling curve, the carrier fluid can be integrated in the four-lump system. So, most of the carrier medium is classified as "Residue", but it also partly contains the boiling cuts of "Spindle oil" and "Light Liquids". This composition is used for the starting conditions of the model. As already mentioned, test campaigns with the pure carrier fluid are performed and the reaction model like shown in Figure 2 is determined.

Subsequently, test runs with a certain plastic type are used to evaluate the thermal degradation of this plastic. The basic approach is the same as for the carrier fluid, but the plastic model needs an adaptation to insert the plastic into the four-lump model. So, an additional lump "Plastic" is linked to the heavy lump. The heavy lump is now called "Wax" (but has the same definition as "Residue"), the other lumps are named the same, "Spindle oil "(SO), "Light Liquids" (LL) and "Gas", but are indexed with p for plastic. The lumped model for the plastic pyrolysis is shown in Figure 3.

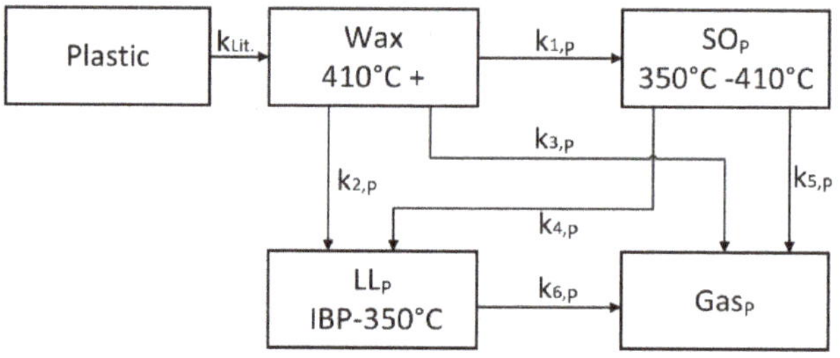

Figure 3. 5-lump model for the plastic pyrolysis with six monomolecular, irreversible, first order reactions and one starting reaction of the plastic degradation.

The six reaction rates $k_{i,P}$ of the five-lump model for plastic can be determined the same way as described before, just with the difference that the fraction of carrier fluid in the mixture is being calculated with the already determined four-lump model of the carrier fluid. The initial reaction $k_{Lit.}$ cannot be evaluated in this pilot plant, because the conversion of the plastic needs to be very high (estimated 90–100%), in order to avoid blockage of the small pipes. Therefore, the already determined kinetics for different plastic types from literature are used to describe this first degradation step. Thermogravimetric analytics are suited well to predict the pyrolysis to gaseous components, so overall first order, degradation reactions are used like that described in [15–17]. In the Lumped Kinetic Model, the plastic is cracked down to a liquid wax phase described by the kinetics derived from TGA experiments ($k_{Lit.}$). Based on the given six pseudo reactions $k_{i,P}$, the lumps react in a way which meet the measured mass fractions of the lumps.

2.2.2. Reactor Model

Based on the lumped kinetic model which represents the chemistry and kinetics of the cracking in the reactor, an overall reactor model is also developed. This model is based on the assumption of a plug flow reactor, whereas the mass balances can be constituted with deployed reaction rates as shown in Equations (2)–(6) for the lump system shown in Figure 3,

$$\frac{dc_{Plastic}}{dz} = \frac{-k_{Lit.} * c_{Plastic}^{n_{Lit.}}}{w}, \tag{2}$$

$$\frac{dc_{Wax}}{dz} = \frac{+k_{Lit.} * c_{Plastic}^{n_{Lit.}} - k_{1,P} * c_{Wax} - k_{2,P} * c_{Wax} - k_{3,P} * c_{Wax}}{w}, \tag{3}$$

$$\frac{dc_{SO}}{dz} = \frac{+k_{1,P} * c_{Wax} - k_{4,P} * c_{SO} - k_{5,P} * c_{SO}}{w}, \tag{4}$$

$$\frac{dc_{LL}}{dz} = \frac{+k_{2,P} * c_{Wax} + k_{4,P} * c_{SO} - k_{6,P} * c_{LL}}{w}, \tag{5}$$

$$\frac{dc_{Gas}}{dz} = \frac{+k_{3,P} * c_{Wax} + k_{5,P} * c_{SO} + k_{6,P} * c_{LL}}{w}, \tag{6}$$

in the mass balances c describes the mass fraction of the respective lump, z is the reactor length, w is the local flow velocity of the whole stream (gas and liquid) and $n_{Lit.}$ is the order of reaction for the used TGA degradation reaction of the plastic. It is assumed that all lumps (gas and liquid) move in the same velocity w, so there is no slip.

Additionally, a heat balance is integrated to evaluate the temperature profile over the reactor length (Equation (7)) because there is just six temperature measurements at the reactor wall,

$$\frac{dT}{dz} = \frac{1}{\rho * c_p} * \left(\frac{-k_{thermal} * 4 * (T - T_W)}{d * w} - H * \rho \right), \tag{7}$$

whereas T is the local temperature, ρ is the density of the local mixture, c_p is the heat capacity of the local mixture, $k_{thermal}$ is the local heat transfer coefficient, T_W is the local wall temperature measured in the sand bath, d is the inner diameter of the reactor pipe and H is defined by Equation (8),

$$H = H_r * \left(\frac{\frac{dc_{Plastic}}{dz}}{MM_{Plastic}} + \frac{\frac{dc_{Wax}}{dz}}{MM_{Wax}} + \frac{\frac{dc_{SO}}{dz}}{MM_{SO}} + \frac{\frac{dc_{LL}}{dz}}{MM_{LL}} + \frac{\frac{dc_{Gas}}{dz}}{MM_{Gas}} \right) + H_V + H_{melt}, \tag{8}$$

in Equation (8), H_r is the standard reaction enthalpy per mole change, MM is the mean molecular mass of the respective lump, H_{melt} is the melting energy of the plastic by passing by the melting temperature and H_V is calculated of the vaporizing fraction (vf) of the species respectively multiplied with the respective heat of evaporation (H_{Vap}) (Equation (9)),

$$H_V = H_{Vap,\ Wax} * \Delta vf_{Wax} + H_{Vap,\ SO} * \Delta vf_{SO} + H_{Vap,\ LL} * \Delta vf_{LL} + H_{Vap,\ Gas} * \Delta vf_{Gas}, \tag{9}$$

H_r has been calculated out of the average paraffinic, olefinic and naphthenic composition of the cracking production. If a big paraffinic molecule split into two molecules, there are two possibilities:

- High Paraffin → Paraffin + Olefin
- High Paraffin → Paraffin + Naphthene

For each reaction, the enthalpies calculated of the heat of formations do not differ very much, independently how big the broken parts are as long the C and H balance is respected. The reaction enthalpies to paraffins and olefins is averaged to 76,232 kJ/mol and these to paraffins and naphthene is averaged to 57,101 kJ/mol. The mean distribution of olefins and naphthenes is 17:3, accordingly H_r is determined by 73,289 kJ/mol.

The last unknown factor of the heat balance is $k_{thermal}$. To evaluate the heat transfer coefficient equations of the "VDI Wärmeatlas" [28] are used. Due to the complex construction of the plant, the reactors have different sections:

- Horizontal pipes
  - One-phase flow [28] (Ga)
  - Two-phase flow [28] (Hbb)
- Vertical up streamed pipes [28] (Hbb)
- Vertical down streamed pipes [28] (Hbb)
- Up streamed coils [28] (GC)
- Down streamed coils [28] (GC)

The length of all these sections is measured for all different reactor setup used. Moreover, in all sections one phase flow (only liquid) or two-phase flow may occur. For all these cases, suitable equations are available in [28] to calculate the heat transfer coefficient.

The coupled differential equation system has to be solved simultaneously, which is performed with Matlab. The solver calculates the reactor with the finite-element method discretely, whereas the minimal step size is one centimeter (Figure 4). Each discrete element is computed at constant conditions and the input is the starting values of the educt or the results of the previous element.

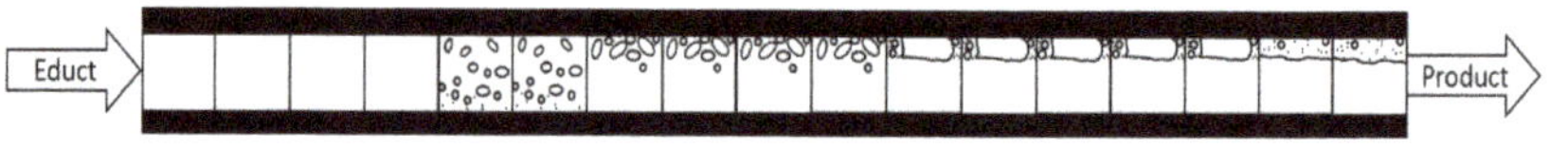

Figure 4. Sketch of the discrete calculation of the tubular reactor with a finite-element method.

Finally, the kinetic data fitted to this model are determined with the nonlinear fitting tool. In this study, Matlab is used, which can calculate the model with the data By using reasonable starting points from previous models or from the literature [15,20–23], the model is calculated with random or algorithmic specified values until a local or global minimum is found by the method of least squares.

3. Results

This section reports the results for the pure carrier fluid, polypropylene, low density and high-density polyethylene, and compares the experimental results with the model. An overview of main test runs is given in Table 2. The main input parameters, such as mass fraction of plastic and carrier fluid, the mean reactor temperature, the pressure and the flow are shown. Additionally, the calculated residence time of the reaction medium above 400 °C and the conversion referring to the lump "Residue" are inserted in Table 2.

Table 2. Exemplary test runs performed in the pilot plant. The total number of test runs was 67.

Test Run	Mass Fraction Carrier Fluid	Feed	Mass Fraction Feed	Mean Temperature	Pressure	Flow	Residence Time >400 °C	Conversion Based on "Residue"
(-)	(%)	(-)	(%)	(°C)	(barg)	(g/h)	(min)	(%)
1	100	-	0	493	5	2478	3.3	50
5	100	-	0	529	5	3219	2.5	29
11	100	-	0	522	5	2533	0.9	56
16	100	-	0	415	15	855	18.9	7
23	100	-	0	430	14	555	35.7	23
27	90	PP	10	479	5	1540	5.3	42
34	90	PP	10	487	5	2265	3.8	42
37	80	PP	20	472	5	1982	4.5	53
38	90	LDPE	10	467	15	1083	9	57
45	80	LDPE	20	435	15	504	44.6	29
49	70	LDPE	30	459	15	874	19.3	46
51	90	HDPE	10	444	15	1339	14.8	26
62	70	HDPE	30	441	15	370	54	25
67	80	HDPE	20	450	15	627	30.1	21

The pressure, reactor length, temperature of the sand baths and the flow rate are the main input data for the model. Additionally, the starting conditions have to be defined at the reactor inlet: The starting temperature, the composition of the carrier fluid and the mass fractions of plastic and carrier fluid. Then, the model calculates each discrete element of the reactor with starting kinetic values. After the calculation, the Optimization Toolbox of Matlab compares the calculated yields and the measured ones and changes the kinetic data for a new calculation. This is done until a minimum mean squared error between measured and modeled yields is found. An overview of the obtained values for the kinetic parameters is shown in Table 3. For the carrier fluid all six reactions take place and need to be formulated. This is due to the fact that the starting composition of the carrier fluid already contains the lumps, "Residue", "Spindle oil" and "Light Liquids" (Figure 2). In order to obtain a reasonable fit, all lumps of the carrier fluid have to react. Contrary, kinetics of the plastics behaves differently. In a first attempt, also six reactions have been assumed, as shown in Figure 3. However, just three reactions are sufficient to describe the system, and to find a global minimum to fit all measured yields. All other reaction rates can be neglected (ki set to zero). In case of LDPE the conversion to products is very low and just the reaction k2 producing small amounts of "Light Liquids" is sufficient to model the measured yields. HDPE behaves like LDPE just with an additional reaction k6 to produce some "Gas" (Table 3).

Table 3. Determined Arrhenius constants "A" and Activation Energies" E_A" of the plastic pyrolysis in the pilot plant.

		$k_{Lit.}$ +	k1	k2	k3	k4	k5	k6
Carrier fluid	A (1/s)	-	4.1×10^{15}	6.1×10^{11}	4.7×10^{14}	9.8×10^{15}	1.8×10^{15}	3.2×10^{10}
	E_A (kJ/mol)	-	270	208	269	271	271	199
PP	A (1/s)	3.2×10^{15}	2.0×10^{02}	1.0×10^{5}	5.0×10^{14}	-	-	-
	E_A (kJ/mol)	244	80	100	249	-	-	-
LDPE	A (1/s)	1.0×10^{15}	-	1.6×10^{47}	-	-	-	-
	E_A (kJ/mol)	241	-	700	-	-	-	-
HDPE	A (1/s)	1.9×10^{13}	-	1.0×10^{47}	-	-	-	1.0×10^{18}
	E_A (kJ/mol)	220	-	700	-	-	-	300

+ $k_{Lit.}$ is taken from [15].

Additionally, parameters like the calculated flow velocity, temperature and vapor phase in any element can be exported from the model. With these data, the residence time of each test run can be obtained and the calculated temperature and concentration trends over length or time can be drawn. The corresponding legend is shown in Figure 5 and the trends are shown in Figures 6 and 7.

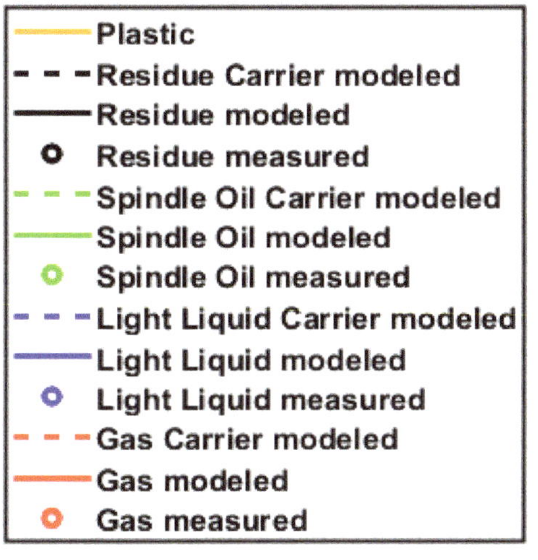

Figure 5. Legend for Figures 6 and 7: The dashed line show the part produced from the carrier fluid and the full lines are the trend of total product, respectively. The points mark the experimental results of this test run.

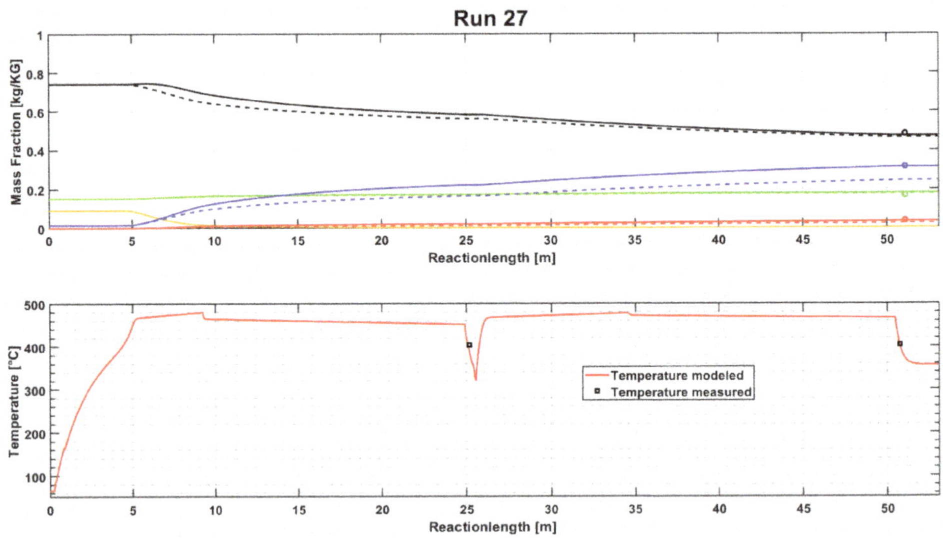

Figure 6. Trends of mass fraction of the lumps and the temperatures over the reactor length for test run 27 with 10% PP.

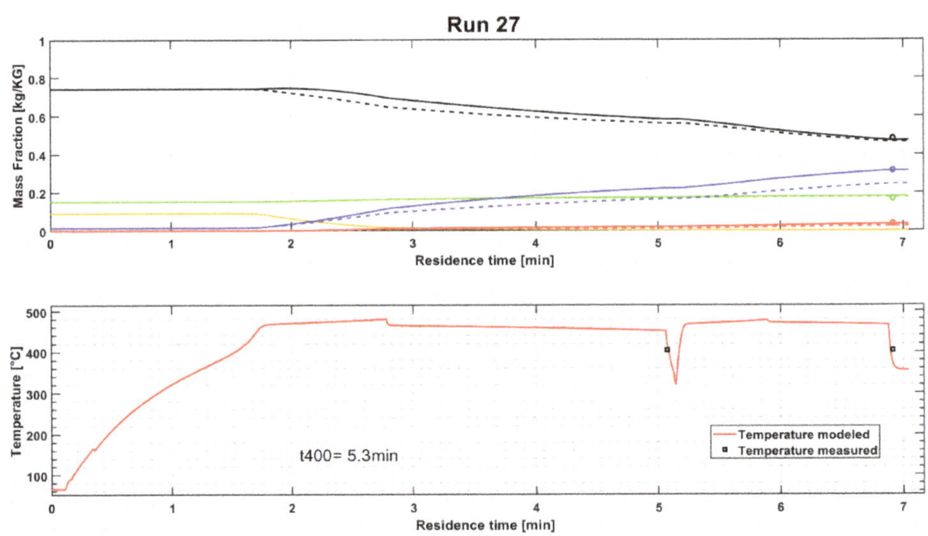

Figure 7. Trends of mass fraction of the lumps and the temperatures over the time for test run 27 with 10% PP. The residence time with temperatures higher than 400 °C is 5.3 min for this test run.

It can be seen in Figures 6 and 7 that PP cracks very fast at a certain temperature and completely converts to "Wax". Then the light product fractions are increasing during the reactor while the "Residue" and the "Wax" are decreasing. The second plot in Figure 6 is the temperature trend. It increases also very fast after entering in the reactor. The medium temperature reaches the temperature of the sand bath nearly after five meter reactor length. After about 25 m, the first sand bath is passed and a short pipe section outside of the sand bath with a temperature measurement follows. A sharp temperature decrease can be seen clearly in the diagram. Then, the second sand bath begins and the temperature reaches again the hot equilibrium temperature.

Both tubular reactors in the sand baths have the same length of about 25 m in test run 27. However, the residence time in both sand baths are different, which can be derived from the temperature trend in Figure 7. The second sand bath has a residence time shorter than two minutes, whereas the first sand bath has more than three minutes residence time. This is due to the decrease of density of the medium, caused by produced gases and other light products, which results in higher flow velocities.

Furthermore, it can be seen that the modelled curves fit the measurement points quite well in test run 27. To determine the accuracy of the model, the next subchapters show deviation plots obtained for the different feedstocks investigated.

3.1. Carrier Fluid

At first the test runs with pure carrier fluids are evaluated. The carrier fluid contains nearly 10% of "Light Liquids" and after the pyrolysis more than 65% of "Light Liquids" can be measured. The deviations of the modeled and measured values are shown in Figure 8. The simple four lump model works very well for the carrier fluid pyrolysis, which can be derived from the low deviations to the diagonal line and also from the uniform distribution around the diagonal. There is just one outlier, which is also the only test run with a very high conversion. Hence, there is the indication that the model predicts well under mild pyrolysis conditions with a "Light Liquid" yield between 10 to 45%.

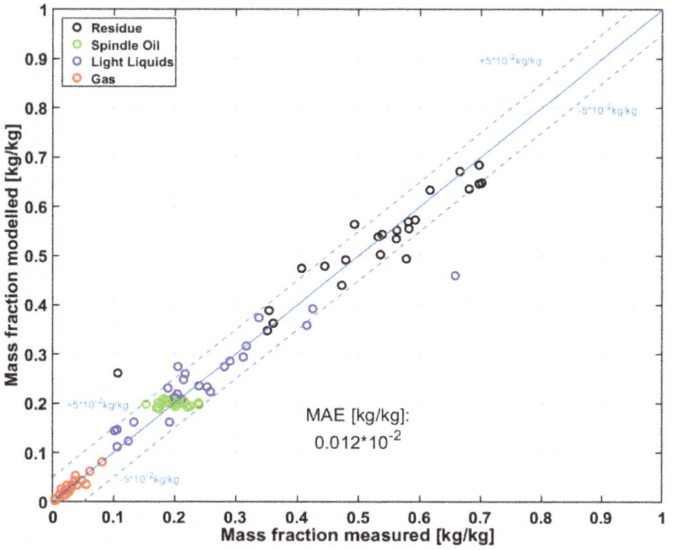

Figure 8. Deviation of the modeled and measured values of the test runs with pure carrier fluid. "MAE" is the mean absolute error of the optimization of 0.01×10^{-2} kg/kg. The dashed lines show the 10% deviation.

3.2. Polypropylene (PP)

Figure 9 shows the deviations between measured and modeled results for the test runs with PP. PP cracks at lower temperatures than the other plastic types studied in this work [8], which is confirmed by yields lower than 50% for the "Residue" fraction, except for one test run. Furthermore, the share of the "Residue" fraction is overpredicted from the model, contrariwise the share of the "Light Liquids" fraction is underpredicted.

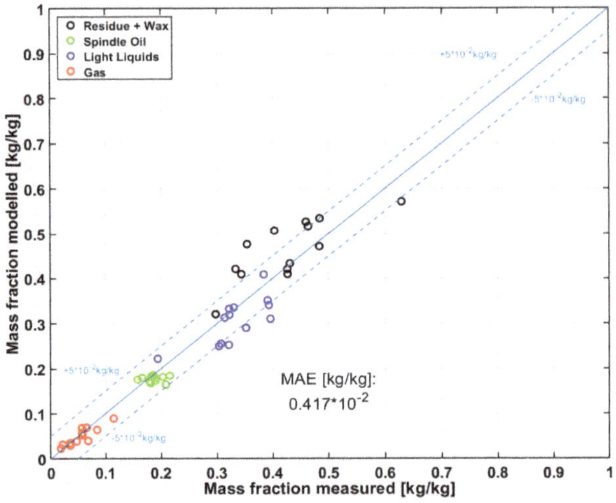

Figure 9. Deviation of the modeled and measured values of the test runs with carrier fluid and PP. "MAE" is the mean absolute error of the optimization of 0.42×10^{-2} kg/kg. The dashed lines show the 10% deviation.

3.3. Low Density Polyethylene (LDPE)

Figure 10 compares the test runs with LDPE with the model. The "Light Liquids" fraction is uniformly distributed which indicates a good fit, However, the "Spindle oil" fraction is overpredicted, even there is no reaction to produce "Spindle oil" (compare with Table 3). Consequently, the "Spindle Oil" fraction originates from the carrier fluid which indicates that there is a strong interaction between carrier fluid and LDPE. Additionally, the "Gas" lump has a uniform, but a high relative variance.

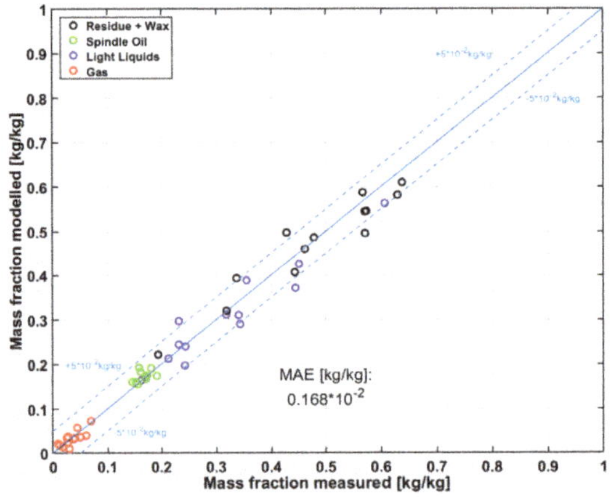

Figure 10. Deviation of the modeled and measured values of the test runs with carrier fluid and LDPE. "MAE" is the mean absolute error of the optimization of 0.17×10^{-2} kg/kg. The dashed lines show the 10% deviation.

3.4. High Density Polyethylene (HDPE)

The determined kinetic data for HDPE shows irregularities. At low conversions, the "Residue" lump is underpredicted, whereas the "Light Liquid" lump is overpredicted as shown in Figure 11. As well as in the LDPE kinetic the "Spindle oil" lump is overpredicted without having a "Spindle oil" producing reaction pathway in the plastic decomposition, and the "Gas" lump is slightly underpredicted.

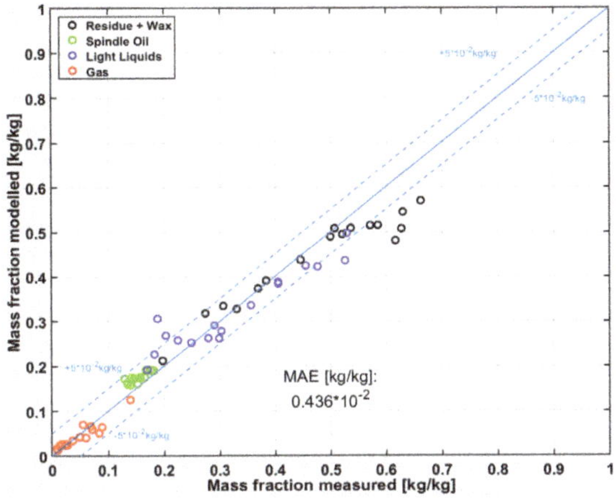

Figure 11. Deviation of the modeled and measured values of the test runs with carrier fluid and HDPE. "MAE" is the mean absolute error of the optimization of 0.44×10^{-2} kg/kg. The dashed lines show the 10% deviation.

4. Discussion

As shown in the previous chapter, the simple lump models provide a good agreement with the measured data from the pilot plant, especially the carrier fluid can be predicted with sufficient accuracy. The carrier fluid has a significant impact on the pyrolysis, although the reaction rates are lower than those of plastics, as shown in Table 4. Furthermore, the expected trends can be well modeled, whereas following order: PP > LDPE > HDPE for the reaction rates are expected from literature [22]. This is confirmed by the reactions rates for a given temperature producing "Light Liquids" via reactions path k2.

Table 4. Reaction rates at 460 °C of all determined reactions and the used plastic decomposition reactions "$k_{Lit.}$" from [15].

k@460 °C (1/s)	$k_{Lit.}$	k1	k2	k3	k4	k5	k6
Carrier fluid	-	2.38×10^{-4}	8.38×10^{-4}	2.97×10^{-5}	4.48×10^{-4}	8.42×10^{-5}	1.83×10^{-4}
PP	1.32×10^{-2}	3.99×10^{-4}	7.51×10^{-3}	8.28×10^{-4}	-	-	-
LDPE	6.76×10^{-3}	-	2.15×10^{-3}	-	-	-	-
HDPE	4.02×10^{-3}	-	1.34×10^{-3}	-	-	-	4.21×10^{-4}

The kinetic values for the plastic degradation reaction $k_{Lit.}$ (Table 3) are taken from [15]. Also, the kinetic values of the carrier fluid, obtained by the model calculation are in a similar order of magnitude, which is also confirmed by the activation energy of around 200 kJ/mol. The calculated activation energies of the reaction paths involved in the PP pyrolysis fit well to literature, the lowest activation energy is 80 kJ/mol for the reaction from "Wax" to "Spindle oil" [5,8,16,17]. However, the calculated activation energy of 700 kJ/mol for reaction path k2 for LDPE and HDPE exceed the range of literature, even to those in comparison to oils [24]. Activation energies of oil cracking are in a frame of 50 to 500 kJ/mol [23]. The reason for this deviation can be assigned to the comparatively low product yields of "Light Liquids", which can only be confirmed in test runs with higher temperatures. Further, test runs with increased temperatures and higher conversions need to be performed to receive more detailed data.

However, some differences can be seen between the kinetic data of PP and HDPE in Figures 9 and 11, respectively. In the case of PP, the "Light Liquids" lump is underpredicted, although the total plastic feed is depolymerized, and its resulting "Wax" is also converted with the other product lumps completely. This indicates that there is an interaction between the carrier fluid and the plastic pyrolysis, which is not yet considered in the model. Hence, PP increases the outcome of products of the co-pyrolysis.

Contrariwise, the kinetic data of HDPE underpredicts the "Residue" lump at low conversions. Again, the HDPE is fully decomposed, and the produced "Light Liquids" lump is not originating 2 "Residue" lump is underpredicted, what indicates an interaction of the co-pyrolysis, in which the degradation of the carrier fluid is inhibited, or the carrier fluid produces a high boiling (410+ °C) product instead of a "Light Liquids" lump together with HDPE.

Additionally, the lump "Spindle oil" remains on the same level in the model, resulting in a horizontal trend in Figures 8 and 9. "Spindle Oil" is also not produced from LDPE and HDPE, but is nevertheless overpredicted. Consequently, the reaction mechanism of producing "Spindle Oil" is not sufficiently considered in the model, but this has just a low effect on the variance.

5. Conclusions

The provided lumping approach reduces the complexity of the chemical reaction system significantly by reducing the interacting components, this drastically lowers the analytical and computational effort for the model calculation. In contrast, in order to the common practice operating mostly with TGA measurements, this model can investigate the decomposition processes in a tubular reactor by adding consecutive reactions. The reactor calculation can describe the effect of evaporation and the resulting flow regimes inside the reactor pipe. Also, the complexity of the reaction model is still low (just four lumps), the kinetic data for a satisfying description of the pyrolysis of PP, LDPE and HDPE are found. Simultaneously to the chemical reactions, the heat transfer is described analogously to flow boiling which results in reasonable temperature trends. Due to the temperatures, evaporation, condensation and chemical reactions the density of the reaction medium changes over the whole reactor length, but the model considers all these mechanisms, and thus, provides a reasonable residence time of reaction mixtures in the reactor. Furthermore, the often-used Arrhenius approach obviously describes the temperature dependency sufficiently.

Under the used mild cracking conditions, polyethylene has imperfect conversion towards the products lumps "Spindle oil", "Light Liquids" and "Gas". In particular, the "Spindle oil" lump seems to have different chemical mechanisms, interacting with the carrier fluid. Also PP shows interactions with the carrier fluid, in which the conversion to "Light Liquids" is enhanced.

To generate a more precise prediction, these interactions of the carrier fluid and the plastics should be introduced. Moreover, the formation of coke should be included, which could be inserted with a C/H balance to the system.

However, this lumped kinetic model achieve a sufficient description of the process and hence it is already used in a plant simulation to provide data for an upscale.

Author Contributions: Conceptualization, A.E.L. and T.S.; methodology, A.E.L.; software, A.E.L.; validation, A.E.L. and T.S.; formal analysis, A.E.L.; investigation, A.E.L.; resources, A.E.L. and W.H.; data curation, A.E.L. and T.S.; writing—original draft preparation, A.E.L.; writing—review and editing, A.E.L.; visualization, A.E.L. and T.S.; supervision, M.L. project administration, W.H. All authors have read and agreed to the published version of the manuscript.

Funding: This research received no external funding.

Data Availability Statement: Restrictions apply to the availability of these data. Data was obtained from OMV Downstream GmbH and are available from the authors with the permission of OMV Downstream GmbH.

Conflicts of Interest: The authors declare no conflict of interest.

References

1. Plasticseurope. Available online: https://www.plasticseurope.org/ (accessed on 20 October 2020).
2. European Commission. Eine Europäische Strategie für Kunststoffe in der Kreislaufwirtschaft. 2018. Available online: https://eur-lex.europa.eu/resource.html?uri=cellar:2df5d1d2-fac7-11e7-b8f5-01aa75ed71a1.0002.02/DOC_3&format=PDF (accessed on 8 August 2018).
3. Europäische Union. Richtlinie (EU) 2018/ des Europäischen Parlaments und des Rates vom 30 Mai 2018 zur Änderung der Richtlinie 2008/98/EG über Abfälle. 2018. Available online: https://eur-lex.europa.eu/legal-content/DE/TXT/?uri=CELEX:32018L0852 (accessed on 8 August 2018).
4. Quicker, P. Evaluation of Recent Developments Regarding Alternative Thermal Waste Treatment with a Focus on Depolymerisation Processes. 2019. Available online: https://www.vivis.de/wp-content/uploads/WM9/2019_WM_359-370_Quicker.pdf (accessed on 25 November 2020).
5. Aguado, J.; Serrano, D.P. *Feedstock Recycling of Plastic Wastes*; Royal Society of Chemistry: Cambridge, UK, 1999.
6. Al-Salem, S.M.; Antelava, A.; Constantinou, A.; Manos, G.; Dutta, A. A review on thermal and catalytic pyrolysis of plastic solid waste (PSW). *J. Environ. Manag.* **2017**, *197*, 177–198. [CrossRef] [PubMed]
7. Anuar Sharuddin, S.D.; Abnisa, F.; Wan Daud, W.M.; Aroua, M.K. A review on pyrolysis of plastic wastes. *Energy Convers. Manag.* **2016**, *115*, 308–326. [CrossRef]
8. Scheirs, J.; Kaminsky, W. (Eds.) *Feedstock recycling and Pyrolysis of Waste Plastics: Converting Waste Plastics into Diesel and Other Fuels*; John Wiley & Sons Ltd.: Chichester, UK, 2006.
9. Vogel, J.; Krüger, F.; Fabian, M. Chemisches Recycling. Hintergrund July 2020. Available online: https://www.umweltbundesamt.de/sites/default/files/medien/1410/publikationen/2020-07-17_hgp_chemisches-recycling_online.pdf (accessed on 24 August 2020).
10. Lederer, C.; Lehner, M. Development of an upscalable process for polyolefins and other post-consumer plastic fractions via solvent-based depolymerization. In Proceedings of the 7th International Symposium on Feedstock Recycling of Polymeric Materials (7th ISFR 2013), New Delhi, India, 23–26 October 2013.
11. Schubert, T.; Lechleitner, A.; Lehner, M.; Hofer, W. 4-Lump kinetic model of the co-pyrolysis of LDPE and a heavy petroleum fraction. *Fuel* **2020**, *262*, 116597. [CrossRef]
12. Brown, M.E. *Introduction to Thermal Analysis: Techniques and Applications*; Springer: Dordrecht, The Netherlands, 2001.
13. Kiang, J.K.Y.; Uden, P.C.; Chien, J.C.W. Polymer Reactions—Part VII: Thermal Pyrolysis of Polypropylene. *Polym. Degrad. Stab.* **1980**, *2*, 113–127. [CrossRef]
14. Bockhorn, H.; Hornung, A.; Hornung, U. Mechanisms and kinetics of thermal decomposition of plastics from isothermal and dynamic measurements. *J. Anal. Appl. Pyrolysis* **1999**, *50*, 77–101. [CrossRef]
15. Westerhout, R.W.; Waanders, J.; Kuipers, J.A.; Swaaij, W.P. Kinetics of the Low Temperature Pyrolysis of Polyethene, Polypropene, and Polystyrene, Modeling, Experimental Determination, and Comparison with Literature Models and Data. *Industiral Eng. Chem. Res.* **1997**, *36*, 1955–1964. [CrossRef]
16. Bockhorn, H.; Hornung, A.; Hornung, U.; Schawaller, D. Kinetic study on the thermal degradation of polypropylene and polyethylene. *J. Anal. Appl. Pyrolysis* **1999**, *48*, 93–109. [CrossRef]
17. Aboulkas, A.; El harfi, K.; El Bouadili, A. Thermal degradation behaviors of polyethylene and polypropylene. Part I: Pyrolysis kinetics and mechanisms. *Energy Convers. Manag.* **2010**, *51*, 1363–1369. [CrossRef]
18. Tsuchiya, Y.; Sumi, K. Thermal decomposition products of polypropylene. *J. Polym. Sci. Part. A-1* **1969**, *7*, 1599–1607. [CrossRef]
19. Buxbaum, L.H. The Degradation of Poly(ethylene terephthalate). *Angew. Chem. Int. Ed. Engl.* **1968**, *7*, 182–190. [CrossRef]
20. Schubert, T.; Lehner, M.; Hofer, W. Experimental and modeling approach of LDPE thermal cracking for feedstock recycling. In Proceedings of the 9th International Symposium on Feedstock Recycling of Polymeric Materials (9th ISFR 2017), Ostrava, Czech Republic, 10–13 July 2017.
21. Schubert, T.; Lehner, M.; Karner, T.; Hofer, W.; Lechleitner, A. Influence of reaction pressure on co-pyrolysis of LDPE and a heavy petroleum fraction. *Fuel Process. Technol.* **2019**, *193*, 204–211. [CrossRef]
22. Singh, J. (Ed.) *Feedstock Recycling and Pyrolysis of Waste Plastics: Converting Waste Plastics into Diesel and Other Fuels*; Wiley & Sons: Chichester, UK, 2006.
23. Singh, J.; Kumar, M.M.; Saxena, A.K.; Kumar, S. Reaction pathways and product yields in mild thermal cracking of vacuum residues: A multi-lump kinetic model. *Chem. Eng. J.* **2005**, *108*, 239–248. [CrossRef]
24. Del Bianco, A.; Panariti, N.; Anelli, M.; Beltrame, P.L.; Carniti, P. Thermal cracking of petroleum residues. *Fuel* **1993**, *72*, 75–80. [CrossRef]
25. Kuo, J.C.; Wei, J. Lumping Analysis in Monomolecular Reaction Systems. Analysis of Approximately Lumpable System. *Ind. Eng. Chem. Fund.* **1969**, *8*, 124–133. [CrossRef]
26. Wei, J.; Kuo, J.C. Lumping Analysis in Monomolecular Reaction Systems. Analysis of the Exactly Lumpable System. *Ind. Eng. Chem. Fund.* **1969**, *8*, 114–123. [CrossRef]
27. Schubert, T. *Experimental Investigation and Modeling of a Polyolefin Pyrolysis Process*; Montanuniversitaet: Leoben, Austria, 2018.
28. Kabelac, S. (Ed.) *VDI-Wärmeatlas: [Berechnungsunterlagen für Druckverlust, Wärme- und Stoffübergang]*; Springer: Berlin/Heidelberg, Germany, 2006.

processes

Article

Chemical Recycling of WEEE Plastics—Production of High Purity Monocyclic Aromatic Chemicals

Tobias Rieger [1], Jessen C. Oey [1], Volodymyr Palchyk [1], Alexander Hofmann [1], Matthias Franke [1,*] and Andreas Hornung [1,2]

1 Fraunhofer Institute for Environmental, Safety and Energy Technology UMSICHT, Institute Branch Sulzbach-Rosenberg, An der Maxhütte 1, 92237 Sulzbach-Rosenberg, Germany; tobias.rieger@umsicht.fraunhofer.de (T.R.); jessen.oey@umsicht.fraunhofer.de (J.C.O.); volodymyr.palchyk@umsicht.fraunhofer.de (V.P.); alexander.hofmann@umsicht.fraunhofer.de (A.H.); andreas.hornung@umsicht.fraunhofer.de (A.H.)

2 School of Chemical Engineering, University of Birmingham, Birmingham B15 2TT, UK

* Correspondence: matthias.franke@umsicht.fraunhofer.de; Tel.: +49-9661-8155-600

Abstract: More than 200 kg real waste electrical and electronic equipment (WEEE) shredder residues from a German dismantling plant were treated at 650 °C in a demonstration scale thermochemical conversion plant. The focus within this work was the generation, purification, and analysis of pyrolysis oil. Subsequent filtration and fractional distillation were combined to yield basic chemicals in high purity. By means of fractional distillation, pure monocyclic aromatic fractions containing benzene, toluene, ethylbenzene, and xylene (BTEX aromatics) as well as styrene and α-methyl styrene were isolated for chemical recycling. Mass balances were determined, and gas chromatography–mass spectrometry (GC-MS) as well as energy dispersive X-ray fluorescence (EDXRF) measurements provided data on the purity and halogen content of each fraction. This work shows that thermochemical conversion and the subsequent refining by fractional distillation is capable of recycling WEEE shredder residues, producing pure BTEX and other monocyclic aromatic fractions. A significant decrease of halogen content (up to 99%) was achieved with the applied methods.

Keywords: WEEE; chemical recycling; pyrolysis; recovery of aromatics; oil upgrading; dehalogenation

Citation: Rieger, T.; Oey, J.C.; Palchyk, V.; Hofmann, A.; Franke, M.; Hornung, A. Chemical Recycling of WEEE Plastics—Production of High Purity Monocyclic Aromatic Chemicals. *Processes* **2021**, *9*, 530. https://doi.org/10.3390/pr9030530

Academic Editor: Daniel Vollprecht

Received: 15 February 2021
Accepted: 9 March 2021
Published: 16 March 2021

1. Introduction

Waste electrical and electronic equipment (WEEE) represents a significant source of almost all precious and critical metals, but their recovery potential is far from being fully exploited as things stand today. At the end of state-of-the-art WEEE treatment processes, one or more output fractions are left behind, which are usually sent to landfills or to energetic utilization in waste incinerators ([1] pp. 131–133), [2,3]. With those fractions, remaining metals get lost, irretrievable for material recovery ([1] pp. 209–212), [4]. At the same time, WEEE and its output fractions contain high-quality plastics like HIPS, ABS, epoxy resins, PS, PE, PP, and PVC [5–7]. However, these plastics show high concentrations of flame retardants (FR) as TBBPA, DDO, HBCD, and DDE [3], resulting in bromine and chlorine concentrations of 0.6–4.0 wt.% [8].

As mechanical recycling recovers plastics in their given polymeric composition, state-of-the-art processes are not able to remove or to eject flame retardants from FR-containing WEEE plastics effectively. The recovery of unpolluted plastics by means of mechanical recycling is thus mostly limited to FR-free fractions ([1] pp. 209–212), [2,3,9]. Against this background, pyrolysis seems to offer a promising solution to complement established mechanical recycling processes, especially regarding highly chlorinated and brominated WEEE-plastics ([1] pp. 131–133), [3,5,8,10].

Due to the importance of effectively recycling and decontaminating WEEE plastics [1–3,9,11–13], numerous researchers have investigated chemical recycling of WEEE or of the flame retarded plastics present in such [14–24]. Most studies on novel recycling approaches regarding WEEE focus on the recovery of metals [4,25–27], whereas investigations on recycling of nonmetal fractions of real WEEE are limited [12].

Different chemical recycling technologies for WEEE plastics like thermal pyrolysis [4,21,22,28–32], co-pyrolysis [33–36], two step pyrolysis [33,37], catalytic pyrolysis [23,28,38–45], microwave-assisted pyrolysis [28,46,47], and combinations thereof [24,44,47,48] have been studied.

A very recent review by Charitopoulou et al. [12] on recycling of WEEE plastics comes to the conclusion that pyrolysis is considered a favorable technology to recycle FR-containing plastics from WEEE. The formation of PXDD/F is theoretically suppressed due to the absence of oxygen. However, the oxygen already present in the nonmetal fraction of WEEE in practice leads to PXDD/F [11,41,48].

Several pretreatment technologies were studied to remove BFR from WEEE plastics prior to pyrolysis [31,49–56]. Examples of common pretreatment technologies are solvent extraction with isopropanol, toluene, or methanol, as well as supercritical fluid technologies with acetone, methanol, or ethanol as media. Such techniques were found to have good dehalogenation effectiveness. However, their high operational cost and energy consumption limit industrial implementation [12].

Patent research revealed several patent publications in the field of pyrolysis reactors or processes to treat mixed plastic wastes [57–64] as well as WEEE in particular [65–69]. Most reactor and process concepts are designed to either decompose WEEE in order to prevent landfilling of large volumes of hazardous wastes, dehalogenation, and/or to produce valuable chemicals or fuels at temperatures between 200 and 800 °C. Patented reactor concepts include screw reactors [58,70], U-shaped reactors [68], and fluidized bed reactors [61]. Most patents address thermal or catalytic pyrolysis. However, methods including vacuum pyrolysis [67] or the addition of other material, e.g., heavy oil or hydrocracking steams to a pyrolysis process [60,63], were published. Furthermore, methods comprising multistep processes are present. Examples are upstream technologies prior to pyrolysis [63,65,66] or downstream purification technologies to enhance product quality [57,59,60,67]. Flame retarded polymers sum up to on average 30% of WEEE plastics [71,72]. A crucial point in the thermochemical processing of FR-containing plastics is, however, the copper catalyzed formation of hazardous polyhalogenated dibenzo-*p*-dioxins and furans (PXDD/F) as precursors for PXDD/F, namely chlorine and bromine, are present in WEEE [2,73]. In order to prevent or to minimize their formation during thermochemical treatment, the critical temperature for PXDD/F formation between 200 and 600 °C [74,75] should be skipped quickly by rapidly heating up the plastics to >450 °C [4].

Consequently, Fraunhofer UMSICHT developed an innovative thermochemical conversion process for treatment of composite materials, including WEEE [70]. This process is based on an innovative auger reactor equipped with a unique heat exchanger (Section 2.1). The so-called iCycle® process (intelligent Composite Recycling) enables the conversion of WEEE fractions at very constant and controlled process conditions (heating up and retention time of feedstock, stability of process temperature). The current contribution deals with the recovery of chemicals from the liquid oil fraction generated during thermochemical treatment of shredder residues from a state-of-the art WEEE dismantling process. By use of iCycle® and downstream processing of the oil, the objectives of the current contribution are as follows:

- to investigate suitability and potential of thermochemical conversion for the generation of intermediate products for chemical recycling;
- to isolate monocyclic aromatic fractions for application in the chemical industry and plastics synthesis by a combination of filtration and fractional distillation;
- to analyze the opportunities and limitations of the applied process combination for the removal of chlorine and bromine in order to provide virgin grade basic chemicals.

2. Materials and Methods

2.1. Thermochemical Conversion

The residual WEEE fraction consisted of IT-appliances (collection group 5) and was provided by a manufacturer as shredder residue with a maximum particle size of 20 mm. This shredder residue contained around 40% metals and inorganics and 60% organics. 231 kg of the feedstock was treated in the continuous thermochemical demonstration plant (load capacity ~70 kg/h) illustrated in Figure 1. Prior to treatment, the plant was flushed with nitrogen overnight and heated up to an operating temperature of 650 °C.

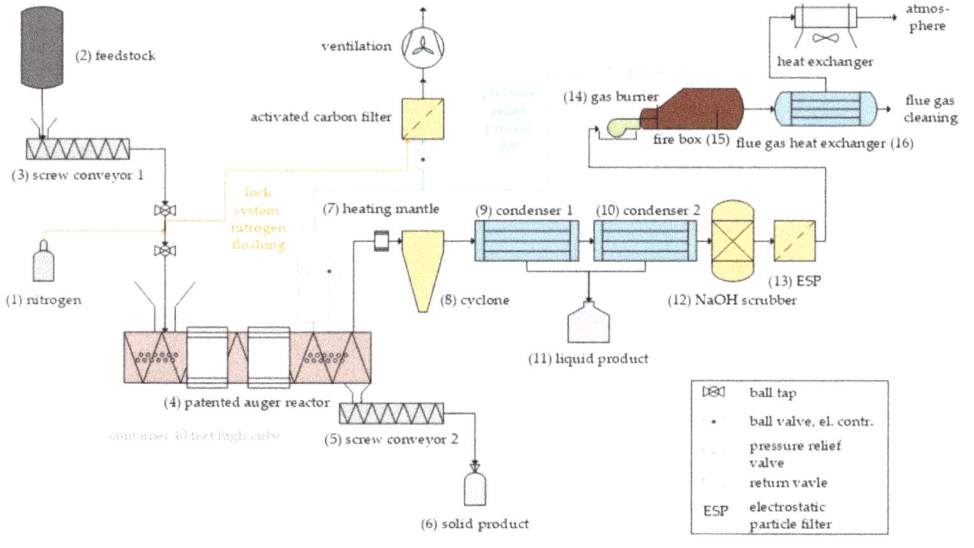

Figure 1. Process flow diagram of the iCycle® pilot plant (thermochemical conversion process).

As seen in Figure 1, throughout the treatment, the feedstock material is fed by a screw conveyor (3) batch-wise from a receiver tank (2) into a lock, where it is flushed with N_2 (1) to ensure that no ambient air enters the reactor. Thermochemical treatment and decomposition of the plastic fraction takes place in the reactor (4) at 650 °C. The reactor is a patented system with an innovative heat exchanger design. By use of an Archimedean screw, internal heating by externally preheated cycled spheres can be achieved. Along their way through the auger reactor, the spheres do not get in contact with the feedstock itself as they are moved forward in the inner section of the Archimedean screw. Thus, clogging of feedstock to the hot spheres is prevented. The system in combination with a relatively high temperature of 650 °C ensures a high heating rate of reaction media to avoid the formation of polyhalogenated dioxins and furans due to the presence of flame retardants [4,70]. In addition to the internal heat supply by cycled spheres, external heating is provided by heating sleeves covering the surface of the auger reactor.

Inside the reactor, the auger unit moves the solid material from the feeding point to the discharge point, where remaining solid matter drops out of the reactor and is transferred by a conveyer screw (5) into a collection reservoir (6). The heated auger reactor has a length of 6000 mm and a diameter of 470 mm. Gaseous and vaporous decomposition products (at 650 °C) leave the reactor through a cyclone (7) and subsequently the cooling and condensation train (9, 10), consisting of two tube bundle heat-exchangers, connection pipes, and a pump that transfers the condensate into a collection reservoir (11). All material found in the cooling and condensation train after the experiment is referred to as "condensate". The remaining gas is cleaned in a NaOH-scrubber unit (12) and an electrostatic particle filter (ESP) (13) and is subsequently burned on sight (14, 15) in accordance with German Federal Emission Control Act (BImSchG).

2.2. Pyrolysis Oil Pretreatment

Prior to fractional distillation, the pretreatment of condensate was performed to remove solids and aqueous phase present. Two pretreatment methods were conducted in this research. The first method was a vacuum filtration process using a filter paper and a Buchner funnel to separate solids from the crude condensate produced throughout the iCycle® process. Filtration was performed within two steps with two different pore sizes of filter paper, i.e., 40 μm and 2 μm. The condensate was then poured through the funnel into a borosilicate flask, and the solids that were larger than the pore size of the filter paper were separated. A vacuum pump (KNF, model: N810FT.18; 100 kPa) was used to initiate the oil suction and enhance the filtration process. The filter paper was frequently replaced, as soon as the filter paper clogs in order to prevent performance decrease. The mass of the total solids that remained in filter papers was measured and prepared for analysis.

Subsequently, a separation of the observable aqueous phase from the filtrate of the second filtration was conducted using a separation funnel. In this way, the lower phase (aqueous) was released by gravitational force via the stopcock (tap) at the bottom of the funnel. The weight of the separated phase was measured and the water phase prepared for analysis.

2.3. Fractional Distillation

For the fractional distillation of 5 kg of the pretreated condensate (oil), a batch distillation system (PILODIST-104) was operated. The plant's column features a stainless-steel wire mesh (20 theoretical stages), arranged with a head temperature sensor and a reflux divider. The column was insulated by a heating mantle, where the temperature was adjusted to maintain at 5 K below the column head temperature to provide an adiabatic condition during the distillation process. The main condenser with a subsequent distillate cooler at the top of the column ensures a sufficient condensation of ascending distillate vapors. The distillation flask at the bottom of the column, surrounded by an insulation mantle, has a volume of 20 L. The heating of the flask was controlled according to the temperature difference (ΔT) between the heating device and the oil temperature inside the heating flask. The reflux ratio was defined by adjusting the off-take time and the reflux time of the distillate fraction. A vacuum pump was connected to the top of the condenser and the fraction collector to maintain the desired pressure. The system was constantly flushed with nitrogen with a flow rate of 0.5 L/min in order to prevent oxidation reactions. A liquid nitrogen cold trap was used to protect the vacuum pump during the distillation with reduced pressure. The following parameters were operated in order to distill the oil generated:

- Operated pressure: 1 atm (RT–85 °C); 100 mbar (>85 °C)
- Temperature difference between heating device and oil in the heating flask: ΔT = 100 K
- Off-take time/reflux time/reflux ratio: 4 s/20 s/5

Based on the boiling points of benzene, toluene, ethylbenzene, and xylene (BTEX aromatics) as well as phenolic compounds present in the operated oil, the temperature intervals and operating pressures depicted in Table 1 were selected to divide the oil into distillate fractions comprising high purities of BTEX-aromatics and monocyclic aromatic substances as styrene, α-methyl styrene, phenol and cresols.

Table 1. Fractional distillation operating parameters.

Fraction	Pressure	Temperature Interval
[-]	[-]	[°C]
1	1 atm	RT–85
2	100 mbar	85–115 (AET [1])
3	100 mbar	115–140 (AET [1])
4	100 mbar	140–150 (AET [1])
5	100 mbar	150–190 (AET [1])
6	100 mbar	190–205 (AET [1])
7	100 mbar	205–225 (AET [1])

[1] Atmospheric equivalent temperature.

2.4. *Analysis of Pyrolysis Oil and Fraction Characterization*

2.4.1. Gas Chromatography–Mass Spectrometry (GC-MS) Analysis

The composition of oil was analyzed on a gas chromatograph coupled with a mass spectrometer Shimadzu GCMS-QP2020. The chromatograph was equipped with a 30 m nonpolar 0.25 mm inner diameter (i. d.), 0.25 µm film thickness DB-5ms, and 2.5 m middle polar 0.15 mm i. d., 0.15 µm film thickness VF-17ms column set from Agilent Technologies. Helium with 5.0 purity was used as carrier gas for all experiments. The injection volume was set to 1 µL. Dilution was conducted with 1 mg of sample in 1 mL of DCM. The measurements were performed at a constant linear velocity 40 cm/min of carrier gas. The temperature of the GC oven was programmed using the settings starting at 40 °C, 3 min hold to 320 °C, 3 min hold at 10 °C/min. The temperatures of the injector, the MS-interface, and MS were set to 250, 280 and 200 °C, respectively. The quadrupole MS detector was operated at scan speed of 5000 Hz using a mass range of 35–500 m/z. Solvent cut time was 3 min, with the MS start at 3.2 min. Total analysis time was 34 min. BTEX, styrene, α-methyl styrene, phenol, cresols and naphthalene were identified using standards solutions of pure chemicals. NIST-17 Mass Spectral Library was used for all other substance identification. Only the substances detected with the similarity index (SI) of more than 70 were identified. The proportion of each substance in the sample was given in percentage area. Only the proportions of the substances of particular interest in this research were shown in the graph, while other substances were not included and were referred to as "others".

2.4.2. Energy Dispersive X-ray Fluorescence (EDXRF) Analysis

A quantitative halogen analysis of all the samples was made with an energy dispersive X-ray fluorescence spectrometer (EDXRF) from Shimadzu (EDX 720). The concentrations for chlorine and bromine in oils were calculated based on the calibration curve and the Cl/Br peak intensities measured by EDXRF. Standard solutions were made with 1,3,5-trichlorobenzene, 98% from Alfa Aesar™ for chlorine and with 1,3,5-tribromobenzene, and 98% from Alfa Aesar™ for bromine. Toluene at 99.8% from Merck was used as a solvent. Each sample was measured three times in order to get statistical data.

2.4.3. Water Content Analysis

Water content in pyrolysis oil and distillate fractions was measured using a Karl-Fischer volumetric titration method. HYDRANAL™—Composite 5 from Honeywell Fluka™ was used as one component reagent titrant and methanol ACS Reagent, while ≥99.8% from Honeywell Riedel-de Haën™ was used as a titration medium. All measurement were performed on Metrohm 915 KF Ti-Touch and leaned on DIN 51777. Same as in case of EDXRF, measurements were performed three times.

3. Results

Real WEEE shredder residues were thermally converted by means of pyrolysis at 650 °C with a residence time of 30 min. Solid, liquid, and gaseous products were yielded from this process; these products will be termed solid residue, condensate, and gas hereinafter, respectively. The obtained condensate underwent two pretreatment steps, namely solids filtration and aqueous phase separation. The obtained product, termed "oil", has thereafter been divided into eight fractions (including residue) by a fractional batch distillation process. The aim of the processing of the WEEE shredder residues was to produce highly enriched monocyclic aromatic mixtures with reduced content of solids, aqueous phase, and halogen concentrations to make them valuable for industrial reuse. The results of the individual process steps are presented in the following section.

3.1. Thermochemical Conversion

The weights of the treated feedstock material, the collected solid product, and the condensate were determined at 231, 74, and 67 kg. The weight of the gaseous products was determined by a difference at 89 kg. Hence, the products yields were approximately 32, 29, and 39 wt.% for the solid product, the condensate, and the gaseous product, respectively. The results are illustrated in Table 2.

Table 2. Mass balance of the thermochemical conversion process.

Material	Mass	Mass Fraction
[-]	[kg]	[wt.%]
Initial	231	100
Solid (solid residue)	74	32
Liquid (condensate)	67	29
Gas	90	39

The GC-MS analysis of the condensate showed a composition of 6.26 area% benzene, 22.05 area% toluene, 8.39 area% ethylbenzene, 0.73 area% xylenes, 26.63 area% styrene, 6.94 area% phenol, 3.90 area% α-methyl styrene, and 1.91 area% cresols. It follows that the condensate consisted of 72.16 area% of monocyclic aromatic substances to be recovered potentially. The composition is also illustrated in Figure 2 in Section 3.3.

3.2. Pyrolysis Oil Pretreatment

The condensate produced by thermochemical conversion underwent pretreatment processes in order to remove the solids and aqueous phase contained. A filtration process using a Buchner funnel with a filter paper with a pore size of 40 μm was first carried out, followed by a second filtration using a filter paper with a pore size of 2 μm. Subsequently, the observable aqueous phase in the condensate was removed using a separation funnel. Hereinafter, the filtrate <2 μm without aqueous phase will be referred to as oil.

The amount and the mass fraction of the condensate, the filtered condensate, the solids and aqueous phase removed are summarized in Table 3. By means of pretreatment, 0.3846 kg solids (>40 μm), 0.6125 kg solids (2–40 μm), and 0.4739 kg aqueous phase were separated from the condensate, which correspond to mass fractions of 2.59 wt.%, 4.12 wt.%, and 3.19 wt.% of the initial crude condensate, respectively.

After the conducted pretreatment steps, the oil contained chlorine, bromine, and water concentrations of 4833 ± 785 ppm, 2251 ± 135 ppm, and 8267 ± 340 ppm, respectively. Halogen and water concentrations are depicted in Figure 3.

Table 3. Mass balance of condensate pretreatment.

Material	Mass	Mass Fraction
[-]	[kg]	[wt.%]
Crude condensate	14.8477	100.00
Filtrate (<40 μm)	13.5219	91.07
Solids (>40 μm)	0.3846	2.59
Loss (filtration 1)	0.9412	6.34
Filtrate (oil) (<2 μm)	11.7514	79.15
Solids (2–40 μm)	0.6125	4.12
Loss (filtration 2)	1.1580	7.80
Aqueous phase	0.4739	3.19

3.3. Fractional Distillation

The pretreated condensate (referred as oil) was successfully divided into its fractions by means of distillation. The initial boiling point of the oil was 67 °C. The highest column head temperature reached was AET 225 °C at 100 mbar. All substances (and mixtures) with boiling points above 225 °C were considered as residue and were not prepared for analysis. Four pure monocyclic aromatic fractions (referred to as main products) including BTEX aromatics and styrene with a total mass fraction of 46.68 wt.% of the initial oil feed were obtained from distillation. Additionally, three fractions enriched in valuable monocyclic aromatic substances as styrene, phenol, α-methyl styrene, cresols, and polycyclic naphthalene with a total mass fraction of 20.89 wt.% were yielded. Along with these, 27.66 wt.% residue and 2.04 wt.% cold trap fractions were produced. Two cold trap fractions were collected during distillation. A mass balance of the distillation process is presented in Table 4, implying a mass loss of 2.73 wt.%, presumably due to mass hold-up in the column.

Table 4. Distillate fractions with temperature intervals and mass composition.

Fraction	Temperature Interval	Mass	Mass Fraction
[-]	[°C]	[kg]	[wt.%]
Initial	-	5.3830	100
1	RT–85	0.4800	8.92
2	85–115	0.7690	14.29
3	115–140	0.3205	5.95
4	140–150	0.9430	17.52
5	150–190	0.5835	10.84
6	190–205	0.3360	6.24
7	205–225	0.2050	3.81
Residue	>225	1.4890	27.66
Cold trap 1	-	0.0540	1.00
Cold trap 2	-	0.0561	1.04
Loss	-	0.1469	2.73

All fractions yielded by means of distillation are depicted in Figure 4. It can be observed that fractions 1 to 4 were colorless and transparent, indicating high chemical stability. Fractions 5 and 6 showed a light brown color, which is typical for oxidized phenolic compounds contained in the fractions. Fraction 7 and the distillation residue comprise dark brown to black opaque appearance.

Fraction 1 (RT–85 °C), with a mass fraction of 8.92 wt.%, solely consisted of benzene and toluene with a proportion of 88.00 area% benzene and 12.00 area% toluene. The chlorine, bromine, and water concentrations were determined to be 221 ± 25 ppm, 20 ± 1 ppm, and 8833 ± 170 ppm, respectively.

Fraction 2 (85–115 °C) contained mostly toluene with a proportion of 96.23 area% as well as benzene (3.20 area%), 113 ± 33 ppm chlorine, 3 ± 1 ppm bromine, and 1333 ± 125 ppm water. The mass fraction represented 14.29 wt.% of the initial mass.

Fraction 3 (115–140 °C) with a mass fraction of 5.95 wt.% contained a mixture of toluene, ethylbenzene, xylenes, and styrene. The GC-MS result shows that ethylbenzene had the highest proportion with 43.15 area%, followed by 26.83 area% toluene, 25.41 area% styrene, and 4.15 area% xylenes. The fraction had a chlorine concentration of 836 ± 93 ppm, a bromine concentration of 6 ± 1 ppm, and a water concentration of 400 ± 216 ppm.

Fraction 4 (140–150 °C) with a mass fraction of 17.52 wt.% mostly consisted of styrene 79.98 area%. The GC/MS analysis also determined 16.83 area% ethylbenzene, 2.26 area% xylenes, and a trace amount of toluene and 1-methylethyl-benzene. Moreover, this fraction contained 150 ± 23 ppm chlorine, 280 ± 3 ppm bromine, and 333 ± 170 ppm water.

Fraction 5 (150–190 °C) consisted of styrene, phenol, α-methyl styrene, and naphthalene, where phenol 35.13 area% is the most present, followed by 23.52 area% styrene, and 10.48 area% α-methyl styrene. It was noted that this fraction also contained 12.74 area% indene (SI 96%). The bromine concentration of fraction 5 was the highest compared to the other fractions (1100 ± 58 ppm), along with 303 ± 27 ppm chlorine and 1233 ± 125 ppm water. The mass fraction of fraction 5 was 10.84 wt.%

Fraction 6 (190–205 °C), with a mass fraction of 6.24 wt.%, consisted of 37.07 area% phenol, 27.16 area% naphthalene, 19.06 area% benzonitrile (SI 98%), 13.98 area% cresols, trace amounts of aniline, 3-phenyl-2-propenal, and derivatives of benzene, indene, and benzofuran. This fraction exhibited chlorine, bromine, and water concentrations of 99 ± 13 ppm, 249 ± 1 ppm, and 833 ± 170 ppm, respectively.

Fraction 7 (205–225 °C) represented 3.81 wt.% of the initial mass, predominantly containing naphthalene with a proportion of 40.45 area%, followed by 31.81 area% cresols, 6.87 area% 1-methyl-naphthalene (SI 97%), 4.52 area% 2-methyl-benzonitrile (SI 96%), 3.56 area% benzonitrile (SI 98%), 1.24 area% phenol, as well as several derivatives of benzonitrile, naphthalene, benzofuran and traces of other compounds. High chlorine (1914 ± 517 ppm) and bromine (405 ± 10 ppm) concentration compared to lower boiling point fractions were determined. This fraction's water concentration exhibited 900 ± 141 ppm.

In the distillation residue (boiling point >225 °C), derivatives of naphthalene, benzene, fluorene, anthracene, and pyrene were identified, containing high concentrations of chlorine (1538 ± 36 ppm) and bromine (4316 ± 159 ppm).

The results from GC-MS, EDXRF, and water concentration analysis are summarized and presented in Figures 2 and 3.

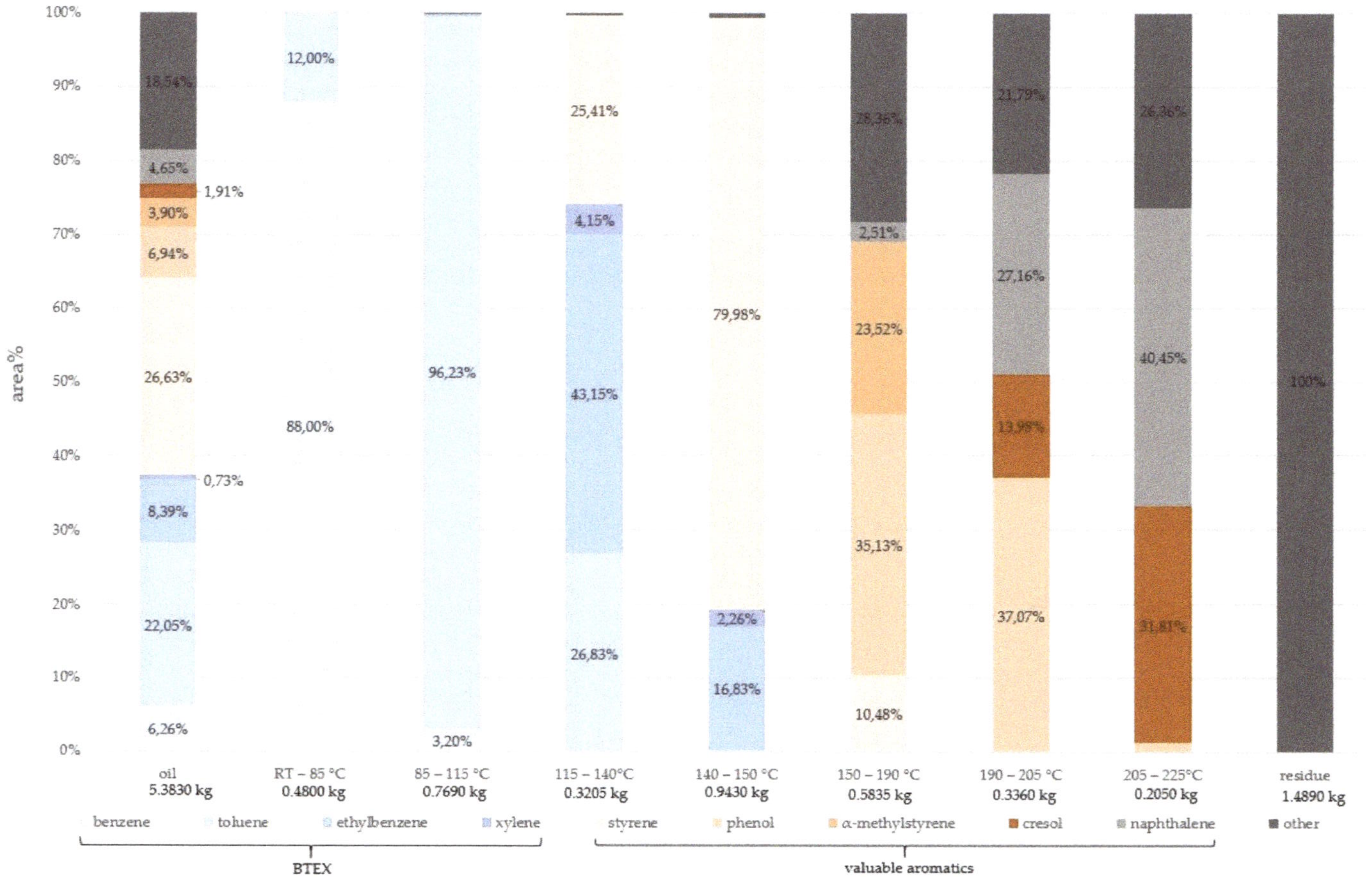

Figure 2. Composition of oil and distillate fractions.

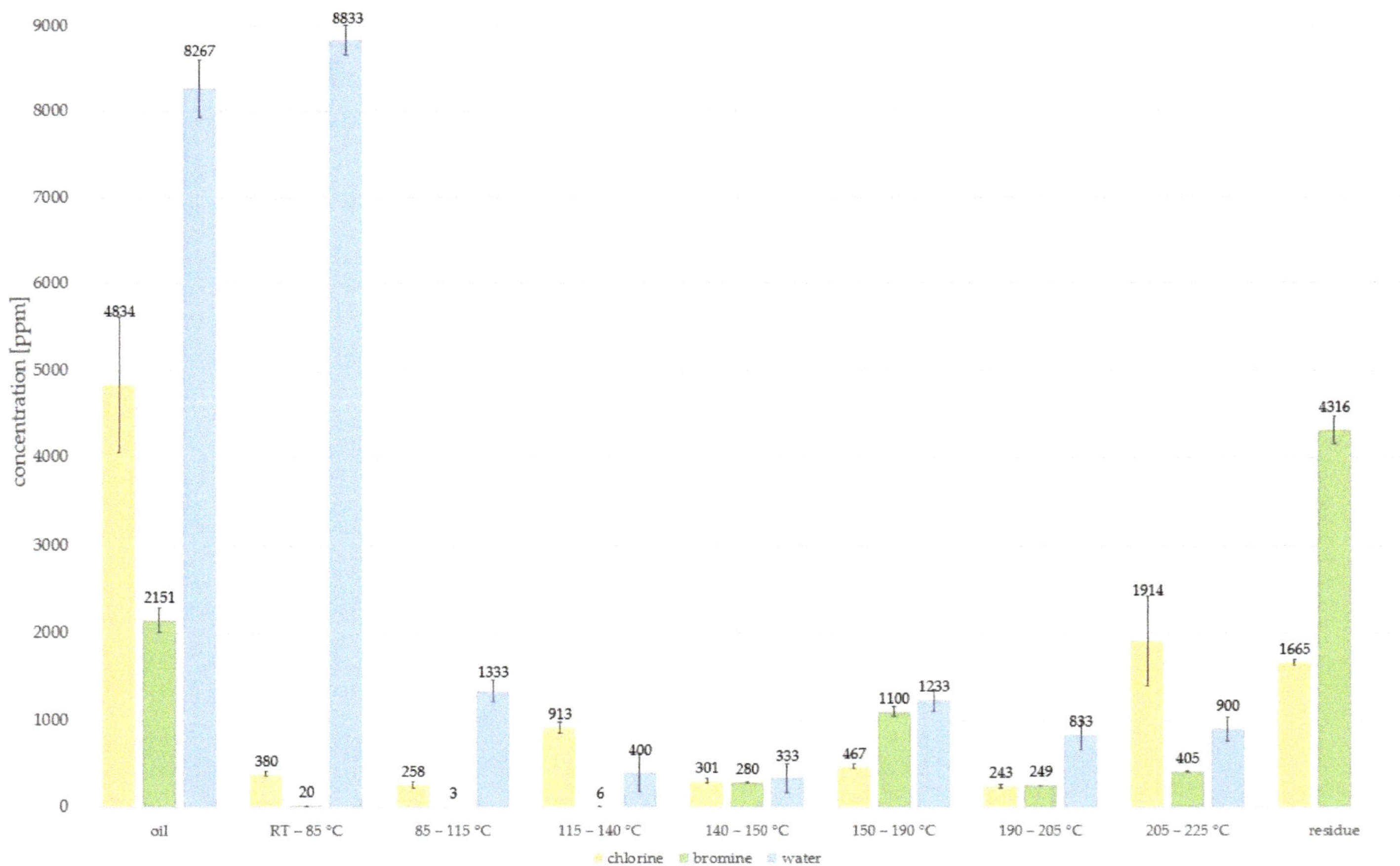

Figure 3. Energy dispersive X-ray fluorescence (EDXRF) results and water concentration analysis of oil and distillate fractions.

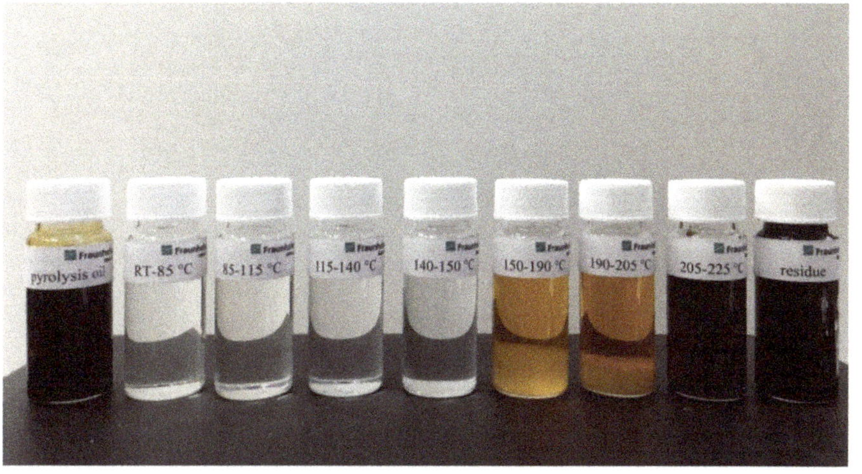

Figure 4. Fractions obtained by fractional distillation.

4. Discussion

The pyrolysis of WEEE or flame-retarded polymers contained in WEEE with and without catalyst has been conducted in a number of recent works [2,6,9,10,15,20–23,30,32,76,77]. Furthermore, mixed WEEE residues can vary widely in terms of composition, strongly depending on their origin [1] (pp. 209–212) [2,3,7,9].

On average, the thermal conversion of WEEE at different temperatures yields 36 wt.% condensate, 39 wt.% gas, and 25 wt.% solid residue [6]. Recent works yielded 71–91 wt.% condensate, 3–21 wt.% solid residue and 2–8 wt.% gas from thermal pyrolysis of real mixed WEEE at 600 °C in a laboratory scale [76,78]. The iCycle® process applied for this work yielded 29 wt.% condensate, 32 wt.% solid residue, and 39 wt.% gas on a demonstration scale plant. The converted WEEE contained around 40% metals and inorganics, which explains the comparatively high amount of solids. Due to the operating temperature of 650 °C, the production of 39 wt.% gas is in an expectable range. However, the distribution of the products is, aside from operational parameters and scale, also highly dependent on the composition of the input material. Similar results (40 wt.% condensate, 30 wt.% solid residue, 12.5 wt.% tar, and 13.5 wt.% gas) were found by Vasile et al. [77] throughout the pyrolysis of mixed WEEE in a temperature range of 430–470 °C in a smaller demonstration scale plant. The produced oil (condensate) comprised a comparable content of 62.75 vol% aromatics as BTEX, styrene, and phenol derivatives. The condensate produced throughout the present work was also very rich in monocyclic aromatic compounds (including BTEX aromatics, styrene, phenol, and cresols), representing 76.81 area% in the pretreated pyrolysis oil. Condensates by Santella et al. [76] contained less than 20 wt.% aromatic compounds and mostly styrene. The pyrolysis oil produced from mixed WEEE by Hall et al. [78] consisted mainly of phenol, isopropyl phenol (30.2 wt.%), and styrene (5.9 wt.%), indicating major differences in the composition of the investigated feedstock. However, Hall et al. [78] found a similar chemical composition of the liquid pyrolysis product from waste refrigerators; bromine and chlorine contents of the condensate were 0.3 % and 0.1 %, respectively.

The current investigation showed that the pretreatment of the condensate in terms of filtration and phase separation is capable of removing roughly 10 wt.% of substances as solids and water that are undesired for further upgrading of the oil, as they are likely to disturb the refinement processes to produce other materials, e.g., bulk chemicals or fuels. Hence, the pretreatment technologies are crucial in terms of condensate processing, and they need to be conducted before further upgrading.

The chlorine and bromine concentrations of the investigated pyrolysis oil amount to roughly 5000 ppm chlorine and 2000 ppm bromine. Other related works found less than 600 ppm chlorine and less than 900 ppm bromine in the pyrolysis oil from mixed WEEE [78]. This confirms the enormous differences in WEEE composition [7,72] and manifests the need of the dehalogenation of pyrolysis oil in order to upgrade them to fuels or chemicals.

By means of fractional distillation, pure BTEX fractions and concentrated monocyclic aromatic fractions were produced:

- Benzene fraction (88 area% benzene, 12 area% toluene)
- Toluene fraction (3 area% benzene, 96 area% toluene)
- BTEX/styrene fraction (27 area% toluene, 43 area% ethylbenzene, 4 area% xylenes, 25 area% styrene)
- Styrene fraction (17 area% ethylbenzene, 2 area % xylenes, 80 area% styrene)
- Phenol fraction (35 area% phenol, not considered as main product)

The results evidenced that the pyrolysis of WEEE shredder residues and subsequent distillation of the condensate might be applied to produce high purity chemical fractions which can be used as a feedstock in the chemical industry and in polymer synthesis to produce virgin grade chemical products and polymers and also to substitute crude oil consumption.

The chlorine and bromine concentrations were significantly reduced by up to 99% in the distillate fractions. The EDXRF analysis showed that bromine accumulates in the distillation residue where chlorine concentration is decreased in all fractions, including the residue. The decrease in chlorine concentration in all fractions promote the assumption that chlorine (in the form of HCl) was dissolved in the oil and released throughout distillation [77].

In the first three fractions (RT–140 °C), bromine concentrations do not exceed 20 ppm, which represents a reduction of 99% in relation to the crude condensate and can already be sufficient debromination for the industrial application of the produced chemicals. Fractions 4–7 comprise bromine concentrations roughly between 200 ppm and 400 ppm, representing a significant reduction from the initial oil (2151 ppm).

Fractions 1, 2, 4, and 6 chlorine concentrations were reduced to roughly 250–470 ppm from 4843 ppm in the initial oil. Fraction 3 and 5 still comprise 913 ppm and 1914 ppm chlorine, respectively. Further dehalogenation is needed in order to reuse such fractions for industrial applications [77].

Beside the chemicals, the solid fraction is also suitable for recycling. It can be supplied to copper smelters, where metals are recovered. The gas can be used for energy recovery after treatment as described in Figure 1. The water phase, however, needs to be treated as hazardous waste as it contains metal compounds, organic, and halogen residues.

Several studies in the field of chemical recycling of WEEE have investigated the production of fuels, which is less valuable as a product compared to high-purity monocyclic aromatic fractions. Furthermore, the chemical recycling to valuable basic chemicals and monomers has a positive environmental impact in terms of climate change and fossil resource depletion compared to the productions of fuels. Thus, the conversion of WEEE to valuable chemicals has great potential to close the loop of the nonmetal fraction of WEEE as it will be mainly kept in the material cycle where the production of fuels entails the release of the nonmetal fraction (especially in the form of carbon dioxide) to the atmosphere [79,80].

5. Conclusions

A significant amount of more than 200 kg real WEEE shredder residues from a German recycling company were treated in a demonstration scale pyrolysis plant and the subsequent fraction was distillated, in order to produce basic chemicals with high purity. Such treatment of liquid pyrolysis products from WEEE shredder residues to produce pure chemical fractions have not been reported yet. Thus, a new approach for recovering valuable chemicals like aromatics from pyrolysis oil was investigated and was successfully proven in a demonstration scale plant.

From the conducted experiments, it can be concluded that:

- Pyrolysis is a promising technology for production of an intermediate oil for chemical upgrading.
- Pretreatment such as filtration and phase separation is capable of removing solids and water, which are undesired for the further upgrading of the oil.
- It was proven that a combination of pyrolysis and subsequent fractional distillation is a suitable method for the isolation of high purity BTEX fractions and concentrated monocyclic aromatic fractions.

Current results revealed that a combination of filtration followed by fractional distillation is suitable for the reduction of halogen content. The halogen content was reduced up to more than 99% in the obtained fractions. Therefore, the recovered aromatics have great potential to be used as feedstock in chemical industry.

Estimations made by Fraunhofer UMSICHT suggest that the pyrolysis of WEEE for the recovery of metals is already economically viable in a scale >250 kg/h. The recovery of aromatics from WEEE nonmetal fraction instead of energetic utilization, however, enables significant improvement of both economic and ecologic aspects of WEEE recycling.

Further research works need to be conducted on higher halogen reduction and are part of our current projects. The development and scale-up of the presented technologies are the next steps to implement the recovery of aromatic compounds from different plastic waste streams in industry.

Author Contributions: Conceptualization, A.H. (Alexander Hofmann); methodology, T.R. and V.P.; software, T.R. and V.P.; formal analysis, V.P.; data curation, J.C.O. and T.R.; writing—original draft preparation, T.R.; writing—review and editing, A.H. (Alexander Hofmann), M.F. and V.P.; visualization, T.R.; supervision, A.H. (Andreas Hornung), A.H. (Alexander Hofmann), V.P. and T.R.; project administration, A.H. (Alexander Hofmann), V.P. All authors have read and agreed to the published version of the manuscript.

Funding: This research received no external funding.

Institutional Review Board Statement: Not applicable.

Informed Consent Statement: Not applicable.

Acknowledgments: The authors are grateful to Jan Grunwald and Martin Nieberl for technical support and discussion.

Conflicts of Interest: The authors declare no conflict of interest.

References

1. Chagnes, A. *WEEE Recycling: Research, Development, and Policies*; Elsevier Science: Amsterdam, The Netherlands, 2017; ISBN 978-0-12-803363-0.
2. Wang, R.; Xu, Z. Recycling of non-metallic fractions from waste electrical and electronic equipment (WEEE): A review. *Waste Manag.* **2014**, *34*, 1455–1469. [CrossRef]
3. Yang, X.; Sun, L.; Xiang, J.; Hu, S.; Su, S. Pyrolysis and dehalogenation of plastics from waste electrical and electronic equipment (WEEE): A review. *Waste Manag.* **2013**, *33*, 462–473. [CrossRef]
4. Hense, P.; Reh, K.; Franke, M.; Aigner, J.; Hornung, A.; Contin, A. Pyrolysis of waste electrical and electronic equipment (WEEE) for recovering meatals and energy: Previous achievements and current approaches. *Environ. Eng. Manag. J. (EEMJ)* **2015**, *14*, 1637–1647. [CrossRef]
5. Bhaskar, T.; Kaneko, J.; Muto, A.; Sakata, Y.; Jakab, E.; Matsui, T.; Uddin, A. Pyrolysis studies of PP/PE/PS/PVC/HIPS-Br plastics mixed with PET and dehalogenation (Br, Cl) of the liquid products. *J. Anal. Appl. Pyrolysis* **2004**, *72*, 27–33. [CrossRef]
6. Alston, S.M.; Clark, A.D.; Arnold, J.C.; Stein, B.K. Environmental Impact of Pyrolysis of Mixed WEEE Plastics Part 1: Experimental Pyrolysis Data. *Environ. Sci. Technol.* **2011**, *45*, 9380–9385. [CrossRef]
7. Martinho, G.; Pires, A.; Saraiva, L.; Ribeiro, R. Composition of plastics from waste electrical and electronic equipment (WEEE) by direct sampling. *Waste Manag.* **2012**, *32*, 1213–1217. [CrossRef]
8. Beccagutti, B.; Cafiero, L.; Pietrantonio, M.; Pucciarmati, S.; Tuffi, R.; Ciprioti, S.V. Characterization of Some Real Mixed Plastics from WEEE: A Focus on Chlorine and Bromine Determination by Different Analytical Methods. *Sustainability* **2016**, *8*, 1107. [CrossRef]
9. Buekens, A.; Yang, J. Recycling of WEEE plastics: A review. *J. Mater. Cycles Waste Manag.* **2014**, *16*, 415–434. [CrossRef]

10. Blazsó, M.; Czégény, Z.; Csoma, C. Pyrolysis and debromination of flame retarded polymers of electronic scrap studied by analytical pyrolysis. *J. Anal. Appl. Pyrolysis* **2002**, *64*, 249–261. [CrossRef]
11. Nnorom, I.C.; Osibanjo, O. Sound management of brominated flame retarded (BFR) plastics from electronic wastes: State of the art and options in Nigeria. *Resour. Conserv. Recycl.* **2008**, *52*, 1362–1372. [CrossRef]
12. Charitopoulou, M.A.; Kalogiannis, K.G.; Lappas, A.A.; Achilias, D.S. Novel trends in the thermo-chemical recycling of plastics from WEEE containing brominated flame retardants. *Environ. Sci. Pollut. Res.* **2020**, 1–24. [CrossRef] [PubMed]
13. Hornung, A.; Balabanovich, A.; Donner, S.; Seifert, H. Detoxification of brominated pyrolysis oils. *J. Anal. Appl. Pyrolysis* **2003**, *70*, 723–733. [CrossRef]
14. Kowalska, E.; Radomska, J.; Konarski, P.; Diduszko, R.; Oszczudłowski, J.; Opalińska, T.; Więch, M.; Duszyc, Z. Thermogravimetric investigation ofwastes from electrical and electronic equipment (WEEE). *J. Therm. Anal. Calorim.* **2006**, *86*, 137–140. [CrossRef]
15. De Marco, I.; Caballero, B.; Chomón, M.; Laresgoiti, M.; Torres, A.; Fernández, G.; Arnaiz, S. Pyrolysis of electrical and electronic wastes. *J. Anal. Appl. Pyrolysis* **2008**, *82*, 179–183. [CrossRef]
16. Moltó, J.; Font, R.; Gálvez, A.; Conesa, J. Pyrolysis and combustion of electronic wastes. *J. Anal. Appl. Pyrolysis* **2009**, *84*, 68–78. [CrossRef]
17. Acomb, J.C.; Nahil, M.A.; Williams, P.T. Thermal processing of plastics from waste electrical and electronic equipment for hydrogen production. *J. Anal. Appl. Pyrolysis* **2013**, *103*, 320–327. [CrossRef]
18. Caballero, B.; De Marco, I.; Adrados, A.; López-Urionabarrenechea, A.; Solar, J.; Gastelu, N. Possibilities and limits of pyrolysis for recycling plastic rich waste streams rejected from phones recycling plants. *Waste Manag.* **2016**, *57*, 226–234. [CrossRef] [PubMed]
19. Dias, P.; Javimczik, S.; Benevit, M.; Veit, H. Recycling WEEE: Polymer characterization and pyrolysis study for waste of crystalline silicon photovoltaic modules. *Waste Manag.* **2017**, *60*, 716–722. [CrossRef]
20. Esposito, L.; Cafiero, L.; De Angelis, D.; Tuffi, R.; Ciprioti, S.V. Valorization of the plastic residue from a WEEE treatment plant by pyrolysis. *Waste Manag.* **2020**, *112*, 1–10. [CrossRef]
21. Evangelopoulos, P.; Kantarelis, E.; Yang, W. Investigation of the thermal decomposition of printed circuit boards (PCBs) via thermogravimetric analysis (TGA) and analytical pyrolysis (Py–GC/MS). *J. Anal. Appl. Pyrolysis* **2015**, *115*, 337–343. [CrossRef]
22. Hall, W.J.; Williams, P.T. Pyrolysis of brominated feedstock plastic in a fluidised bed reactor. *J. Anal. Appl. Pyrolysis* **2006**, *77*, 75–82. [CrossRef]
23. Hall, W.J.; Williams, P.T. Removal of organobromine compounds from the pyrolysis oils of flame retarded plastics using zeolite catalysts. *J. Anal. Appl. Pyrolysis* **2008**, *81*, 139–147. [CrossRef]
24. Ma, C.; Yu, J.; Wang, B.; Song, Z.; Xiang, J.; Hu, S.; Su, S.; Sun, L. Chemical recycling of brominated flame retarded plastics from e-waste for clean fuels production: A review. *Renew. Sustain. Energy Rev.* **2016**, *61*, 433–450. [CrossRef]
25. Hao, J.; Wang, Y.; Wu, Y.; Guo, F. Metal recovery from waste printed circuit boards: A review for current status and perspectives. *Resour. Conserv. Recycl.* **2020**, *157*, 104787. [CrossRef]
26. Zhang, K.; Wu, Y.; Wang, W.; Li, B.; Zhang, Y.; Zuo, T. Recycling indium from waste LCDs: A review. *Resour. Conserv. Recycl.* **2015**, *104*, 276–290. [CrossRef]
27. Zhang, L.; Xu, Z. A review of current progress of recycling technologies for metals from waste electrical and electronic equipment. *J. Clean. Prod.* **2016**, *127*, 19–36. [CrossRef]
28. Vinu, R.; Ojha, D.; Nair, V. Polymer Pyrolysis for Resource Recovery. In *Reference Module in Chemistry, Molecular Sciences and Chemical Engineering*; Elsevier BV: Amsterdam, The Netherlands, 2016.
29. Evangelopoulos, P.; Arato, S.; Persson, H.; Kantarelis, E.; Yang, W. Reduction of brominated flame retardants (BFRs) in plastics from waste electrical and electronic equipment (WEEE) by solvent extraction and the influence on their thermal decomposition. *Waste Manag.* **2019**, *94*, 165–171. [CrossRef] [PubMed]
30. Evangelopoulos, P.; Persson, H.; Kantarelis, E.; Yang, W. Performance analysis and fate of bromine in a single screw reactor for pyrolysis of waste electrical and electronic equipment (WEEE). *Process. Saf. Environ. Prot.* **2020**, *143*, 313–321. [CrossRef]
31. Shen, Y.; Zhao, R.; Wang, J.; Chen, X.; Ge, X.; Chen, M. Waste-to-energy: Dehalogenation of plastic-containing wastes. *Waste Manag.* **2016**, *49*, 287–303. [CrossRef] [PubMed]
32. Hall, W.J.; Williams, P.T. Fast Pyrolysis of Halogenated Plastics Recovered from Waste Computers. *Energy Fuels* **2006**, *20*, 1536–1549. [CrossRef]
33. Ma, C.; Sun, L.; Jin, L.; Zhou, C.; Xiang, J.; Hu, S.; Su, S. Effect of polypropylene on the pyrolysis of flame retarded high impact polystyrene. *Fuel Process. Technol.* **2015**, *135*, 150–156. [CrossRef]
34. Terakado, O.; Ohhashi, R.; Hirasawa, M. Bromine fixation by metal oxide in pyrolysis of printed circuit board containing brominated flame retardant. *J. Anal. Appl. Pyrolysis* **2013**, *103*, 216–221. [CrossRef]
35. Hlaing, Z.; Wajima, T.; Uchiyama, S.; Nakagome, H. Reduction of Bromine Compounds in the Pyrolysis Oil of Computer Casing Plastics Using Shell, $Ca(OH)^2$ and NaOH. *APCBEE Procedia* **2014**, *10*, 193–197. [CrossRef]
36. Jung, S.-H.; Kim, S.-J.; Kim, J.-S. Fast pyrolysis of a waste fraction of high impact polystyrene (HIPS) containing brominated flame retardants in a fluidized bed reactor: The effects of various Ca-based additives (CaO, $Ca(OH)^2$ and oyster shells) on the removal of bromine. *Fuel* **2012**, *95*, 514–520. [CrossRef]
37. Bhaskar, T.; Hall, W.J.; Mitan, N.M.M.; Muto, A.; Williams, P.T.; Sakata, Y. Controlled pyrolysis of polyethylene/polypropylene/polystyrene mixed plastics with high impact polystyrene containing flame retardant: Effect of decabromo diphenylethane (DDE). *Polym. Degrad. Stab.* **2007**, *92*, 211–221. [CrossRef]

38. Miandad, R.; Barakat, M.; Aburiazaiza, A.S.; Rehan, M.; Nizami, A. Catalytic pyrolysis of plastic waste: A review. *Process. Saf. Environ. Prot.* **2016**, *102*, 822–838. [CrossRef]
39. Panda, A.K.; Singh, R.; Mishra, D. Thermolysis of waste plastics to liquid fuelA suitable method for plastic waste management and manufacture of value added products—A world prospective. *Renew. Sustain. Energy Rev.* **2010**, *14*, 233–248. [CrossRef]
40. Wang, Y.; Sun, S.; Yang, F.; Li, S.; Wu, J.; Liu, J.; Zhong, S.; Zeng, J. The effects of activated Al_2O_3 on the recycling of light oil from the catalytic pyrolysis of waste printed circuit boards. *Process. Saf. Environ. Prot.* **2015**, *98*, 276–284. [CrossRef]
41. Brebu, M.; Bhaskar, T.; Murai, K.; Muto, A.; Sakata, Y.; Uddin, A. Thermal degradation of PE and PS mixed with ABS-Br and debromination of pyrolysis oil by Fe- and Ca-based catalysts. *Polym. Degrad. Stab.* **2004**, *84*, 459–467. [CrossRef]
42. Bhaskar, T.; Matsui, T.; Uddin, A.; Kaneko, J.; Muto, A.; Sakata, Y. Effect of Sb_2O_3 in brominated heating impact polystyrene (HIPS-Br) on thermal degradation and debromination by iron oxide carbon composite catalyst (Fe-C). *Appl. Catal. B Environ.* **2003**, *43*, 229–241. [CrossRef]
43. Wu, H.; Shen, Y.; Harada, N.; An, Q.; Yoshikawa, K. Production of Pyrolysis Oil with Low Bromine and Antimony Contents from Plastic Material Containing Brominated Flame Retardants and Antimony Trioxide. *Energy Environ. Res.* **2014**, *4*, 105. [CrossRef]
44. Ma, C.; Kamo, T. Enhanced debromination by Fe particles during the catalytic pyrolysis of non-metallic fractions of printed circuit boards over ZSM-5 and Ni/SiO_2-Al_2O_3 catalyst. *J. Anal. Appl. Pyrolysis* **2019**, *138*, 170–177. [CrossRef]
45. Park, Y.-K.; Han, T.U.; Jeong, J.; Kim, Y.-M. Debrominated high quality oil production by the two-step catalytic pyrolysis of phenolic printed circuit boards (PPCB) using natural clays and HY. *J. Hazard. Mater.* **2019**, *367*, 50–58. [CrossRef]
46. Lam, S.S.; Chase, H.A. A Review on Waste to Energy Processes Using Microwave Pyrolysis. *Energies* **2012**, *5*, 4209–4232. [CrossRef]
47. Sharuddin, S.D.A.; Abnisa, F.; Daud, W.M.A.W.; Aroua, M.K. A review on pyrolysis of plastic wastes. *Energy Convers. Manag.* **2016**, *115*, 308–326. [CrossRef]
48. Liu, W.-J.; Tian, K.; Jiang, H.; Yu, H.-Q. Lab-scale thermal analysis of electronic waste plastics. *J. Hazard. Mater.* **2016**, *310*, 217–225. [CrossRef]
49. Vilaplana, F.; Ribes-Greus, A.; Karlsson, S. Microwave-assisted extraction for qualitative and quantitative determination of brominated flame retardants in styrenic plastic fractions from waste electrical and electronic equipment (WEEE). *Talanta* **2009**, *78*, 33–39. [CrossRef]
50. Zhang, C.-C.; Zhang, F.-S. Removal of brominated flame retardant from electrical and electronic waste plastic by solvothermal technique. *J. Hazard. Mater.* **2012**, *221–222*, 193–198. [CrossRef] [PubMed]
51. Zhong, Y.; Peng, P.; Yu, Z.; Deng, H. Effects of metals on the transformation of hexabromocyclododecane (HBCD) in solvents: Implications for solvent-based recycling of brominated flame retardants. *Chemosphere* **2010**, *81*, 72–78. [CrossRef] [PubMed]
52. Wang, Y.; Zhang, F.-S. Degradation of brominated flame retardant in computer housing plastic by supercritical fluids. *J. Hazard. Mater.* **2012**, *205–206*, 156–163. [CrossRef] [PubMed]
53. Onwudili, J.A.; Williams, P.T. Degradation of brominated flame-retarded plastics (Br-ABS and Br-HIPS) in supercritical water. *J. Supercrit. Fluids* **2009**, *49*, 356–368. [CrossRef]
54. Xiu, F.-R.; Qi, Y.; Wang, S.; Nie, W.; Weng, H.; Chen, M. Application of critical water-alcohol composite medium to treat waste printed circuit boards: Oil phase products characteristic and debromination. *J. Hazard. Mater.* **2018**, *344*, 333–342. [CrossRef] [PubMed]
55. Xiu, F.-R.; Li, Y.; Qi, Y.; Yu, X.; He, J.; Lu, Y.; Gao, X.; Deng, Y.; Song, Z. A novel treatment of waste printed circuit boards by low-temperature near-critical aqueous ammonia: Debromination and preparation of nitrogen-containing fine chemicals. *Waste Manag.* **2019**, *84*, 355–363. [CrossRef]
56. Grause, G.; Fonseca, J.D.; Tanaka, H.; Bhaskar, T.; Kameda, T.; Yoshioka, T. A novel process for the removal of bromine from styrene polymers containing brominated flame retardant. *Polym. Degrad. Stab.* **2015**, *112*, 86–93. [CrossRef]
57. Mcnamara, D. Conversion of Waste Plastics Material to Fuel. International Patent 2011,077,419, 30 June 2011.
58. Yamaguchi, K. Pyrolysis Apparatus for Waste Plastic. Japanese Patent 2004,300,186, 28 October 2004.
59. Peltekis, K.; Kumble, B.; Astill, C. Plant and Process for Pyrolysis of Mixed Plastic Waste. International Patent 2018,000,050, 4 January 2018.
60. Ward, A.M.; Oprins, A.J.M.; Narayanasway, R. Process for Converting Mixed Waste Plastic (mwp) into Valuable Petrochemicals. European Patent 3,110,912, 4 January 2017.
61. Stankevitch, V. Process for the Conversion of Waste Plastics to Produce Hydrocarbon Oils. U.S. Patent 6,534,689, 13 March 2003.
62. Pajala, T. Pyrolysis Apparatus. International Patent 2017,198,896, 23 November 2017.
63. Csokai, V.; Szinay, Z.; Boday, A. Process for Termical Degradation of pvc and Other Wastes Containing Halogen-Containing Polymer Waste. International Patent 2012,025,771, 1 March 2012.
64. Riedewald, F. Process and System for Whole Tyres and Plastic Composites Pyrolysis to Fuel Conversion and Compound Recovery. U.S. Patent 2015,0,184,079, 2 July 2015.
65. Thomas, C.; Menuet, J.; Vanhelle, G. Method for Recovering Metals from Electronic Waste Containing Plastics Materials. U.S. Patent 8,800,775, 15 September 2014.
66. Menad, N.-E.; Guignot, S.; Göklap, I.; Bostyn, S.; Graz, Y.; Poirier, J. Method for Recycling Waste Electrical and Electronic Equipment. U.S. Patent 15,539,570, 21 December 2017.
67. Sun, S.; Long, L.; Zhong, S. The Separation of Each Component Material and Recovery Method in a Kind of Waste and Old Printed Circuit Board. Chinese Patent 1,01,612,628, 30 December 2009.

68. Riedewald, F. Process for the Recycling of Waste Batteries and Waste Printed Circuit Boards in Molten Salts or Molten Metals. International Patent 2014,167,139, 16 October 2015.
69. Brandhorst, H.W.; Engel, U.H., Jr.; Ludwig, C.T.; Zavoral, E.J. Multistage Thermolysis Method for Safe and Efficient Conversion of E-Waste Materials. U.S. Patent 9,850,433, 6 July 2017.
70. Hornung, A.; Hense, P.; Aigner, J.; Reh, K.; Franke, M. Rohrofen und Verfahren zur Chemischen Umsetzung International Patent 2016,189,138, 1 December 2017.
71. Bientinesi, M.; Petarca, L. Comparative environmental analysis of waste brominated plastic thermal treatments. *Waste Manag.* **2009**, *29*, 1095–1102. [CrossRef] [PubMed]
72. Alaee, M. An overview of commercially used brominated flame retardants, their applications, their use patterns in different countries/regions and possible modes of release. *Environ. Int.* **2003**, *29*, 683–689. [CrossRef]
73. Birnbaum, L.S. Health effects of polybrominated dibenzo-p-dioxins (PBDDs) and dibenzofurans (PBDFs). *Environ. Int.* **2003**, *29*, 855–860. [CrossRef]
74. Tan, P.; Neuschütz, D. Study on polychlorinated dibenzo-p-dioxin/furan formation in iron ore sintering process. *Metallurg. Mater. Transact. B.* **2004**, *35*, 983–991. [CrossRef]
75. Mei, J.; Wang, X.; Xiao, X.; Cai, Y.; Tang, Y.; Chen, P. Characterization and inventory of PBDD/F emissions from deca-BDE, polyethylene (PE) and metal blends during the pyrolysis process. *Waste Manag.* **2017**, *62*, 84–90. [CrossRef]
76. Santella, C.; Cafiero, L.; De Angelis, D.; La Marca, F.; Tuffi, R.; Cipriotti, S.V. Thermal and catalytic pyrolysis of a mixture of plastics from small waste electrical and electronic equipment (WEEE). *Waste Manag.* **2016**, *54*, 143–152. [CrossRef]
77. Vasile, C.; Brebu, M.A.; Karayildirim, T.; Yanik, J.; Darie, H. Feedstock recycling from plastic and thermoset fractions of used computers (I): Pyrolysis. *J. Mater. Cycles Waste Manag.* **2006**, *8*, 99–108. [CrossRef]
78. Hall, W.J.; Williams, P.T. Analysis of products from the pyrolysis of plastics recovered from the commercial scale recycling of waste electrical and electronic equipment. *J. Anal. Appl. Pyrolysis* **2007**, *79*, 375–386. [CrossRef]
79. Alston, S.M.; Arnold, J.C. Environmental Impact of Pyrolysis of Mixed WEEE Plastics Part 2: Life Cycle Assessment. *Environ. Sci. Technol.* **2011**, *45*, 9386–9392. [CrossRef]
80. Meys, R.; Frick, F.; Westhues, S.; Sternberg, A.; Klankermayer, J.; Bardow, A. Towards a circular economy for plastic packaging wastes–the environmental potential of chemical recycling. *Resour. Conserv. Recycl.* **2020**, *162*, 105010. [CrossRef]

processes

MDPI

Article

Comparative Analysis of the Behaviour of Marine Litter in Thermochemical Waste Treatment Processes

Johann Hee [1,*], Kai Schlögel [1], Simone Lechthaler [2,3], Jacqueline Plaster [4], Kristina Bitter [5], Lars Mathias Blank [5] and Peter Quicker [1]

1 Unit of Technology of Fuels, RWTH Aachen University, 52056 Aachen, Germany; schloegel@teer.rwth-aachen.de (K.S.); quicker@teer.rwth-aachen.de (P.Q.)
2 Institute of Hydraulic Engineering and Water Resources Management, RWTH Aachen University, 52056 Aachen, Germany; lechthaler@iww.rwth-aachen.de
3 Department of Geography, Chair of Physical Geography and Geoecology, RWTH Aachen University, 52056 Aachen, Germany
4 PGS Pacific Garbage Screening GmbH (Future Everwave), 52070 Aachen, Germany; plaster@everwave.de
5 Institute of Applied Microbiology, Aachen Biology and Biotechnology, RWTH Aachen University, 52056 Aachen, Germany; kristina.bitter@rwth-aachen.de (K.B.); lars.blank@rwth-aachen.de (L.M.B.)
* Correspondence: hee@teer.rwth-aachen.de

Abstract: Plastic in the ocean, especially plastic on the ocean surface is not only researched intensively but also photos and reports rise awareness of the challenge in the general public. While research is concerned with the fate of marine litter in the environment, recycling of these materials after collection is rarely addressed, mainly because there is neither considerable data on composition nor a suggested process to do so. This study is the first to analyse and evaluate chemical recycling (pyrolysis, gasification) and energy recovery (incineration) of marine litter. Two heterogenous marine litter samples from Sylt and Norderney, North Sea, Germany, were analysed, consisting of six different material groups. Agricultural mulch foil was used as reference material. The thermochemical treatment processes were reproduced by thermogravimetric analysis. Furthermore, pyrolysis trials on a semi-technical scale were conducted and the residues were analysed by proximate, ultimate and X-ray fluorescence analysis. The results indicate that heterogeneous and weathered material mixtures can be treated by thermochemical processes. Finally, the pyrolysis condensates are discussed as substrate for biotechnological upcycling. In summary, we present a comprehensive approach from the material characterisation of marine litter to the analysis of three different thermochemical treatment processes and the possibility to use the generated pyrolysis condensate for subsequent upcycling. The data collected form the basis for the evaluation and application of possible treatment options for the collected marine litter.

Keywords: marine litter; waste treatment; plastic waste; pyrolysis; gasification; incineration; thermogravimetric analysis; biotechnological upcycling; plastics recycling

Citation: Hee, J.; Schlögel, K.; Lechthaler, S.; Plaster, J.; Bitter, K.; Blank, L.M.; Quicker, P. Comparative Analysis of the Behaviour of Marine Litter in Thermochemical Waste Treatment Processes. *Processes* **2021**, *9*, 13. https://dx.doi.org/10.3390/pr9010013

Received: 30 November 2020
Accepted: 18 December 2020
Published: 23 December 2020

Publisher's Note: MDPI stays neutral with regard to jurisdictional claims in published maps and institutional affiliations.

1. Introduction

Plastics are a versatilely and controversially discussed topic. Due to the many advantages of the various materials concerning durability, low-costs in production, light weight and the possibility to fulfil numerous requirements (formability, heat resistance, insulation, etc.) it is ubiquitous [1]. Historic production is estimated at 8300 million metric tons (Mt) of virgin plastics with a waste generated of 6300 Mt [2].

While the materials themselves have a high added value, the problem of their material characteristics arises when getting into the environment. Here, there are different ways of entry which can be divided into land-based including river transport and ocean-based including oil and gas platforms [3]. Causes are littering, industries, natural storm events, wear and tear and defective waste management. The plastic leakage of land-based, mismanaged waste which is entering the oceans from coastlines is calculated to an annual amount of 4.8 to 12.7 Mt taking into account solid waste, economic status, population density and coastal access [4]. After entering the environment, plastics are transported mainly in riverine systems, temporarily accumulate on sediments, in soils and on water surfaces and eventually deposit on the sea floor as a final sink [3].

Based on an oceanographic model which was additionally calibrated with data from numerous expeditions, Eriksen et al. [5] estimated the total count of plastic pieces and its weight floating in the oceans worldwide (Table 1). Thereby, the results showed a total count of 525.0 × 10^{10} plastic particles with a total weight of 2689.4 × 10^2 tons. The results do not consider the plastics that sank to the ground.

Table 1. Estimated count of plastic pieces (all size classes) and its weight floating in the oceans worldwide: North Pacific (NP), North Atlantic (NA), South Pacific (SP), South Atlantic (SA), Indian Ocean (IO), Mediterranean Sea (MED) (data from Eriksen et al. [5]).

	NP	NA	SP	SA	IO	MED	Total
Count [$n \times 10^{10}$]	199.0	93.0	49.1	29.7	130.0	24.7	525.0
Weight [$t \times 10^2$]	964.0	564.7	210.2	127.8	591.3	231.5	2689.4

Research on plastic in the environment started in 1972 [6]. Since then, many terms and definitions have been used describing this contamination focusing different aspects, such as size (nano-, micro-, meso- and macroplastic, macro litter), shape (plastic debris) or origin (anthropogenic litter, marine litter, marine plastic, plastic litter) [3]. These different terms make it particularly difficult to compare results of already published studies and thus to derive further prevention and waste treatment strategies.

A frequently used term also applied on a political level (United Nations, European Commission, Joint Group of Experts on the Scientific Aspects of Marine Environmental Protection) is 'marine litter' [7–9]. According to the United Nations Environment Programme marine litter is defined as 'any persistent, manufactured or processed solid material discarded, disposed of or abandoned in the marine and coastal environment' and thus includes not only plastics but also other possible waste fractions and emphasises the heterogeneity of the material [7].

The ocean surface can be seen as temporary sink within the life cycle of marine litter, as there are already different technical solutions to collect these waste items: extraction by vacuum, skimmer, air barrier, sand filters, drones and robots, waterway litter traps, detection aids, boats and wheels, river booms, large-scale booms and miscellaneous capture [10]. After the collection of marine litter, a further application in terms of recycling is required. In general, there are different treatment possibilities for waste plastics (Figure 1).

Although these methods are proven as treatment processes for plastics, especially the recycling routes are mainly restricted to pre-processed, sorted and separated polymer fractions. Considering marine litter in particular, additional points of conflict occur since the material is not only very heterogenous but also strongly weathered by various environmental influences. These external influences include mechanical degradation based on wave movement and sandy shores as well as degradation due to UV radiation, oxidation and the general process of biofouling [11]. Leaching of chemical additives (e.g., plasticisers) within the plastic fraction due to external influences additionally changes the material properties [12].

There are no standardised treatment processes for marine litter to date caused by missing recovery and recycling strategies. Furthermore, research on the applicability of treatment and recycling processes for marine litter is still lacking. This study is the first to analyse thermochemical waste treatment processes in application for marine litter comparing the methods of pyrolysis, gasification and incineration on a laboratory scale (Figure 1).

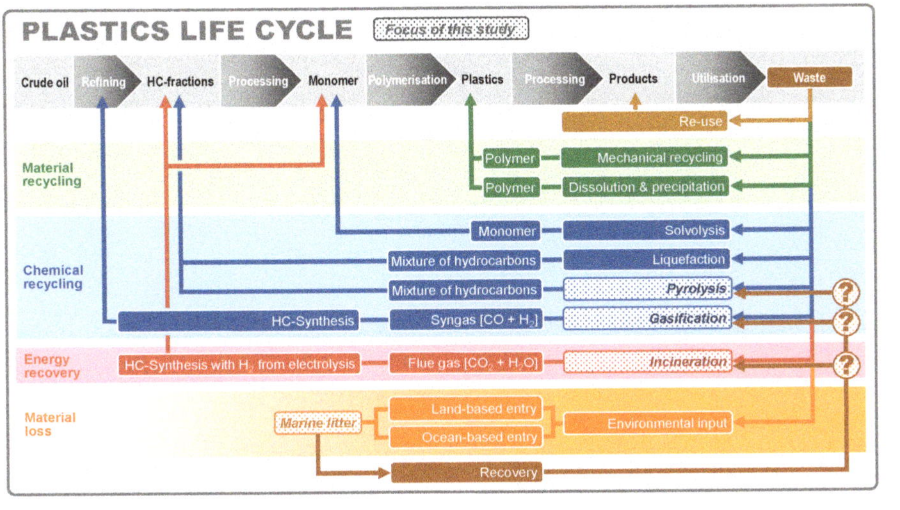

Figure 1. Life cycle of plastics with focus on the investigations of this study. (Figure: Peter Quicker) In terms of environmental input, the figure does not represent the entire life cycle of plastics, which is shown in Reference [3].

2. Materials and Methods

The material used in this study consists of three batches. The batches one and two were procured by Pacific Garbage Screening and originate from the North Sea where it was collected on Sylt and Norderney (Table 2). The sampling sites are shown in Figure 2. The materials were washed up on the beach and due to their location, origin and appearance the materials meet categorisation criteria for marine litter (ML). Manual collection was carried out randomly. Therefore, the material cannot be considered representative for the total waste load of the respective beach. The third batch used as reference material was commercial mulch foil for agricultural usage purchased from a hardware store.

Table 2. Material characteristics of the two batches of marine litter and mulch foil.

	ML Sylt	ML Norderney	Mulch Foil
Characteristics	Marine litter washed up on the beach and mostly collected after high tide	Marine litter washed up on the beach.	Virgin grade LD-PE mulch foil (black)
Localization	North Sea Germany, Sylt, Wadden Sea National Park Sylt, West Beach and East Beach	North Sea Germany, Norderney, Wadden Sea National Park Norderney, Norderney East	-
Coordinates	55.014167° N 8.370444° E	53.715902° N 7.283360° E	-
Weight	5478 g	4522 g	975 g
Volume	0.08 m^3	0.162 m^3	-
Date of collection	22 April–15 May 2020	17 April 2020	-

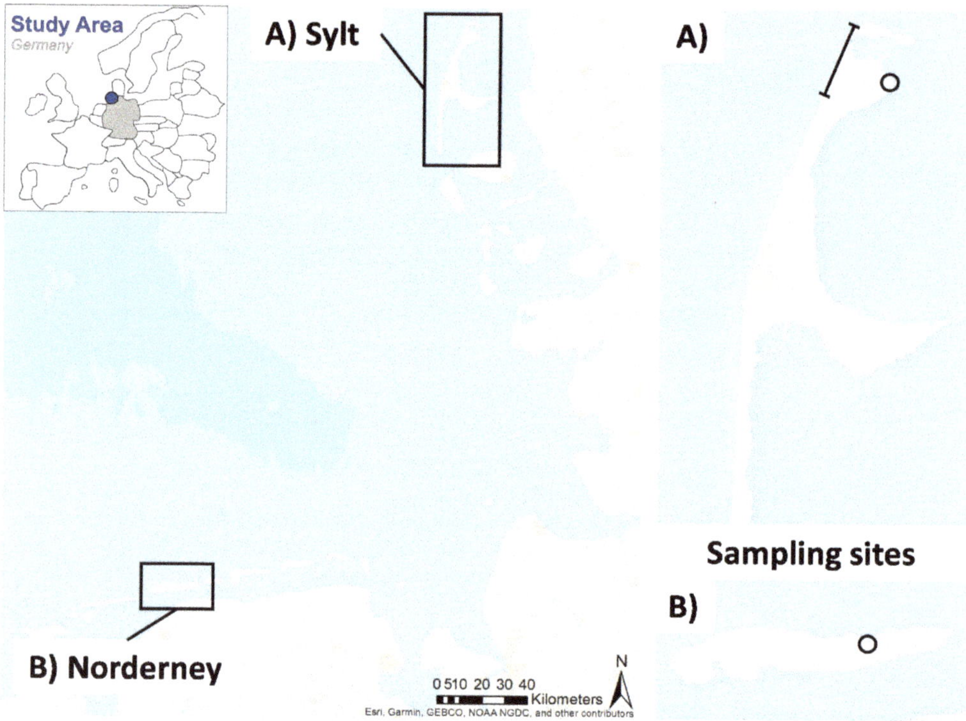

Figure 2. Overview map of the sampling sites (marked black). (**A**) Norderney: the material was samples within a radius of approx. 250 m in a chain of dunes south of the "Schlopp" (garbage hotspot), (**B**) Sylt: the sampled section reached from „List Weststrand Strandsauna" to "List Westfeuer am Ellenbogen" and additional material was sampled at "List Oststrand Wattenmeer." (Figure: Simone Lechthaler, Johann Hee).

2.1. Material Composition According to Material Group Catalogue

The sorting of the samples revealed a mixture of various types of material. They were sorted by hand according to a material group catalogue consisting of the following: 3D plastics, films, metals, nets, rubber, foamed plastics and others (e.g., glass and unidentifiable items). After sorting, the fractions were weighed to determine gravimetric composition (Table 3). Subsequently, the type of plastic of the individual materials was determined using a laboratory scale near-infrared spectroscopy (NIR) analyser.

Table 3. Composition of the marine litter (ML) samples used in this study [wt%].

Material Group	ML Norderney	ML Sylt
3D plastics	88 wt%	15 wt%
Films	10 wt%	6 wt%
Metals	1 wt%	2 wt%
Nets	-	55 wt%
Rubber & Elastomers	-	8 wt%
Foamed plastics	<<1 wt%	5 wt%
Others	1 wt%	9 wt%

2.2. ML Sylt

In the ML Sylt sample, the nets fraction made up the highest proportion with 55 wt% (Table 3). As shown in Figure 3D, the nets consisted of different types of plastic such as polypropylene (PP) and high-density polyethylene (HDPE). 3D plastics made up 15 wt% of the total weight. In contrast to ML Norderney, there were no bottles within the sampled material. This fraction mostly comprised fragments of boxes, caps and canisters made of polystyrene (PS), PP, polyvinyl chloride (PVC), HDPE and acrylonitrile butadiene styrene (ABS) (Figure 3A). 8 wt% were rubber and elastomers, like the glove shown in Figure 3E and 6 wt% films consisting of PP, PVC, PA, HDPE and PS. Foamed plastics made up 5 wt% of the total weight including polyethylene terephthalate (PET), PS, HDPE and polyurethane (PUR) foams. Non-polymer materials of the sample made up 9 wt% and consisted of glass, textiles, bones or feathers. Metals, like the spray can shown in Figure 3C, represented 2 wt% of the sample.

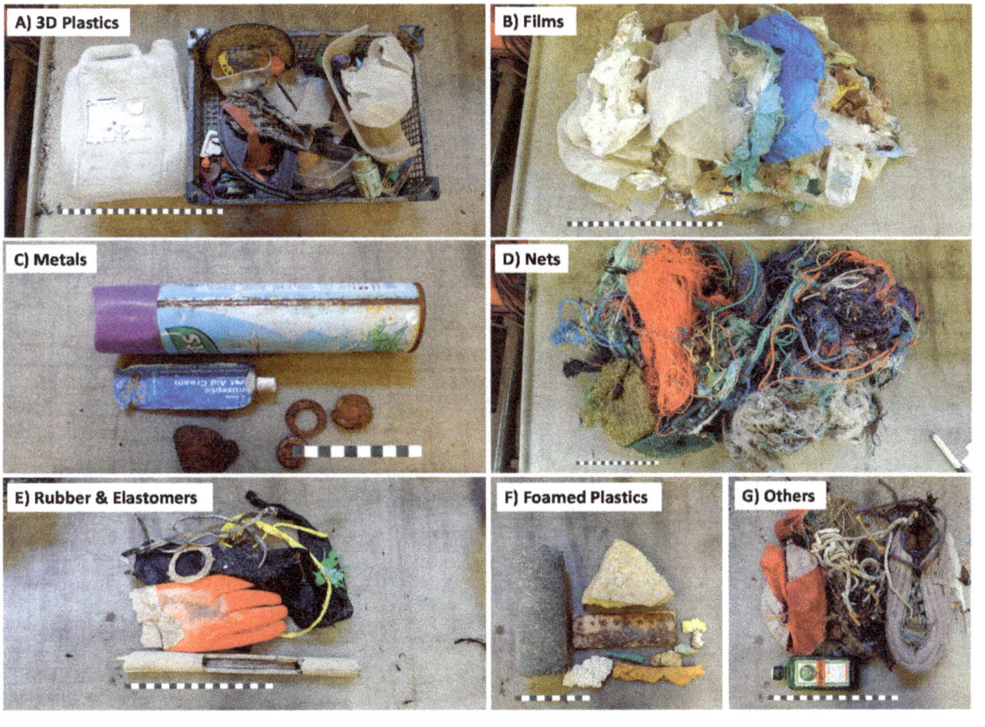

Figure 3. ML fractions Sylt: (**A**) 3D Plastics; (**B**) Films; (**C**) Metals; (**D**) Nets; (**E**) Rubbers & Elastomers; (**F**) Foamed Plastics and (**G**) Others. The scale amounts to 1 cm per section (Picture: Jacqueline Plaster, Johann Hee).

2.3. ML Norderney

For the ML Norderney sample, the 3D fraction made up the highest proportion with 88 wt%, comprised of plastic bottles (Table 3 and Figure 4A). The materials from this fraction consist mainly of PET, PP and HPDE but also PS and PUR. Films made up 10 wt% of the total quantity including PVC, HDPE, PS, PP, ABS (Figure 4). The remaining portion (1 wt%) consisted of metals, others (1 wt%) and foamed plastics with less than 1 wt% in relation to the total weight. In contrast to the material from Sylt, no nets were found.

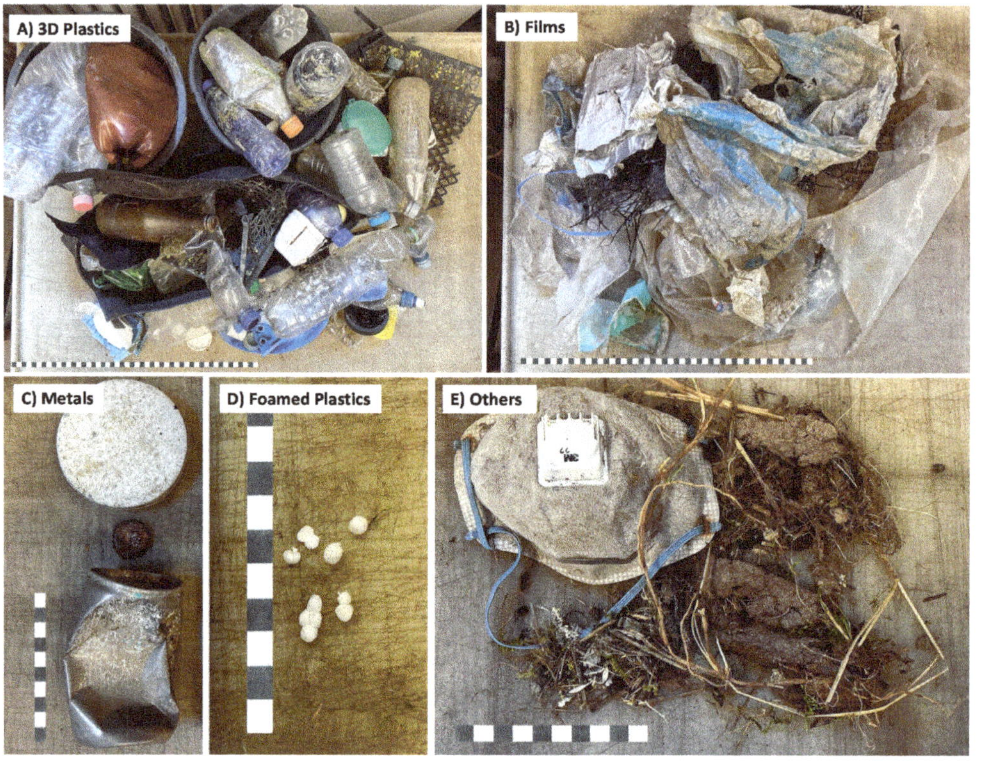

Figure 4. ML fractions Norderney: (**A**) 3D Plastics; (**B**) Films; (**C**) Metals; (**D**) Foamed Plastics and (**E**) Others. The scale amounts to 1 cm per section (Picture: Jacqueline Plaster, Johann Hee).

2.4. Selection of A Comparative Material

For the thermochemical conversion experiments investigating the behaviour of ML in thermal waste treatment processes, mulch foil was chosen as a reference material for the two ML fractions. Mulch foil is often used in agriculture for vegetable production to enhance the soil temperature and to protect crops from unfavourable growing conditions, such as weed growth, evaporation of soil moisture and the spread of soil-borne diseases. The material usually consists of PE from fossil resources which is hardly biodegradable. Due to its benefits and leading to a higher harvest yield the use of mulch foil increased rapidly over the last decades. At seasons end the material requires collection and disposal. Instead, because of wear and tear, photodegradation, lack of waste collection at seasons end and its durability the material continuously accumulates in the environment and thus becomes an increasing contamination [13,14].

Recent developments in recycling processes focus on the biotechnological degradation of PE and subsequent enzymatical conversion of decomposition products into precursor chemicals. The most promising results in precursor yields are currently observed at microbial fermentation experiments on pyrolysis condensate from PE. [15] A possible aim for future so called biotechnological upcycling is to include a pyrolytic pre-treatment to provide access to a suitable carbon source. Consequently, the mulch foils are a suitable reference material for evaluating the behaviour of a homogeneous fraction in thermal waste treatment processes in comparison to heterogenous ML.

2.5. Chemical Composition of Material Samples

For the analytical determination of their chemical composition and thermogravimetric analysis (TGA) trials the respective material fractions were mixed again, shredded and sampled in accordance with LAGA PN 98 [13]. After separation of metallic fractions, the samples were milled to <1 mm grain size.

Table 4 depicts the results of laboratory analyses of the raw materials used in thermochemical conversion trials. The two types of marine litter ML Sylt and ML Norderney show similar results both in proximate and elemental composition. Apart from the ash contents of 13.52 and 10.55 wt%, they mainly consist of volatile matter of 84.6 and 91.0 wt% respectively. Both present a similar relation of carbon and hydrogen while the ML Sylt sample additionally shows significant amounts of nitrogen, sulphur and chlorine as potential pollutant precursors. Comparatively the virgin nature of the mulch foil reference sample becomes apparent, containing no detectable pollutant precursors and negligible amounts of ash. Subsequently it delivers the highest values for C, H and energy content.

Table 4. Characteristics of ML fractions and mulch foils (as received) (* n = 2; ** n = 3).

	ML Sylt	ML Nordenery	Mulch Foil
Proximate analysis * (wt%)			
Moisture	<0.2	<0.2	<0.2
Ash (550 °C)	13.52	10.55	0.83
Volatile matter	84.6	91.0	96.8
Ultimate analysis ** (wt%)			
C	68.2	66.8	85.6
H	9.2	9.5	13.6
N	0.7	<0.1	<0.1
S	0.6	<0.1	<0.1
Cl	1.3	0.5	<0.1
Calorific value * (kJ/kg)	33,430	35,970	42,920

2.6. Experimental Setup

The thermochemical experiments were conducted in a Linseis STA PT1600 thermobalance. The furnace chamber can be purged with different gases and thus different processing atmospheres can be set. The sample material was placed in a crucible, which rested on a thermocouple measuring the sample temperature through the ceramic bottom. All experiments were conducted with a heating rate of 30 °C/min up to 850 °C sample temperature. To establish the atmospheric conditions of pyrolysis, gasification and incineration a mass-flow-controller for nitrogen and synthetic air was used. During the experiments, the weighing system of the thermobalance was protected by nitrogen as purge gas flow at 75 mL/min, not interfering with the conversion process. The sample gas, which determines the thermochemical process, was injected separately and flushed the sample at 50 mL/min. Pyrolysis was performed in nitrogen atmosphere while synthetic air was used for incineration. For gasification, an air ratio of 0.3 was calculated based on data from fuel analysis and nitrogen and synthetic air flow set accordingly.

The sample materials created for laboratory analysis were also used as sample material in the TGA. The sample weight was 100 mg per trial and at least three repetitions (n) were carried out.

Pyrolysis trials were repeated for all three materials in a retort furnace with an increased sample size of 0.5 kg. As shown in Figure 5, the setup consisted of a modified Thermconcept KM 70/13 chamber furnace containing a stainless steel retort as fixed-bed reactor.

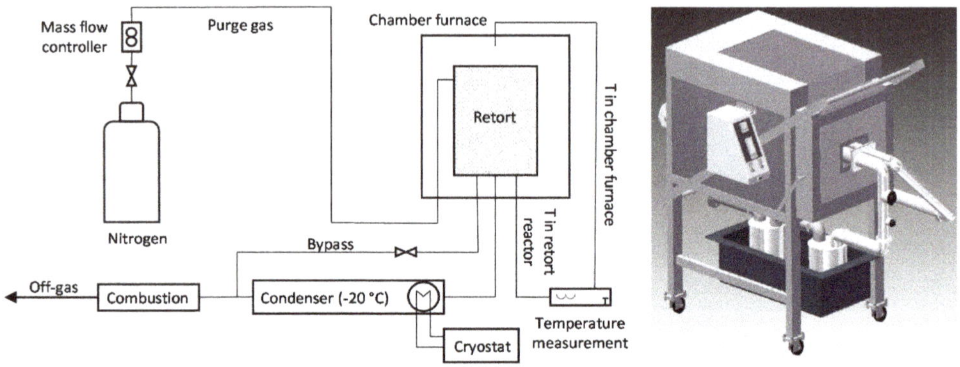

Figure 5. Retort furnace for pyrolysis (Figure: Kai Schlögel).

To ensure oxygen exclusion and thus pyrolytic conditions, the retort was continuously flushed with nitrogen. The furnace was heated up at a rate of ca. 8–10 K/min. To minimise the risk of sudden pressure rises the heating program involved a 30 min plateau at 250 °C before heating up to the target temperature of 700 °C and holding for another 30 min. The pyrolysis gas produced in the retort was led through steel containers tempered to −20 °C to collect all condensable components. In case of blockage this condenser can be bypassed through a bursting disc which was not necessary in the presented trials. The remaining permanent gas fraction was burned in a post-combustion chamber to minimise harmful gas emissions. After each trial, char and condensable product fractions were collected from the retort and the condenser respectively and analysed for proximate and elemental composition. As mentioned above, the condensate was additionally used for pre-trials in biotechnological processing. On this basis, an outlook is given for its potential to produce precursors for chemical industries.

2.7. Biotechnological Upcycling

Microorganisms can colonise various habitats by metabolising a variety of molecules. In biotechnology, this potential is applied by converting inexpensive substrates into economically interesting molecules. For this purpose, microbes are isolated from diverse habitats. Hydrocarbon-degrading microbes can be found for example in soils and sediments and even enriched in the event of an oil-spill [16]. Lignolytic microbes often have the properties to degrade polycyclic aromatic hydrocarbons [17]. The idea of biotechnological upcycling is to use those abilities to metabolize pyrolysis condensate components and convert them into molecules that can be further used in the sense of a circular economy [18].

The approach to utilise pyrolysis condensate as feedstock has already been demonstrated for pure fractions. Ward et al. [19] were able to produce the biodegradable plastic polyhydroxyalkanoate (PHA) by feeding *Pseudomonas putida* CA-3 with pyrolysis oil from polystyrene (PS). For the pure fraction of polyethylene (PE), a pyrolysis condensate consisting of 99% alkanes and alkenes was also converted to PHA with *Pseudomonas aeruginosa* PAO-1 [20]. Pure polyethylene terephthalate (PET) pyrolysis condensate, which mainly consisted of oligomers of terephthalic acid, was added to a sodium hydroxide solution, resulting in 97% terephthalic acid of the solid fraction. Kenny et al. [21] were able to isolate several bacteria, also of the genus *Pseudomonas*, which metabolise terephthalic acid and thus also capable produce PHA from the PET pyrolysis condensate.

These studies each used a pure plastic fraction as starting material. In the case of mixed polymer fractions, the composition of the pyrolysis condensate can differ considerably from the pyrolysate of the pure fraction. It can consist of a mixture of all the substrates described above and additional molecules. The organism growing on this mixture must be able to ideally convert a wide range of substrates and tolerate the presence of potentially toxic substances. Furthermore, the following requirements apply when using mixed-fraction pyrolysis condensate as a substrate:

1. The hydrophobic condensate must be introduced into the microbial liquid culture in such a way that the microbes are able to access and use it as carbon source. Some microbes naturally produce biosurfactants to gain access to hydrophobic substrates. To improve substrate uptake, bioavailability can be increased by the targeted application of biosurfactants such as rhamnolipids [20]. With increased organic phase surface, the mass transfer into the water phase or directly to microbes increases, allowing increased conversion rates.
2. In the next step, the microbe must be able to transfer the molecule into the cell, while the cell wall normally serves as a barrier for hydrophobic compounds. Therefore, the microbe must possess transporters supporting this transfer such as an *alk*L pore in the outer cell membrane of some alkane-degrading *Pseudomonas* [22].
3. After uptake of the hydrophobic molecules into the cell, the microbe must be equipped with biochemical pathways for their degradation. Here, the metabolic funnel of microbes allows ideally the co-consumption of alternative carbon and energy sources [23], while the use of synthetic microbial mixed-cultures are discussed, in which microbial members take-up individual degradation tasks, converting the mixed substrate source to a single product [24].

3. Results and Discussion

3.1. Thermogravimetric Analysis

In total 39 marine litter trials were conducted with TGA to simulate pyrolysis (n = 14), gasification (n = 11) and incineration (n = 14).

The curves of the TGA (Figure 6A, Figure 7A, Figure 8A) and first derivative of TG (DTG) (Figure 6B, Figure 7B, Figure 8B) show similar trends for pyrolysis, gasification and incineration. As expected, the curves for the homogeneous reference material mulch foils are significantly smoother in all trials. However, the curves for the individual materials differ in relation to the respective thermochemical process. Pyrolysis mass losses were determined for ML Sylt at 79 wt%, Norderney at 88 wt% and mulch foils at 91 wt%. For all materials, the TGA and DTG signal show gradual progress of further mass loss above 500 °C until the end of the programme, which indicates a migration of fixed carbon to volatile phase.

For ML Sylt, mass loss for gasification (Figure 6) increases (compared to pyrolysis) insignificantly up to 80 wt%. The increase in mass loss of the other two materials is larger and an indication that the char remaining after pyrolysis is gasified. In the case of incineration, the analysis results of the ash content listed in Table 4 differ from those of the TGA for ML. As shown in Figure 7, the ash content can be determined for ML Sylt at 14 wt%, ML Norderney at 5 wt% and mulch foil at 0.1 wt%. Reason for this difference is the heterogeneity of the collected sample or the calcination of contained carbonates at temperatures above 550 °C in TGA.

TGA measurements can vary greatly depending on the sampling. Figure 9 shows a comparison of determined sample mass constancy of all conducted TGA trials. In order to evaluate the significance of the TGA trials independently of the heterogeneity of the sample, the deviation from the reference material mulch foil can be considered. The complexity of the investigated thermochemical processes pyrolysis (A), gasification (B) and incineration (C) are clearly visible. Further oxidation decreases deviation in total.

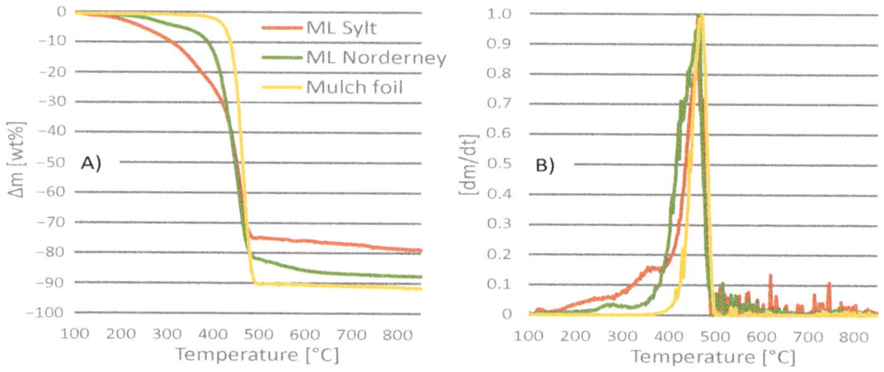

Figure 6. (**A**) shows the thermogravimetric analysis (TGA) mean value of pyrolysis. (**B**) shows the standardised derivative of TG (DTG) signal. (ML Sylt n = 5, ML Norderney n = 5, Mulch foil n = 4).

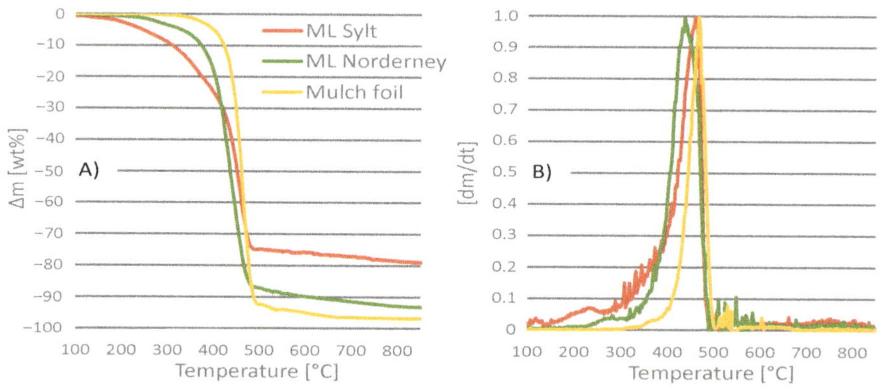

Figure 7. (**A**) shows the TGA mean value of gasification. (**B**) shows the standardised DTG signal. (ML Sylt n = 4, ML Norderney n = 4, Mulch foil n = 3).

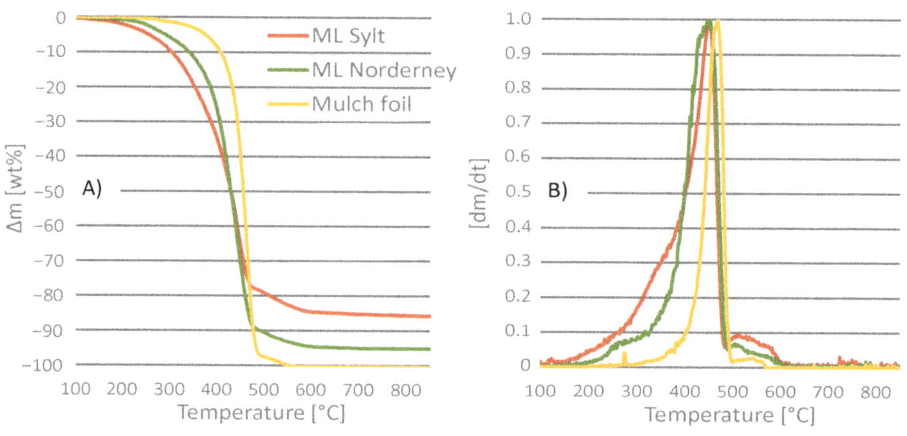

Figure 8. (**A**) shows the TGA mean value of incineration. (**B**) shows the standardised DTG signal. (ML Sylt n = 5, ML Norderney n = 5, Mulch foil n = 4).

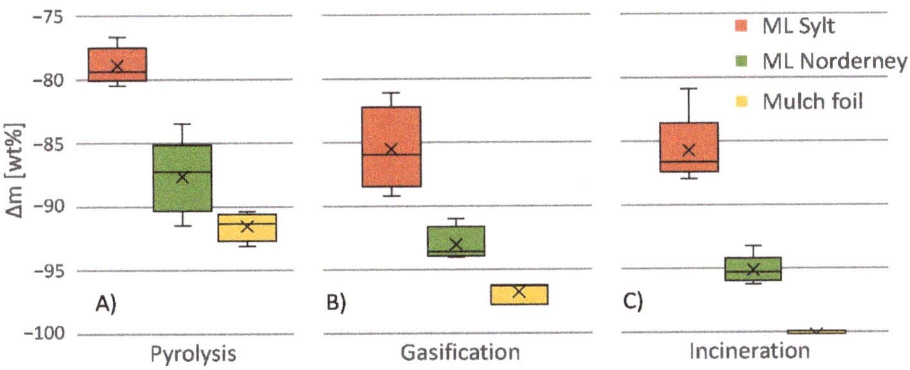

Figure 9. Comparison of determined sample mass constancy of all conducted TGA trials. (**A**) shows deviation for pyrolysis: ML Sylt n = 5, ML Norderney n = 5, Mulch foil n = 4. (**B**) shows deviation for gasification, ML Sylt n = 4, ML Norderney n = 4, Mulch foil n = 3. (**C**) shows deviation for incineration: ML Sylt n = 5, ML Norderney n = 5, Mulch foil n = 4.

Table 5 shows the measured reaction temperatures and calculated standard deviation, respectively. T_i indicates the initial temperature when 1 wt% mass loss is quantified while T_c shows the end of the reaction and a 99 wt% mass constancy. T_p marks the temperature of the highest relative mass loss as calculated by the derivation of the TGA signal. Although no moisture could be quantified in the sample of ML Sylt, initial mass loss can be observed at temperatures below 150 °C. The temperature of the reaction peak ranges from 438–467 °C. T_p differs slightly between the ML fractions, while T_p for mulch foil is very constant. Based on T_c, it can be observed that devolatilization of ML during pyrolysis is significantly slower and occurs up to high temperatures. The reaction with oxygen during incineration is completed earlier that is, at lower temperatures. In conclusion, it can be stated that the standard deviation for the results of the ML fractions is consistently higher than for the mulch foils.

Table 5. Overview of the specific reaction temperatures and calculated standard deviation. T_i indicates the initial temperature when 1 wt% mass loss is quantified while T_c shows the end of the reaction and a 99 wt% mass constancy. T_p marks the temperature of the highest relative mass loss as calculated by the derivation of the TGA signal.

		ML Sylt	ML Norderney	Mulch Foil
T_i	[°C]			
	Pyrolysis	137.6 ± 37.1	187.6 ± 23.7	384.7 ± 5.2
	Gasification	120.5 ± 9.6	226.0 ± 14.0	343.3 ± 5.2
	Incineration	148.4 ± 31.4	212.8 ± 16.9	285.5 ± 7.0
T_p	[°C]			
	Pyrolysis	458.8 ± 1.6	456.8 ± 3.2	466.5 ± 2.2
	Gasification	459.5 ± 1.7	437.5 ± 1.7	466.7 ± 0.9
	Incineration	448.4 ± 4.5	440.0 ± 12.0	465.5 ± 3.8
T_c	[°C]			
	Pyrolysis	751.2 ± 20.6	666.0 ± 28.0	497.0 ± 7.3
	Gasification	782.0 ± 24.0	747.5 ± 5.0	634.7 ± 10.5
	Incineration	633.2 ± 36.2	580.8 ± 12.9	543.0 ± 6.6

3.2. Retort Furnace

The pyrolysis trials in the retort furnace allow separate capture of char and condensate products and subsequent analysis of samples. The results are presented in Table 6. The trials resulted in char yields ranging from 9.3–18.2 wt% and condensate yields of 21.5–

32.3 wt%. As suggested by raw material analysis the char fractions from marine litter educts contain higher amounts of ash and thus lower carbon. Elementary composition in general reflects the results of ultimate analyses as shown in Table 4 with heteroatoms of hydrocarbons concentrating in the char fractions up to 2.0 wt% of sulphur and 4.4 wt% of chlorine. Ultimate analysis shows high contents of carbon and hydrogen for marine litter condensates likely deriving form organic hydrocarbons while mulch foil condensate leaves a significant gap in the balance of detectable elements. This does not match its comparatively high calorific value suggesting an imprecision in elementary analysis. For all educts calorific values of condensate exceed their respective char fractions, demonstrating the release of energy-rich compounds in the pyrolysis process. These results appear plausible considering the raw materials consisting mainly of ash and volatile matter.

Table 6. Characteristics of pyrolysis products derived from different plastic wastes (as received) (* n = 2; ** n = 3).

	Sylt Char	Norderney Char	Mulch Foil Char	Sylt Condensate	Norderney Condensate	Mulch Foil Condensate
Product fraction (wt%)	18.2	14.8	9.3	24.1	21.5	32.3
Proximate analysis * (wt%)						
Moisture	<0.2	<0.2	<0.2			
Ash (550 °C)	84.89	52.75	27.57			
Volatile matter	7.3	4.1	4.3			
Ultimate analysis ** (wt%)						
C	20.3	49.7	72.2	77.0	72.7	56.6
H	0.2	1.0	0.3	11.0	10.3	8.8
N	0.7	0.3	<0.1	0.4	<0.1	<0.1
S	2.0	0.2	0.3	0.1	<0.1	<0.1
Cl	4.4	3.0	<0.1	0.2	0.9	<0.1
Calorific value *(kJ/kg)	8590	18,440	25,080	41,500	37,210	42,170

For further elemental analysis of the pyrolysis products an X-ray fluorescence (XRF) analysis was performed and the detected elements with the highest mass fractions were plotted as shown in Figures 10A, 11A and 12A. Furthermore, the content of heavy metals mentioned in the German ordinance for waste incineration (17. BImSchV) was determined in the samples and plotted on the right side (B). Due to the methodical detection limits of the XRF analyser, sodium (Na) and magnesium (Ma) were not included for condensate products. The XRF analysis of the ML raw materials shows significantly increased contents of sodium, silicon and calcium compared to the reference material (Figure 10A). These elements are used in plastics manufacturing as additives with the function of for example, filler and reinforcing material [25]. However, the cause of the increased concentration in ML could be explained by the origin from sand and soil before collection. As shown in Figure 12 almost all elements are concentrated in the pyrolysis char. Amounts of potential harmful substances were found in low quantities in all the samples, peaking at 0.015 wt% lead for ML Sylt raw material and 0.35 wt% in the char, which had been expected higher for material containing fishing gear. The lead is partly built into the ropes as a weighting for the nets to improve sinking in the water. Previous studies showed lead contents of up to 0.75 wt% in the raw material and 2.25 wt% in the char of ML [26]. Norderney condensate and mulch foil char evince unexpected measurements, as contents of iron, chromium and nickel show the same ratio, which indicates cross-contamination with stainless steel in the sample pre-treatment.

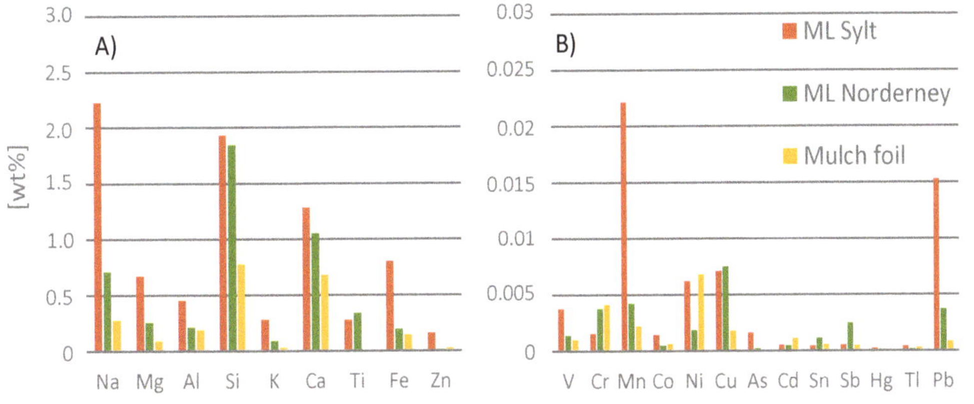

Figure 10. Mass fraction of selected elements in raw material determined by X-ray fluorescence (XRF) (n = 2). (**A**) shows detected elements with the highest mass fractions. (**B**) shows heavy metals mentioned in the German ordinance for waste incineration (17. BImSchV).

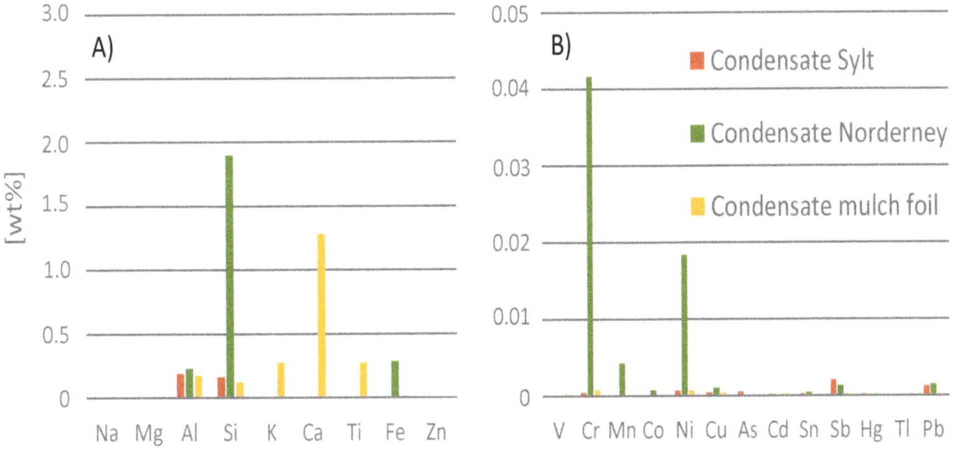

Figure 11. Mass fraction of selected elements in pyrolysis condensate determined by XRF (n = 2). (**A**) shows detected elements with the highest mass fractions. (**B**) shows heavy metals mentioned in the German ordinance for waste incineration (17. BImSchV).

Combining the results of pyrolysis product analysis with those of gas measurements allows examining the distribution of selected elements. Carbon and hydrogen were balanced according to their recovery rate in the product fractions, results are presented in Figure 13. Comparing elementary content in raw material with the accumulated content in pyrolysis products shows minor deviations in relation to the methodically inevitable imprecisions especially in pyrolytic processes. Carbon distributes across all three product fractions while hydrogen almost completely releases from the solid phase. In both elemental balances the gas fraction represents by far the highest fraction, due to slow cooling rates and thus high residence time in the gas phase allowing further cracking of condensable components into permanent gases. The distribution of elements across the product fractions is similar in all trials conducted.

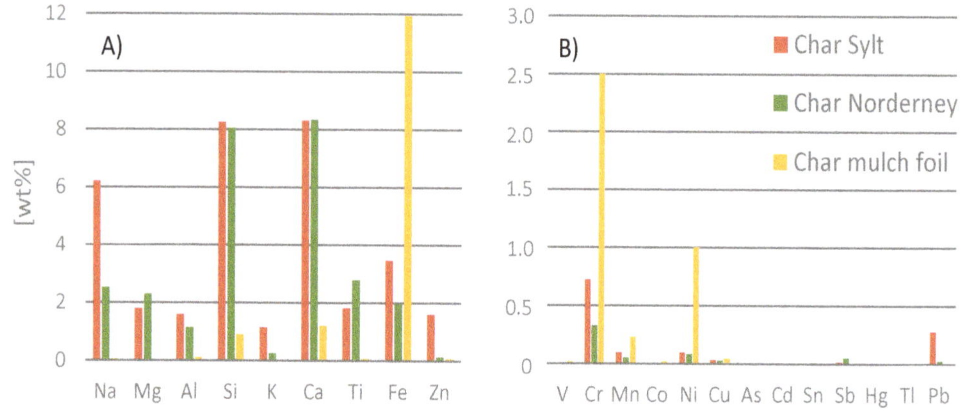

Figure 12. Mass fraction of selected elements in pyrolysis char determined by XRF (n = 2). (**A**) shows detected elements with the highest mass fractions. (**B**) shows heavy metals mentioned in the German ordinance for waste incineration (17. BImSchV).

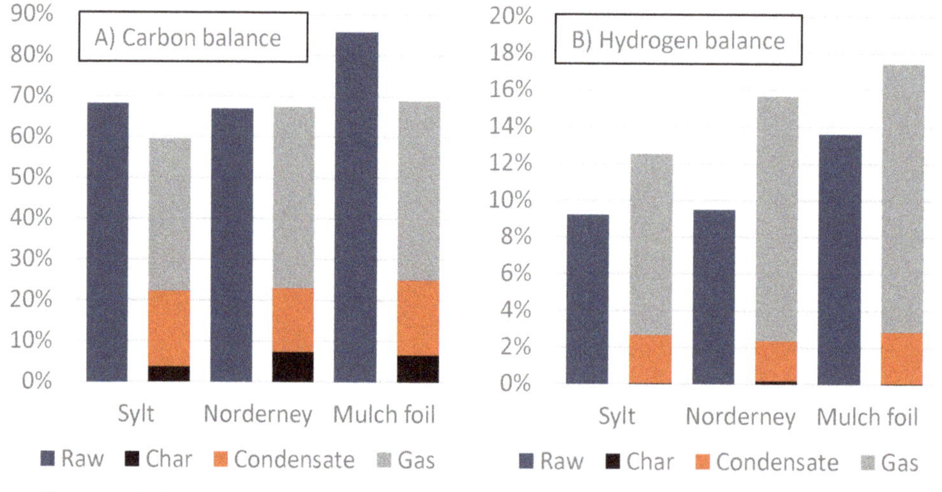

Figure 13. Elementary balances of pyrolysis products compared to analysed content in raw material. (**A**) carbon distribution across product fractions in wt%. (**B**) hydrogen distribution across product fractions in wt%.

To rank the transferability of results from the different experimental setups the distribution of pyrolysis products from the retort furnace are compared to the residual mass from the TGA setup. The latter one represents the char fraction while the mass loss is ascribed to the volatile compounds as a sum parameter consisting of condensate and gas as they correspond to the retort furnace. As presented in Figure 14 the char yields of the materials show very similar results in retort furnace and TGA setup. Both experiments used low heating rates of 8–10 °C/min and 30 °C/min thus returning virtually equal amounts of char. Target temperature (700 °C for retort furnace, 850 °C for TGA) appears to have an insignificant influence. Overall results validate both experimental setups used, suiting them for general material characterisation.

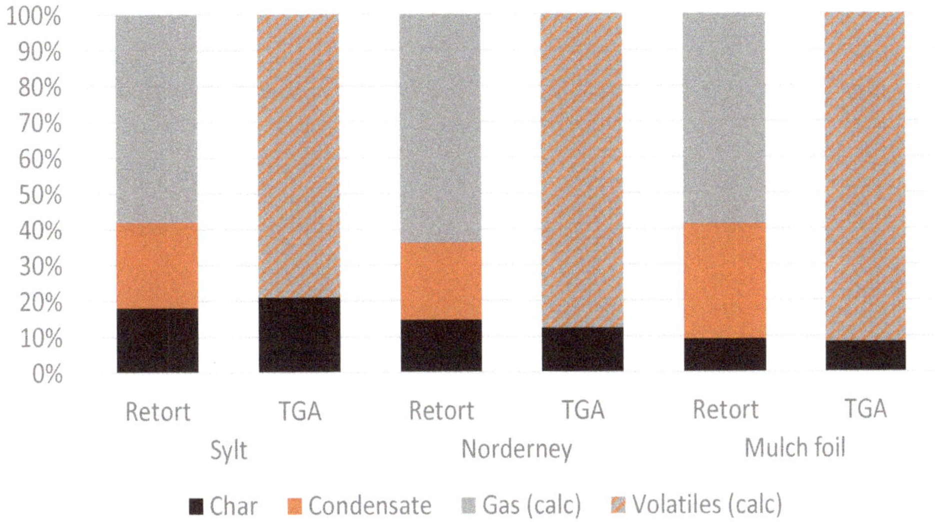

Figure 14. Comparison of product yields from pyrolysis trials in retort furnace and TGA in wt%.

The three pyrolysis condensates of this study are unexpectedly of wax-like consistency. Previous studies on ML pyrolysis in the same retort furnace produced liquid condensates of honey-like consistency for the temperature investigated [26]. The current focus of the upcycling-efforts therefore is on requirement 1 (Section 2.7) how to make the condensate bioavailable by introduction into the liquid culture. Solvents and biosurfactants are tested for the ability to dissolve the condensate and their biocompatibility. As shown in Figure 15C,D, simply supplementing the liquid medium with the condensate and rhamnolipids, as in the study of Guzik et al. [20], will not be sufficient with the condensates of this study. One promising solvent is ethyldecanoate (Figure 15A), which has recently been shown to not affect the growth of *Pseudomonas putida* KT2440 but to be suitable for biotechnological application [27].

Figure 15. Pyrolysis condensate in the solvents (**A**) ethyldecanoate; (**B**) hexadecane; and water supplemented with each 10% of the emulsifiers; (**C**) rhamnolipids and (**D**) Tween80 (Picture: Kristina Bitter).

Biocompatibility is tested with several strains of *Pseudomonas* but also other metabolic versatile organisms with affinity towards hydrocarbons as *Rhodococcus opacus* who harbours genes for the *Alk*B monooxygenase [28] and known alkane-degrader *Alcanivorax borkumensis* [29] fulfilling requirements (2) and (3) (Section 2.7). In the following step, potential growth of microbes on pyrolysis condensate will be monitored via CO_2-development. When bacterial growth on the mixed-plastic pyrolysis condensate can be confirmed, the production of valuable compounds will be investigated.

4. Conclusions

The investigations performed and results presented refer to the collected material of the ML Norderney and ML Sylt. Since the collection was carried out randomly and without standardized conditions, it cannot be considered representative of all the waste washed up on the beach on Norderney and Sylt or other locations. The ML presented in this study rather shows the randomness of the composition of waste samples.

The results presented in this study demonstrate the principal possibility of thermochemical treatment of ML. The determined fluctuations in reaction temperatures, ash content or pollutants do not pose any problems for thermal treatment plants with state of the art process and emission control systems. Process selection depends on legislative incentives and possible cascaded recycling of intermediate and end products.

Conclusively, the necessity of tackling the pollution of the World Ocean with ML becomes obvious. Pyrolysis as part of a chemical recycling process needs extensive pretreatment to remove impurities, for example, by crushing and sorting. The obtained condensable fractions present a valuable product for material recovery. Combustion of pyrolysis gases supplies the necessary process heat. The pyrolysis char is contaminated with pollutants such as heavy metals and must be post-processed safely in waste incineration plants for energy recovery and discharge of pollutants.

The use of condensable fractions for biotechnological upcycling presents an innovative approach to access the material potential of plastics unsuitable for mechanical recycling sustainably. Biotechnological use of pyrolysis condensate poses many challenges but versatile microbes as *Pseudomonas* might be able to meet them. They offer opportunities to produce valuable molecules for further industrial applications, thus reintegrating plastic waste fractions into a circular economy.

Author Contributions: J.H.: Project Administration; Conceptualization, Methodology, Writing—Original Draft, Review and Editing; K.S.: Conceptualization, Methodology, Writing—Original Draft, Review and Editing; S.L.: Writing—Original Draft, Review and Editing; J.P.: Writing—Original Draft, Review and Editing; K.B.: Writing—Original Draft, Review and Editing; L.M.B.: Supervision; P.Q.: Supervision, Writing—Review and Editing. All authors have read and agreed to the published version of the manuscript.

Funding: The research conducted by Institute of Applied Microbiology, Aachen Biology and Biotechnology was funded by iMULCH (FKZ: EFRE-0801169). Pacific Garbage Screening GmbH has received funding from the European Union's Horizon 2020 research and innovation programme under grant agreement No 870294.

Acknowledgments: The authors gratefully acknowledge the volunteers of the Wadden Sea World UNESCO World Natural Heritage Wadden Sea Visitor Centre Norderney and the Centre of Experience Natural Forces Sylt gGmbH who collected the material and provided the material to Pacific Garbage Screening GmbH (Future everwave) for research purposes.

Conflicts of Interest: The authors declare no conflict of interests.

References

1. Hopmann, C.; Michaeli, W. *Einführung in die Kunststoffverarbeitung, 7., Aktualisierte Aufl.*; Hanser: München, Germany, 2015; ISBN 978-3-446-44627-4.
2. Geyer, R.; Jambeck, J.R.; Law, K.L. Production, use, and fate of all plastics ever made. *Sci. Adv.* **2017**, *3*, e1700782. [CrossRef]
3. Lechthaler, S.E.; Waldschläger, K.; Stauch, G.; Schüttrumpf, H. The Way of Macroplastic through the Environment. *Environment* **2020**, *7*, 73. [CrossRef]
4. Jambeck, J.R.; Geyer, R.; Wilcox, C.; Siegler, T.R.; Perryman, M.; Andrady, A.; Narayan, R.; Law, K.L. Plastic waste inputs from land into the ocean. *Science* **2015**, *347*, 768–771. [CrossRef]

5. Eriksen, M.; Lebreton, L.C.M.; Carson, H.S.; Thiel, M.; Moore, C.J.; Borerro, J.C.; Galgani, F.; Ryan, P.G.; Reisser, J. Plastic Pollution in the World's Oceans: More than 5 Trillion Plastic Pieces Weighing over 250,000 Tons Afloat at Sea. *PLoS ONE* **2014**, *9*, e111913. [CrossRef]
6. Carpenter, E.J.; Smith, K.L. Plastics on the Sargasso Sea Surface. *Science* **1972**, *175*, 1240–1241. [CrossRef] [PubMed]
7. UNEP. *Marine Litter, an Analytical Overview*; UNEP: Nairobi, Kenya, 2005.
8. European Commission. Our Oceans, Seas and Coasts: Descriptor 10: Marine Litter. Available online: https://ec.europa.eu/environment/marine/good-environmental-status/descriptor-10/index_en.htm (accessed on 16 November 2020).
9. GESAMP. *Guidelines for the Monitoring and Assessment of Plastic Litter and Microplastics in the Ocean*; GESAMP: London, UK, 2019.
10. Schmaltz, E.; Melvin, E.C.; Diana, Z.; Gunady, E.F.; Rittschof, D.; Somarelli, J.A.; Virdin, J.; Dunphy-Daly, M.M. Plastic pollution solutions: Emerging technologies to prevent and collect marine plastic pollution. *Environ. Int.* **2020**, *144*, 106067. [CrossRef]
11. Chubarenko, I.; Esiukova, E.; Bagaev, A.; Isachenko, I.; Demchenko, N.; Zobkov, M.; Efimova, I.; Bagaeva, M.; Khatmullina, L. Behavior of Microplastics in Coastal Zones. In *Microplastic Contamination in Aquatic Environments*; Elsevier: Amsterdam, The Netherlands, 2018; pp. 175–223.
12. Hahladakis, J.N.; Velis, C.A.; Weber, R.; Iacovidou, E.; Purnell, P. An overview of chemical additives present in plastics: Migration, release, fate and environmental impact during their use, disposal and recycling. *J. Hazard. Mater.* **2018**, *344*, 179–199. [CrossRef]
13. Kasirajan, S.; Ngouajio, M. Polyethylene and biodegradable mulches for agricultural applications: A review. *Agron. Sustain. Dev.* **2012**, *32*, 501–529. [CrossRef]
14. Steinmetz, Z.; Wollmann, C.; Schaefer, M.; Buchmann, C.; David, J.; Tröger, J.; Muñoz, K.; Frör, O.; Schaumann, G.E. Plastic mulching in agriculture. Trading short-term agronomic benefits for long-term soil degradation? *Sci. Total Environ.* **2016**, *550*, 690–705. [CrossRef]
15. Wei, R.; Tiso, T.; Bertling, J.; O'Connor, K.; Blank, L.M.; Bornscheuer, U.T. Possibilities and limitations of biotechnological plastic degradation and recycling. *Nat. Catal.* **2020**, *3*, 867–871. [CrossRef]
16. Rojo, F. Degradation of alkanes by bacteria. *Environ. Microbiol.* **2009**, *11*, 2477–2490. [CrossRef] [PubMed]
17. Haritash, A.; Kaushik, C. Biodegradation aspects of Polycyclic Aromatic Hydrocarbons (PAHs): A review. *J. Hazard. Mater.* **2009**, *169*, 1–15. [CrossRef] [PubMed]
18. Blank, L.M.; Narancic, T.; Mampel, J.; Tiso, T.; O'Connor, K. Biotechnological upcycling of plastic waste and other non-conventional feedstocks in a circular economy. *Curr. Opin. Biotechnol.* **2020**, *62*, 212–219. [CrossRef]
19. Ward, P.G.; Goff, M.; Donner, M.; Kaminsky, W.; O'Connor, K.E. A Two Step Chemo-biotechnological Conversion of Polystyrene to a Biodegradable Thermoplastic. *Environ. Sci. Technol.* **2006**, *40*, 2433–2437. [CrossRef]
20. Guzik, M.W.; Kenny, S.T.; Duane, G.F.; Casey, E.; Woods, T.; Babu, R.P.; Nikodinovic-Runic, J.; Murray, M.; O'Connor, K. Conversion of post consumer polyethylene to the biodegradable polymer polyhydroxyalkanoate. *Appl. Microbiol. Biotechnol.* **2014**, *98*, 4223–4232. [CrossRef]
21. Kenny, S.T.; Runic, J.N.; Kaminsky, W.; Woods, T.; Babu, R.P.; Keely, C.M.; Blau, W.; O'Connor, K.E. Up-Cycling of PET (Polyethylene Terephthalate) to the Biodegradable Plastic PHA (Polyhydroxyalkanoate). *Environ. Sci. Technol.* **2008**, *42*, 7696–7701. [CrossRef]
22. Julsing, M.K.; Schrewe, M.; Cornelissen, S.; Hermann, I.; Schmid, A.; Bühler, B. Outer Membrane Protein AlkL Boosts Biocatalytic Oxyfunctionalization of Hydrophobic Substrates in Escherichia coli. *Appl. Environ. Microbiol.* **2012**, *78*, 5724–5733. [CrossRef]
23. Sudarsan, S.; Dethlefsen, S.; Blank, L.M.; Siemann-Herzberg, M.; Schmid, A. The Functional Structure of Central Carbon Metabolism in Pseudomonas putida KT2440. *Appl. Environ. Microbiol.* **2014**, *80*, 5292–5303. [CrossRef]
24. Utomo, R.N.C.; Li, W.-J.; Tiso, T.; Eberlein, C.; Doeker, M.; Heipieper, H.J.; Jupke, A.; Wierckx, N.; Blank, L.M. Defined Microbial Mixed Culture for Utilization of Polyurethane Monomers. *ACS Sustain. Chem. Eng.* **2020**, *8*, 17466–17474. [CrossRef]
25. Baur, E.; Brinkmann, S.; Osswald, T.A.; Rudolph, N.; Schmachtenberg, E.; Saechtling, H. *Saechtling Kunststoff Taschenbuch, 31. Ausgabe, [komplett überarb., aktualisiert und zum ersten Mal in Farbe]*; Hanser: München, Germany, 2013; ISBN 978-3-446-43442-4.
26. Hee, J.; Schloegel, K.; Lechthaler, S.; Quicker, P. Plastics in the Oceans: Recycling, WtE or landfilling? In Proceedings of the 11th ISWA Beacon Conference on Waste-to-Energy Malmö, Malmö, Sweden, 29–30 October 2019.
27. Demling, P.; Von Campenhausen, M.; Grütering, C.; Tiso, T.; Jupke, A.; Blank, L.M. Selection of a recyclable in situ liquid–liquid extraction solvent for foam-free synthesis of rhamnolipids in a two-phase fermentation. *Green Chem.* **2020**. [CrossRef]
28. Cappelletti, M.; Fedi, S.; Zannoni, D. Degradation of Alkanes in Rhodococcus. In *Beneficial Microorganisms in Food and Nutraceuticals*; Springer: Berlin, Germany, 2019; pp. 137–171.
29. Schneiker, S.; Santos, V.A.P.M.D.; Bartels, D.; Bekel, T.; Brecht, M.; Buhrmester, J.; Chernikova, T.N.; Denaro, R.; Ferrer, M.; Gertler, C.; et al. Genome sequence of the ubiquitous hydrocarbon-degrading marine bacterium Alcanivorax borkumensis. *Nat. Biotechnol.* **2006**, *24*, 997–1004. [CrossRef]

processes

Article

Pyrometallurgical Lithium-Ion-Battery Recycling: Approach to Limiting Lithium Slagging with the InduRed Reactor Concept

Stefan Windisch-Kern *, Alexandra Holzer, Christoph Ponak and Harald Raupenstrauch

Chair of Thermal Processing Technology, Montanuniversität Leoben, Franz-Josef-Strasse 18, 8700 Leoben, Austria; alexandra.holzer@unileoben.ac.at (A.H.); christoph.ponak@unileoben.ac.at (C.P.); harald.raupenstrauch@unileoben.ac.at (H.R.)
* Correspondence: stefan.windisch-kern@unileoben.ac.at

Abstract: The complexity of the waste stream of spent lithium-ion batteries poses numerous challenges on the recycling industry. Pyrometallurgical recycling processes have a lot of benefits but are not able to recover lithium from the black matter since lithium is slagged due to its high oxygen affinity. The presented InduRed reactor concept might be a promising novel approach, since it does not have this disadvantage and is very flexible concerning the chemical composition of the input material. To prove its basic suitability for black matter processing, heating microscope experiments, thermogravimetric analysis and differential scanning calorimetry have been conducted to characterize the behavior of nickel rich cathode materials ($LiNi_{0.8}Co_{0.15}Al_{0.05}O_2$ and $LiNi_{0.33}Mn_{0.33}Co_{0.33}O_2$) as well as black matter from a pretreatment process under reducing conditions. Another experimental series in a lab scale InduRed reactor was further used to investigate achievable transfer coefficients for the metals of interest. The promising results show technically feasible reaction temperatures of 800 °C to 1000 °C and high recovery potentials for nickel, cobalt and manganese. Furthermore, the slagging of lithium was largely prevented and a lithium removal rate of up to 90% of its initial mass was achieved.

Keywords: lithium-ion-batteries; pyrometallurgical recycling; carbothermal reduction

Citation: Windisch-Kern, S.; Holzer, A.; Ponak, C.; Raupenstrauch, H. Pyrometallurgical Lithium-Ion-Battery Recycling: Approach to Limiting Lithium Slagging with the InduRed Reactor Concept. *Processes* **2021**, *9*, 84. https://doi.org/10.3390/pr9010084

Received: 26 November 2020
Accepted: 28 December 2020
Published: 2 January 2021

1. Introduction

Since 1979, when Goodenough et al. finally tested $LiCoO_2$ (short: LCO) as a cathode material, the development and commercialization of electrochemical energy storage based on the lithium-ion technology has been steadily pushed forward [1,2]. Lithium-ion-batteries (LIBs) basically consist of the same components such as anode, cathode, separator or electrolyte as can be found in other battery technologies. This basic principle has not changed since 1979 and therefore also applies to modern LIBs. However, the initially used LCO cathode material is now just one option on a long list of alternatives like NCA ($LiNi_xCo_yAl_zO_2$), NMC ($LiNi_xMn_yCo_zO_2$) or LFP ($LiFePO_4$) materials [3]. The variety of cathode materials is not only based on the fields of possible applications that reach from mobile electronics to e-mobility or stationary storages and their respective demand for performance (energy and power density) or safety aspects, but also on factors like raw material prices, supply risks or social and ecological sustainability. Concerning the development of the LIB market numerous publications can be found. Especially the electric automotive sector will benefit from decreasing costs made possible by mass production and optimized cell chemistry. Berckmans et al. [4] states that by 2030, the cut of fully electric or hybrid vehicles will rise to 25% of the total vehicles sold. In view of the high amount of valuable metals that are contained in LIBs, especially in their cathode materials, and the predicted market demand [5], an efficient recycling process in order to recover the mentioned valuable metals is absolutely necessary.

In general, the recycling of LIBs can be divided into three processing steps, namely pretreatment, metal extraction and metal refining. The recycling chain of LIBs usually starts with pre-treatment processing which aims to separate battery components like the battery management system or the housing and the corresponding materials such as plastics or iron from the active materials of the battery electrodes. For pre-treatment, various processes can be found which differ more or less from each other. Basically, steps during pre-treatment are sorting, dismantling deactivation and mechanical processing and separating. Said active materials–after pre-treatment they are also known as black matter–mainly consist of lithium metal oxides or lithium iron phosphate, carbon and residues from the electrode conductor foils. Umicore's Valéas process, for example, is an exception since it does not need a usual pre-treatment but uses the batteries directly in their pyrometallurgical process. [2,6–12].

While the obtained metal and plastic scrap can be recycled directly, the produced black matter that contains the valuable metals needs to be further treated in a metal extraction step to recover Li, Ni, Co and Mn, at best in a quality that is suitable for closed loop recycling. Therefore, pyro-, bio- and hydrometallurgical methods can be used. Biometallurgical processes like bioleaching are considered as environmentally friendly and low cost alternatives to conventional hydrometallurgy, capable of reaching recovery rates of more than 98% for Ni and Co and more than 80% for Li but suffering from low kinetics and resulting poor throughput rates [13–19].

Typical hydrometallurgical procedures, used to recover metals from black matter, are leaching, solvent extraction, chemical precipitation or electrochemical deposition, with which a high selectivity and therefore product purity can be achieved [20]. The possible recovery rates for Ni, Mn, Co and Li, as for example reported by He et al. [21], can be close to or even higher than 99%. The obtained salts or concentrates can usually be directly used for the production of new cathode materials as it is the case for the Duesenfeld process described by Elwert and Frank [22]. An indication of the importance of hydrometallurgical recycling of spent LIBs is, among other things, the high intensity of research activities in this field. According to Huang et al. [23], more than half of the recycling processes that are currently under investigation are related to hydrometallurgical processing.

Pyrometallurgical approaches use high temperatures, usually above 1400 °C, and reducing conditions to recover valuable metals as a metal alloy. The advantages lie in the experience with and the properties of conventional pyrometallurgical units which are less complex and less vulnerable, e.g., to organic impurities in the black matter, than their hydrometallurgical counterparts. The decisive factor in this regard is the oxygen potential of the contained metals, which is for example low for Ni and Co, leading to a relatively low-effort recovery. On the other hand, the similarity of the oxygen potential between Ni and Co reduce the selectivity of pyrometallurgical processes since they cannot be recovered separately but only as an alloy. The oxygen potential is also responsible for one of the biggest disadvantages of pyrometallurgy. Lithium, which has a much higher oxygen affinity, cannot be recovered as part of the metal alloy but is bound as an oxide in the slag instead [2,20,24–26].

The refining step is usually based on hydrometallurgical methods and aims for a closed loop recycling. Hence, it mainly applies on the metal alloy and slag from pyrometallurgical processing, which without further treatment, cannot be used for the production of new LIBs. The treatment of the metal alloy aims for a separation of the contained metals, while the slag treatment's goal is to recover Li, which is often technically but not economically feasible due to the low Li content in the slag. [2,20,22,27]

However, it can be summarized that there are still a lot of uncertainties in the LIB recycling chain. Not only the development of the waste stream itself, also the number and diversity of pre-treatment processes lead to varying black matter compositions and qualities. For pyrometallurgy, the lack of Li recovery options is a major problem that is not yet solved, but however, gives the desired novel approach with the InduRed reactor a good opportunity to establish itself as an alternative to conventional processes.

The mentioned InduRed reactor might be a possibility to achieve a simultaneous recovery of Ni, Co, Mn as well as Li with a pyrometallurgical process. The existing pilot-scale reactor concept, shown in Figure 1a,b, consists of a packed bed of graphite pieces that is inductively heated by surrounding copper coils. The input material is fed continuously with up to 10 kg/h from the top onto the hot graphite bed. The uppermost induction coil powers the upper third of the reactor where the input material melts and forms a thin molten layer that moves downwards. A second induction coil, placed half way down the reactor, induces enough power so that reduction reactions can take place. Gaseous reaction products are then removed from the reactor via a flue gas pipe whereas the liquid products move further down. The third induction coil makes sure that the temperature within the reactor can be maintained well above the melting temperature of the mixture and enables a continuous flow out of the bottom of the reactor. The advantages of the reactor are the low oxygen partial pressure, the possibility to control different temperature zones, and the big reaction surface due to the graphite bed. Furthermore, the contact time and intensity between gaseous reaction products and the molten phase can be limited because they only need to pass a thin layer or droplets instead of a molten bath like in conventional pyrometallurgical furnaces. Originally, the reactor concept was developed for the recovery of phosphorus from sewage sludge ashes, which is described by Schönberg [28]. The concept was later also adapted by Ponak et al. [29,30] to treat basic oxygen furnace slag with limited iron phosphide formation.

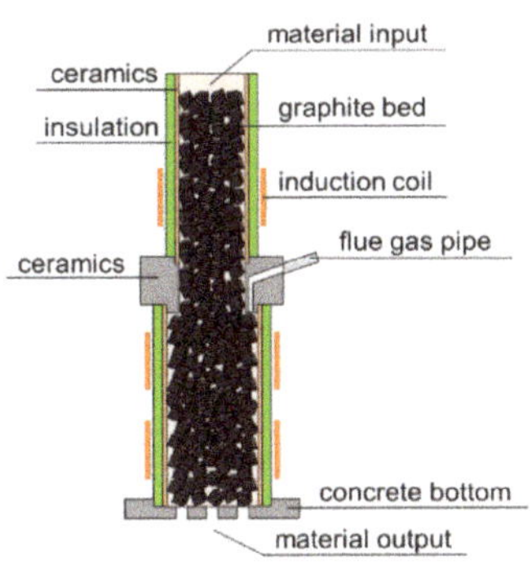

(a)

(b)

Figure 1. (**a**) Schematic illustration of the so called InduRed reactor and (**b**) said reactor operating at a test series for metal recovery from basic oxygen furnace slag. [29,30].

The aim of this work is to investigate if said reactor concept can potentially also provide a solution for LIB black matter recycling.

For the determination of the basic suitability of black matter as an input material for the InduRed reactor, thus its melting and reaction behaviour, heating microscope experiments, thermogravimetric analysis (TGA) and differential scanning calorimetry (DSC) were carried out. Since black matter can have different properties and contents of impurities depending on the pre-treatment procedure, the influence of which on the properties being investigated is difficult to assess, the investigations are also carried out with pure cathode materials. The ability of the InduRed reactor concept to eliminate one of the biggest disadvantages of pyrometallurgical LIB recycling, namely lithium slagging, is finally evaluated by experiments in a lab-scaled batch reactor, which is based on the InduRed concept. The results, in particular the required reaction temperatures and the Li removal rate via the gas phase from the reactor, form the basis on which a decision is made about the fundamental suitability of the reactor to be part of the LIB recycling chain.

2. Materials and Methods

The cathode materials ($LiNi_{0.8}Co_{0.15}Al_{0.05}O_2$, sample abbreviation: NCA and $LiNi_{0.33}Mn_{0.33}Co_{0.33}O_2$, sample abbreviation: NMC) which were used for the experiments have been produced by Gelon Energy Corporation in Linyi, China, while the black matter (sample abbreviation: AM) was provided by a LIB recycling facility operated by Redux GmbH in Bremerhaven, Germany. The chemical composition of said materials is summarized in Table 1 below.

Table 1. Chemical composition of used materials. (mass fraction, w/%).

Species	C	Li	Ni	Co	Mn	Al	P	Fe	Cu	Zn	Pb
AM [1]	29.5	2.4	20.9	4.2	1.1	5.8	0.4	0.6	5.7	0.8	0.1
NCA [2]	-	7.2	48.9	9.2		1.4					
NMC [2]	-	7.2	20.3	20.4	19.0	-					

[1] Data from ICP-MS analysis. [2] Calculated from the molar composition of the cathode materials.

In experiments with NCA and NMC, where reducing conditions were desired (sample abbreviation: NCA_C, NMC_C), fine powdered coke was used as a reducing agent. Since AM already contains 29.5 w/% carbon there was no need to further add a reducing agent.

In order to investigate the general behavior of the cathode materials at high temperatures and under reducing conditions, the work started with two preliminary experimental series. First, heating microscope experiments were conducted in a Hesse Instruments EM 201 with an HR18-1750/30 furnace (Hesse Instruments, Osterode am Harz, Germany) to investigate at which temperatures reactions or transformations in the sample occur. In the heating microscope experiments, black matter and the cathode materials with and without carbon addition were tested at least twice to check the reproducibility of the results. In the reduction experiments, carbon was added in extents of 10 w/% to the NCA and NMC materials. An argon purge with a flow rate of approximately 2.5 l/min was used to inhibit oxidization reactions of the materials. The settings for the heating rate (80 °C/min until 1350 °C, 50 °C/min until 1550 °C and 10 °C/min until 1650 °C with a holding time of 5 min at 1650 °C), the used Al_2O_3 sample plates and the sample size of approximately 0.1 g of powder, pressed in a cylindrical shape, were the same for all experiments.

The second series were simultaneous thermal analyses (STA), more precisely thermogravimetric analysis (TG) and differential scanning calorimetry (DSC), which have been conducted in a Setaram Setsys Evo 2400 at the Chair of Physical Metallurgy and Metallic Materials at the Montanuniversitaet Leoben. The aim of the STA was to confirm the temperature zones in which changes of the materials could be observed in the heating microscope and to further characterize the underlying reaction mechanisms. An argon purge was used to inhibit oxidization reactions of the materials. For carrying out the analyses, graphite crucibles and a carbon addition of 25 w/% were used in order to prevent damages to the analysis hardware. The need for this is due to reactions between standard Al_2O_3 crucibles and the produced metal alloy when carbon is added to the mixture. The reactions lead to a destruction of the Al_2O_3 crucible and the thermocouple underneath gets destroyed. The higher amount of carbon in the STA experiments is needed to prevent reactions between the cathode material and the graphite crucible, which would take part as a reductant.

To simulate the conditions of the InduRed reactor and check its suitability, a third set of experiments has been performed in the so-called InduMelt plant (sample abbreviation starts with: IM_). The InduMelt plant is a single coil induction furnace that is modeled on the InduRed reactor concept and used to perform preliminary experiments. This is due to the fact that the InduMelt plant is easier to use and requires less effort compared to the continuous InduRed reactor but still provides the same reaction conditions. The crucible concept used for these experiments is therefore based on the InduRed reactor and consists of a bed of packed graphite cubes (25 mm edge length) within an Al_2O_3 ceramics ring (70 mm radius, 100 mm height) and is shown in Figure 2a. In Figure 2b, the setup of the InduMelt plant is presented.

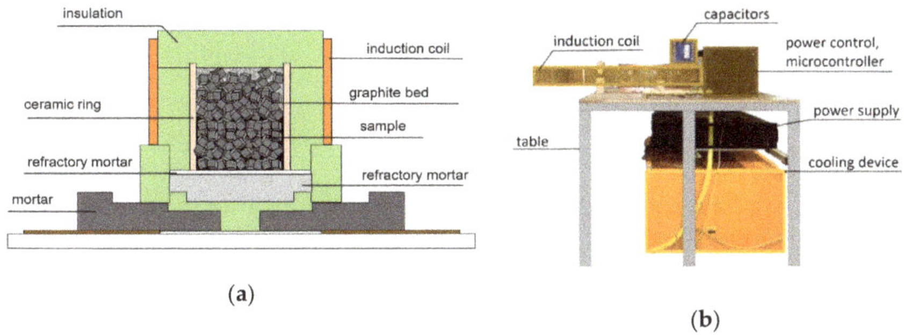

Figure 2. (**a**) Schematic illustration of the crucible concept used in the InduMelt experiments [29] and (**b**) setup of the experimental InduMelt plant.

During the preparation of the experiments, the ceramic ring is fixed on a mortar plate with refractory mortar and alternatingly filled with graphite cubes and input material. In the conducted experiments, the initial sample mass was 552.3 g for IM_NMC_C, 520.0 g for IM_NCA_C and 561.9 g for IM_AM. The filled crucible is then insulated, using 20 mm thick Cerachrome fiber wool with a classification temperature of 1500 °C and placed within the induction coil. The inductive energy input is controlled in such a way that the temperature increases at a maximum rate of 200 °C/h. For the measurement of temperatures of the reactor, two k-type thermocouples are used inside of the reactor to control temperatures up to 1200 °C. To keep track of the temperature after the k-type couples fail due to the high temperatures, two separate s-type couples are mounted on the outer wall of the Al_2O_3 ceramics ring. The temperature distribution in the reactor is known from previous experiments with other waste streams and can show a gradient of several 100 °C towards the end of the experiment, with the highest temperatures occurring at the top of the reactor. The s-type thermocouples are therefore placed at the lower third of the reactor in order to reach the necessary temperatures in the area in which the material is supposed to accumulate.

After the experiments, the reactor needs to cool down for at least 24 h before the sampling can start. Hereby, every graphite cube was picked from the reactor one after another and checked for any metal or slag depositions, which, if present, were removed from the cube's surface and collected. The difficulty to collect every little metal deposition and its influence on the overall mass balance of each experiment is discussed in the results section of this work.

However, representative samples were taken from the collected products and the content of species of interest was examined using inductively coupled plasma mass spectrometry (ICP-MS). For all ICP-MS measurements, which were conducted at the Chair of Waste Processing Technology and Waste Management at the Montanuniversitaet Leoben, the sample preparation was done by aqua regia digestion according to the ÖNORM EN 13657 standard. The measurement of the respective species was carried out according to the ÖNORM EN ISO 17294-2 standard.

3. Results

3.1. Heating Microscope

In the heating microscope experiments, the relative cross-sectional area (CSA) of the sample, thus the trend of cross sectional area of the sample cylinder during heating in relation to its initial value, was observed to investigate at which temperatures changes in the material occur. In Figure 3a, where the results of the test series with NMC are shown, one can see a significant difference between the graphs of NMC_1 and NMC_2 without carbon addition and, respectively, NMC_C_1 and NMC_C_2 in which carbon was added. In this case, the first change of the CSA for NMC_C_1 and NMC_C_2 can be observed at approx. 800 °C, which is almost 200 °C lower than in the tests without carbon addition.

Moreover, the extent to which the change occurs is significantly higher in experiments with carbon addition. The steep decline of NMC_C_1 and NMC_C_2 at approx. 1500 °C was also observed with other cathode materials and can be explained by the melting point of the contained metals. The difference in the trends of the CSA with and without carbon addition can be explained by the origin of the changes. Mao et al. [31] and Kwon and Sohn [32] investigated and described the reaction behaviour of LCO ($LiCoO_2$) with and without carbon addition. According to their findings and due to the fact, that NCA and NMC are structurally identical to LCO, we assume that the changes in experiments without carbon addition are caused by thermal decomposition of the lithium metal oxides, while in experiments with carbon addition, reduction reactions with Li_2O formation led to the observed changes. About the reproducibility it can be said that in the repeated attempts the characteristic changes of the CSA appear at the same temperatures to about the same extent.

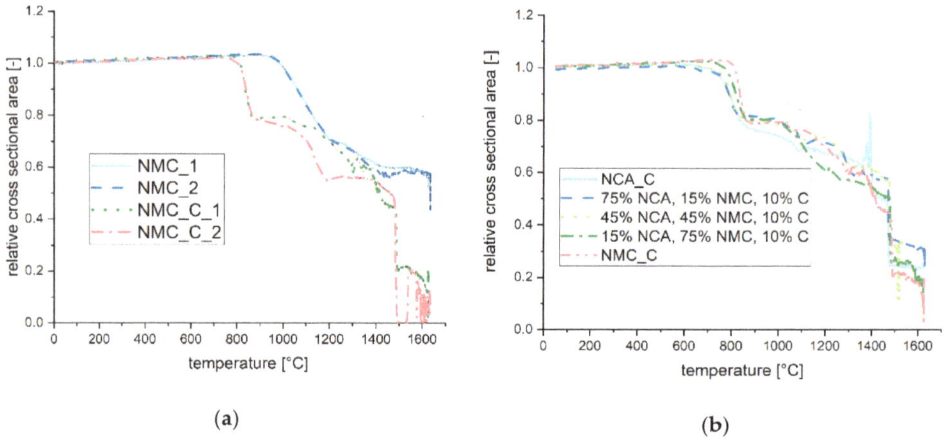

Figure 3. (**a**) Comparison of the cross sectional area of NMC with and without carbon addition in the heating microscope; (**b**) Comparison of the cross sectional area of different mixtures of NMC and NCA, each with carbon addition.

The results, mainly temperature zones and the extent of the correspondence of changes of the CSA, for NCA and NCA_C are very similar to those for NMC and NMC_C. However, since future waste streams are likely to consist of mixtures of different cathode materials, another set of experiments was performed in which NCA and NMC in different compositions and carbon were mixed to investigate if the materials influence each other. In Figure 3b, where the changes of the CSA of NCA_C, NMC_C and mixtures with varying composition are shown, no direct influence can be seen. The following Figure 4a,b show the NMC_C sample before and after the heating microscope experiment. In Figure 4b a perfectly molten metal sphere, indicated by the change of the CSA at approx. 1500 °C, and a fine white crystalline structure can be seen. The blue colour of the Al_2O_3 ceramic is most likely caused by reactions with cobalt and was also observed in all other experiments, especially in those with carbon addition.

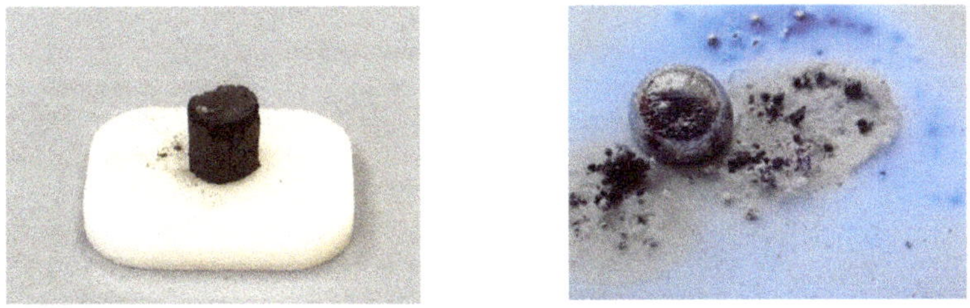

(a) (b)

Figure 4. (**a**) NMC_C sample before and (**b**) after heating to 1600 °C in the heating microscope.

In contrast, the black matter material (AM) showed a completely different behavior, as can be seen in Figure 5a,b in which its CSA does not decrease during heating but increase to almost 120% of its initial value. The lack of the first change of the AMs CSA as well as the absence of any sign of melting at temperatures around 1500 °C indicates that pre-treatment might have a big influence on basic thermophysical properties of the produced black matter.

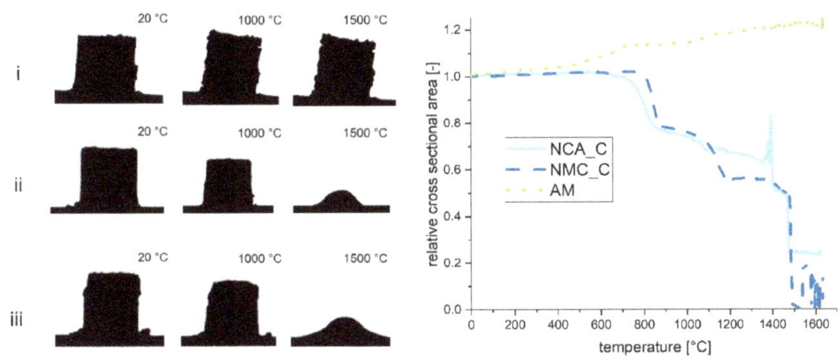

(a) (b)

Figure 5. (**a**) Images of the samples AM (i), NMC_C (ii) and NCA_C (iii) at temperatures of 20 °C, 1000 °C and 1500 °C taken during the heating microscope experiments; (**b**) Trend of the cross sectional area of the samples AM, NMC_C and NCA_C during heating in the heating microscope.

Reasons for the deviating behavior of AM compared to NMC_C and NCA_C could lie in impurities, thus residues from the mechanical processing and separation step during pre-treatment, like Cu and Al from conductor foils. A closer look at the chemical composition of AM in Table 1 reveals that the mass content of Cu and Al with almost 6% each is much higher than anticipated. Moreover, the carbon content is much higher than would be stoichiometrically necessary for the reduction reactions. An example of a disruptive reaction could be the formation of aluminum oxide which, in the appropriate amount, could form a supporting structure and thereby reduce the informative value of the CSA. On the other hand, it is also possible that the anode graphite has a lower reactivity than the fine powdered coke which is used in NMC_C and NCA_C.

The origin of AM, a pre-treatment process that uses thermal deactivation before mechanical shredding, could also cause the observed differences, since some of the reactions might already have taken place if certain temperatures are overcome during this step. By this, the layered structure of the lithium metal oxides could probably have been changed, e.g., due to thermal decomposition which, as can be seen in Figure 3, occurs at approx. 1000 °C and could change the materials properties permanently. However, reliable information about these thermal processes is difficult to access. In our opinion, however, it is quite possible that at certain points in such a process, temperatures above 1000 °C can occur and that therefore the possibility of influencing the material must not be excluded.

3.2. Simultaneous Thermal Analysis

The experiments in the heating microscope gave some first impressions on how NMC, NCA and AM behave at high temperatures and under reducing conditions. For further characterization of the underlaying reactions that cause respective changes in the materials and to create a basis for a kinetics model in the long term, thermogravimetric analysis and differential scanning calorimetry was conducted. The results of the STA are summarized in Figure 6a, showing the trends of the relative mass of the samples, and Figure 6b, which shows the corresponding trends of the heat flow. The evaluation of the measurements, which also includes a correction of the data by reference measurements, was carried out in MATLAB.

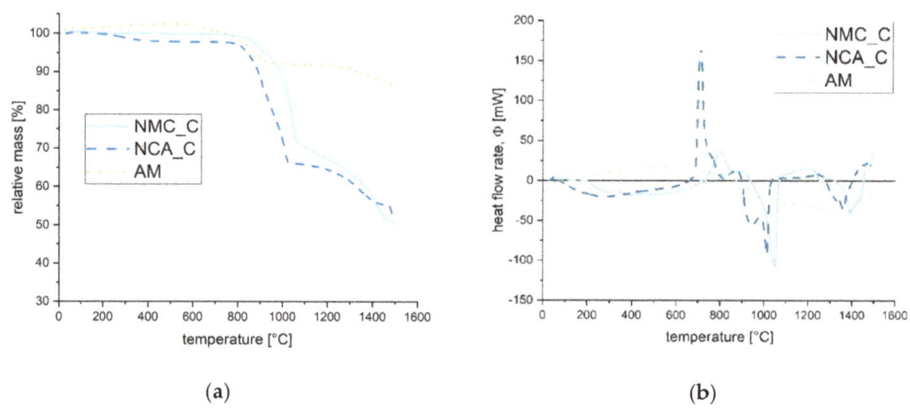

Figure 6. Results of the simultaneous thermal analyses of NMC_C and NCA_C with a heating rate of 40 K/min. (**a**) Trend of the relative mass of NMC_C and NCA_C during heating. (**b**) Trend of the heat flow of NMC_C and NCA_C during heating.

In Figure 6a the beginning of the mass loss at approximately 800 °C matches the observations from the heating microscope experiments. The first mass loss first declines slowly before it becomes steeper around 1000 °C and stops at approximately 70 % of the initial mass which was 40.1 mg for NCA_C and 39.8 mg for NMC_C. At the end of the thermogravimetric curve, the relative mass is about 55% of the initial mass. This means, that additionally to carbon, which had an initial mass content of 25 w/%, also components of the lithium metal oxide, most likely O_2 and Li, had been removed from the sample. Another indication for the presence of reduction reactions between 800 °C and 1000 °C is the trend of the heat flow, shown in Figure 6b. In both samples, the heat flow between 800 °C and approximately 1050 °C is endothermic with a negative peak around 1000 °C where also the biggest slope of the sample mass occurs. The outstanding exothermic peak in the NCA_C at 700 °C heat flow trend could be the result from Al_2O_3 formation whereby a significant amount of heat could be released. In order to confirm this, the samples must be heated in a controlled manner to or just above this temperature and analysed using XRD analysis, which is planned within the further scope of the research project.

As in the heating microscope experiments, the behaviour of the sample AM differs greatly from that of NCA_C and NMC_C. The overall mass loss only accumulates to around 10% and there are no sharp peaks in the heat flow trend. The lower mass loss is on the one hand due to the comparatively lower lithium metal oxide content (<60 w/%) compared to NCA_C and NMC_C (75 w/%) and the resulting decreased ability for CO or CO_2 generation. Since the heating rate was the same in all experiments, the less steep mass loss between 800 °C and 1000 °C and the absence of significant peaks in the heat flow trend indicate a lower reactivity of AM in general. The suspicion from the heating microscope experiments that certain reactions already took place during the thermal deactivation step has gotten stronger.

Finally, the results from the heating microscope experiments and the STA are summarized in Figure 7.

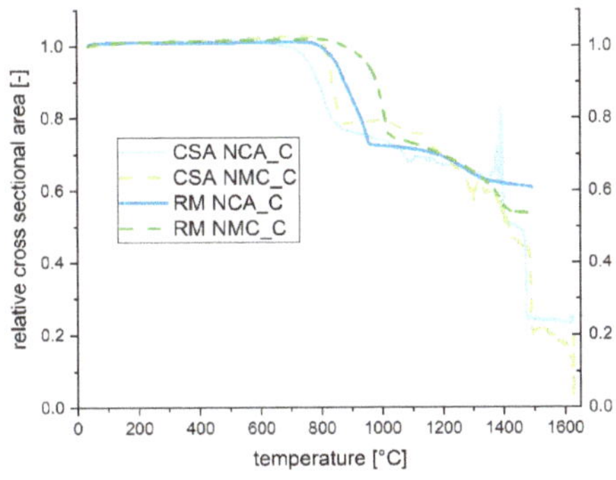

Figure 7. Trends of the cross-sectional area and the relative mass of NMC_C and NCA_C during heating in the heating microscope experiments and the simultaneous thermal analysis.

3.3. *InduMelt Experimients*

The last experimental series was conducted in the presented InduMelt reactor (Figure 2) to investigate the achievable transfer coefficients for Li, Ni, Co and Mn under the particular conditions of the reactor. The trend of the measured temperatures in- and outside of the reactor during one of the experiments is presented in Figure 8a. As explained in the materials and methods section, the slope of the outer s-type couples is used to control the temperature inside of the reactor after the operating temperature of the k-type couples is exceeded.

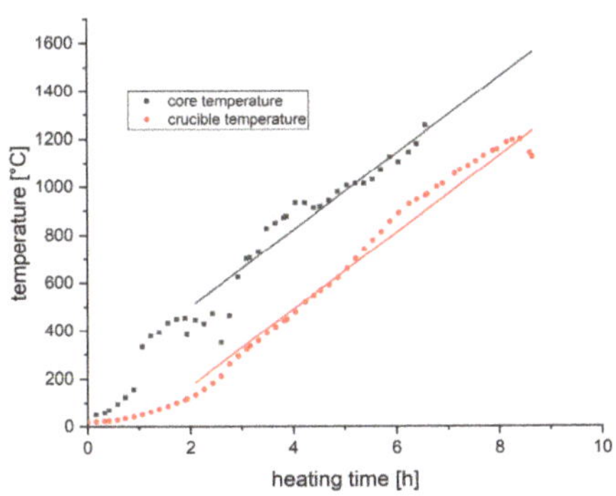

(**a**)

(**b**)

Figure 8. (**a**) Trend of reactor temperatures during IM_NMC_C. (**b**) Picture showing the crucible and the packed bed of graphite cubes with metal depositions after IM_NMC_C.

In Table 2 the compositions of the input mixtures for the InduMelt experiments are shown. For NMC_C and NCA_C the composition matches the stochiometric proportion of the used cathode materials (NCA, $LiNi_{0.8}Co_{0.15}Al_{0.05}O_2$; NMC, $LiNi_{0.33}Mn_{0.33}Co_{0.33}O_2$) with carbon addition of 10 w/%. If the whole added carbon is used and all oxides are removed the mass loss should accumulate to 40–46% of the input mass depending on the amount of Li that can be removed. For AM, which is a mixture of different cathode materials from LIBs and considering its composition most likely also other battery types, the volatile components also accumulate to around 42 w/%.

Table 2. Chemical composition of the input mixtures for the InduMelt experiments. (w/%).

Sample/Element	Li	Ni	Mn	Co	Al	O_2	C	Sum
AM	2.42	20.90	1.08	4.19	5.83	11.16 [2]	29.50	75.08 [1]
NCA_C	6.50	43.98	0.00	8.28	1.26	29.97	10.00	100.00
NMC_C	6.48	18.25	17.09	18.33	0.00	29.85	10.00	100.00

[1] Cu and other impurities are not specified here since they are not in focus of the experimental series. [2] Calculated on basis of the stochiometric Li-O_2 ratio.

Because the aim of the experimental series is to investigate possible recovery and removal rates for certain metals contained in the cathode materials, Cu and other impurities of the sample AM are not further analyzed.

For the first InduMelt experiments with LIB cathode materials and black matter a maximum temperature of approx. 1550 °C was chosen. At this temperature, no further changes of the CSA or mass during the STA and heating microscope were observed and the expected metal alloy's melting point is also some ten degrees lower. This temperature was then held for approx. one hour before the heat input was stopped.

In Figure 8b the reactor after the experiment is shown. All graphite cubes were removed and cleaned from metal and slag deposits which were subsequently weighed. The individual mass of input material and product phases for each experiment can be seen from Table 3.

Table 3. Masses of the input sample and the obtained products in InduMelt experiments. (g).

Experiment/Product	Input	Metal phase	Slag Phase	Powder	Product Sum
IM_NMC_C	552.3	244.2	37.7	11.6	293.5
IM_NCA_C	520.0	267.3	21.6	15.6	304.5
IM_AM [1]	561.9	-	-	396.1	396.1

[1] Neither metal accumulations nor slag depositions could be found.

The obtained product phases are subdivided into metal phase, slag phase and powder. On the first look at Table 3, one can see that the product distribution differs greatly between the experiments IM_NMC_C, IM_NCA_C and the experiment IM_AM. Therefore, the results are presented and discussed separately.

For IM_NCA_C and IM_NMC_C the metal and slag phase accumulates at the bottom of the reactor or can be found as depositions on the graphite cubes and the crucible. To achieve the best mass balance possible, the depositions have been rubbed of the graphite cubes and the metal particles were magnetically separated. By this, 244.2 g respectively 267.3 g of a metal product, which—if we assume that the metal phase only consists of Ni, Co and Mn—accounts for 81% and, respectively, 91% of the said metals in the input material of IM_NMC_C and IM_NCA_C. According to the oxygen potentials of the metals, the slag phase should mainly consist of Li_2O and Al_2O_3. With 37.7 g and 21.6 g of obtained slag for IM_NMC_C and IM_NCA_C compared to an input of approximately 36 g of pure Li alone one can say that this result looks promising, since the amount of oxygen—and of course Al—must also be taken into account. Furthermore, the refractory mortar and the crucible material also consist of Al_2O_3 and can take part in the reactions causing slag formation. Because this discussion is more complex than for the metal phase it will be continued later together with the chemical analyses of the phases. The powder phase of IM_NMC_C and IM_NCA_C is caused by abrasion during the removal of the small metal particles from the graphite cubes and therefore mainly consists of carbon. Summarized, the overall weight loss of IM_NMC_C and IM_NCA_C is 46.8% and, respectively, 41.4% of the input mass. If we assume that Li, O and C are the only volatile components in the input material a maximum weight loss of 47.2% for IM_NMC_C and 50.2% for IM_NCA_C is achievable. For IM_NMC_C, the obtained slag phase is shown in Figure 9a, the metal accumulation in Figure 9b.

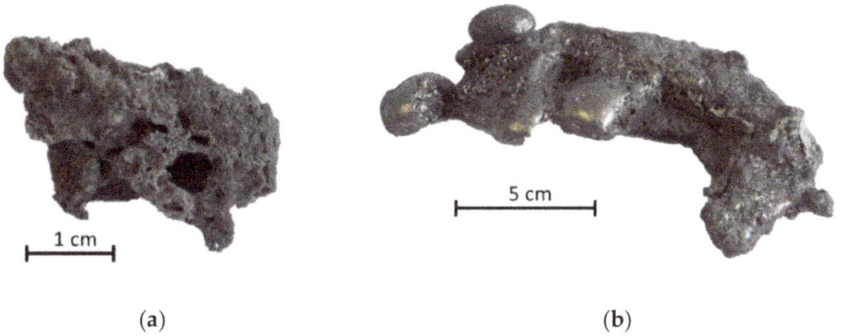

Figure 9. Obtained slag (**a**) and metal sample (**b**) from the experiment IM_NMC_C.

As can be seen, the separation of the metal and slag phase in IM_NMC_C for further chemical analysis was relatively easy since large specimens without fusions could be found. In contrast, the obtained products from IM_NCA_C were harder to separate as Figure 10a–d shows. Therefore, the ICP-MS analysis was performed for both, samples with and without inclusions, and the results weighted during data evaluation.

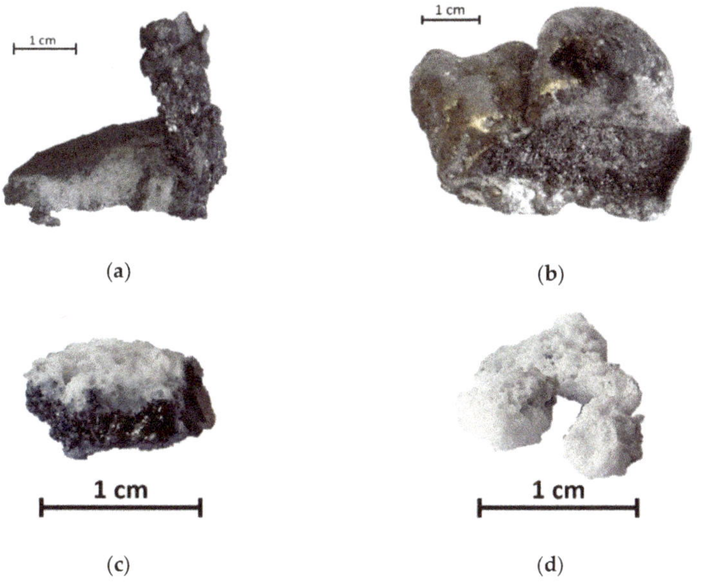

Figure 10. Obtained metal and slag samples from the IM_NCA_C experiment. (**a**) Metal sample 1 which is strongly fused with the produced slag. (**b**) Metal sample 2 with very little slag inclusions. (**c**) Slag sample 1 with metal depositions. (**d**) Slag sample 2 without inclusions or depositions.

To intensify this discussion, we need to look at the results of the chemical analysis, which were achieved by ICP-MS analysis. The discussion starts with the obtained metal phase from the experiments IM_NMC_C and IM_NCA_C for which the results are contained in Table 4.

Table 4. Mass fractions of certain metals in the obtained metal phases. (w/%).

Experiment/Species	Li	Ni	Co	Mn	Al
IM_NMC_C	0.09	38.40	36.10	28.10	0.01
IM_NCA_C_1 [2]	1.31	74.70	7.80	[1]	3.70
IM_NCA_C_2 [3]	0.05	92.20	9.89	[1]	[1]

[1] Species was not analyzed in this experiment. [2] Small inclusions of slag in the metal matrix need to be considered. [3] Slightly over-determined due to weighted consideration of residuals from the aqua regia digestion.

For IM_NMC_C the metal composition mostly matches the expected result. There is almost no Li and Al present in the metal alloy but Ni, Co and Mn. What is noticeable, however, is the significantly lower Mn content compared to Ni and Co. With an equal stoichiometric proportion and similar molecular weight—Mn is a little lighter—the difference should not be that high, which indicates that Mn also accumulates somewhere else than in the metal alloy.

As already explained, the sampling of NCA_C was not trivial due to small slag inclusions within the metal particles. In order to increase the informative value, metal samples with (IM_NCA_C_1) and without (IM_NCA_C_2) small slag particles were analyzed. By this it can be stated that also for IM_NCA_C there was hardly an accumulation of Li and Al in the metal alloy that mainly consists of Ni and Co.

A complete mass balance is hardly feasible due to the difficult collection of the small metal particles. In future experiments and respective analyses, ICP-OES as well as XRD analysis methods will be used to balance all the elements included in greater detail. Nevertheless, compared to the initial amount in the input material it was possible to find around 90% of Ni and Co and 76% of Mn in the metal phase of IM_NMC_C as well as more than 90% of Ni and Co in the metal phase of IM_NCA_C.

In order to investigate the whereabouts of Mn, to clarify whether Ni and Co can also be found in the slag and to finally check the question of whether Li removal from the reactor could be achieved or not we now look at the slag analysis shown in Table 5.

Table 5. Mass fractions of certain metals in the obtained slag phase. (w/%).

Experiment/Species	Li	Ni	Co	Mn	Al
IM_NMC_C	8.22	0.13	3.06	0.15	6.08
IM_NCA_C_1	9.85	1.31	[1]	[1]	7.45
IM_NCA_C_2	4.52	0.24	0.03	[1]	2.48

Species was not analyzed in this experiment.

Beginning with IM_NMC_C it can be said that Ni does hardly accumulate in the slag while a significantly higher but still low amount of Co could be found. For Mn, from which only 76% of its initial input were found in the metal phase, can also not be found in the slag phase. Since Mn is very reactive and has several oxidation states it is likely that parts of it were removed from the reactor via the gas phase. For IM_NMC_C, analogous to the metal phase results, there are again two samples, IM_NCA_1 with metal particles and IM_NCA_2 without metal particles. The data shows that only a small amount of Ni and Co is found in the slag while Li and Al accumulate to higher extents.

If we now compare the amount of Li that was initially inserted in the experiments, which was approx. 36 g for IM_NCA_C and IM_NMC_C with the amount of Li that was found in the metal and slag phase, a lithium removal of 96.72 w/% for IM_NCA_C and 90.76 w/% for IM_NMC_C was achieved.

Before these results are finally summarized, we have to take a look at IM_AM, which, as mentioned at the beginning, behaved differently than IM_NCA_C and IM_NMC_C. As can be seen in Table 3, neither a metal nor a slag accumulation was found but only a fine powder that was optically identical to the input material. The weight loss of 29,5% matches the initial carbon content exactly, which at first sight suggests that only the included carbon was burned in the reactor. However, analysis of the carbon content of the resulting powder revealed a mass content of still 22.6%, which indicates that also in IM_AM reduction reactions occurred. In the thermogravimetric analyses only a decrease in mass of 10% was achieved. This could be an indication that certain reactions proceed more slowly in AM and that longer holding times in the preliminary experiments would have provided better results, which is going to be investigated in the further course of the project. Furthermore, an increase of the average particle size was found that indicates at least an agglomeration of particles even if there was no molten phase. Because there was no slag or metal phase in IM_AM, the results are discussed by a comparison of the chemical composition before and after the InduMelt experiment, which is shown in Table 6.

Table 6. Chemical composition of AM before and after the InduMelt experiment. (w/%).

Experiment/Species	Li	Ni	Co	Mn	Al
AM before IM [1]	2.42	20.90	4.19	1.08	5.83
AM after IM [2]	0.77	35.00	7.00	1.76	4.12

[1] Total mass of input material: 561.9 g. [2] Total mass of product: 396.1 g.

The mass content of Ni, Co and Mn has risen by about 65% each which can only be caused by the mass loss of the sample. A statement about a possible discharge of Mn via the gas phase, as it was observed in IM_NMC_C, should not be made due to the already low concentration in IM_AM. Lithium had an input mass of 13.59 g and was reduced to 3.04 g in the product powder, which corresponds to a decrease of 77.6 w/%. This value is significantly lower than with pure cathode materials but in the light of the different behavior of AM compared to NCA_C and NMC_C in all experimental series still a promising result.

To finally summarize the InduMelt experiments, one must notice that the difficulties to achieve a complete mass balance and the absence of an off-gas analysis lead to the fact that the absolute numbers should only be considered to a limited extent. However, it is not the claim of this work to precisely define transfer coefficients for all species in cathode materials respectively black matter, but to evaluate the magnitude of possible recovery rates for the valuable metals Ni, Co, Mn and Li by using the InduRed reactor technology. In view of this, these tendencies are summarized in Figure 11.

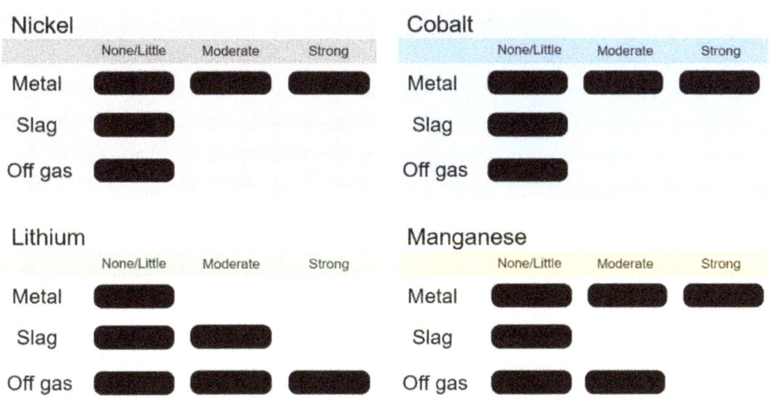

Figure 11. Qualitative consideration of the accumulation of Ni, Co, Mn and Li in the product phases obtained from the InduMelt experiments.

4. Conclusions

The literature research clearly shows that the possibility of simultaneous lithium recovery with a pyrometallurgical process would close a large gap in the recycling chain.

To evaluate if the presented InduRed technology can potentially provide a solution to this problem, a series of experiments have been conducted. By heating microscope experiments and simultaneous thermal analysis, the behavior of NCA and NMC cathode materials as well as black matter (AM) at high temperatures and under reducing conditions was investigated. The results showed that the significant reduction reaction between the lithium metal oxides and carbon take place between 800 °C and 1000 °C and that the produced metal alloy melts at approximately 1500 °C, which are technically feasible temperatures for the desired process.

Experiments, conducted in the InduMelt plant, a lab scale reactor modeled on the InduRed concept, were used to evaluate the transfer coefficients of Ni, Co, Mn and Li in qualitative terms. It was shown that Ni and Co seem to be fully recoverable by this technology while parts of manganese are removed from the reactor via the gas phase. For Li, which is considered to be the bottleneck of pyrometallurgical LIB recycling approaches, very promising results have been seen. In the InduMelt experiments with NCA and NMC more than 90%, respectively more than 75% in the experiment with black matter, of the initial Li were removed from the reactor. The fact that Li does neither accumulate in the slag nor in the metal phase indicates a high potential of the technology to enable new possibilities for Li recovery from the LIB waste stream. If Li is not obtained in small amounts in a slag phase, as in other processes, but can be collected in a separate material flow, its recovery from there can potentially be achieved with less effort and therefore represented more economically.

In order to better examine the removal of Li and Mn from the reactor, the experiments are going to be repeated using a gas vent with gas scrubbing. This should clarify in which form the Li can be obtained from the exhaust gas and how its recovery from there could be achieved. Furthermore, new cathode materials like NMC in other configurations (811, 622, 532 instead of 111) as well as LFP (lithium iron phosphate) are planned to be investigated regarding their suitability for treatment in the InduRed reactor.

The results from experiments with black matter (AM) showed some significant differences, which could partly be attributed to residues from the pre-treatment or excessively high temperatures during the thermal deactivation. Since the contrary behavior of AM in all experimental series cannot be fully elucidated with the available data, further research and experiments are necessary. In addition to that, it is planned to investigate black matter from different pre-treatment processes and the influence of interfering species like Cu or Al in general.

Author Contributions: Conceptualization, S.W.-K. and A.H.; methodology, S.W.-K.; investigation, S.W.-K., A.H. and C.P.; resources, S.W.-K., A.H. and C.P.; writing—original draft preparation, S.W.-K. writing—review and editing, S.W.-K., A.H., C.P. and H.R.; visualization, S.W.-K.; supervision, C.P. and H.R.; project administration, C.P.; funding acquisition, H.R. All authors have read and agreed to the published version of the manuscript.

Funding: This research received no external funding.

Institutional Review Board Statement: Not applicable.

Informed Consent Statement: Not applicable.

Data Availability Statement: The data presented in this study are available on request from the corresponding author.

Conflicts of Interest: The authors declare no conflict of interest.

References

1. Mizushima, K.; Jones, P.C. $LixCoO_2$ (0 <x <−1): A new cathode material for batteries of high energy density. *Mater. Res. Bull.* **1980**, *6*, 783–789.
2. Li, L.; Zhang, X. The Recycling of Spent Lithium-Ion Batteries: A Review of Current Processes and Technologies. *Electrochem. Energy Rev.* **2018**, *4*, 461–482. [CrossRef]
3. Ding, Y.; Cano, Z.P. Automotive Li-Ion Batteries: Current Status and Future Perspectives. *Electrochem. Energy Rev.* **2019**, *1*, 1–28. [CrossRef]

4. Berckmans, G.; Messagie, M. Cost Projection of State of the Art Lithium-Ion Batteries for Electric Vehicles Up to 2030. *Energies* **2017**, *10*, 1314. [CrossRef]
5. Arambarri, J.; Hayden, J. Lithium ion car batteries: Present analysis and future predictions. *Environ. Eng. Res.* **2019**, *4*, 699–710. [CrossRef]
6. Redux Smart Battery Recycling. Available online: https://www.redux-recycling.com/de (accessed on 28 December 2020).
7. Umicore Battery Recycling. Available online: https://csm.umicore.com/en/battery-recycling/e-mobility (accessed on 28 December 2020).
8. Lithium Batterie Recycling. Available online: https://accurec.de/lithium?lang=de (accessed on 28 December 2020).
9. Battery Recycling Datasheet. Available online: https://accurec.de/wp-content/uploads/2018/04/Li-ion-RE_2018.pdf (accessed on 28 December 2020).
10. Arnberger, A.; Coskun, E. Recycling von Lithium-Ionen-Batterien. In *Recycling und Rohstoffe*; Thiel, S., Thomé-Kozmiensky, E., Eds.; TK Verlag: Nietwerder, Germany, 2018; pp. 583–599.
11. Werner, D.; Peuker, U.A. Recycling Chain for Spent Lithium-Ion Batteries. *Metals* **2020**, *3*, 316. [CrossRef]
12. Liu, C.; Lin, J. Recycling of spent lithium-ion batteries in view of lithium recovery: A critical review. *J. Clean. Prod.* **2019**, *228*, 801–813. [CrossRef]
13. Mishra, D.; Kim, D.-J. Bioleaching of metals from spent lithium ion secondary batteries using *Acidithiobacillus ferrooxidans*. *J. Waste Manag.* **2008**, *28*, 333–338. [CrossRef]
14. Xin, Y.; Guo, X. Bioleaching of valuable metals Li, Co, Ni and Mn from spent electric vehicle Li-ion batteries for the purpose of recovery. *J. Clean. Prod.* **2016**, *116*, 249–258. [CrossRef]
15. Zeng, G.; Deng, X. A copper-catalyzed bioleaching process for enhancement of cobalt dissolution from spent lithium-ion batteries. *J. Hazard. Mater.* **2012**, *199*, 164–169. [CrossRef]
16. Zeng, G.; Luo, S. Influence of silver ions on bioleaching of cobalt from spent lithium batteries. *Miner. Eng.* **2013**, *49*, 40–44. [CrossRef]
17. Bahaloo-Horeh, N.; Mousavi, S.M. Use of adapted metal tolerant Aspergillus niger to enhance bioleaching efficiency of valuable metals from spent lithium-ion mobile phone batteries. *J. Clean. Prod.* **2018**, *197*, 1546–1557. [CrossRef]
18. Ghassa, S.; Farzanegan, A. Novel bioleaching of waste lithium ion batteries by mixed moderate thermophilic microorganisms, using iron scrap as energy source and reducing agent. *Hydrometallurgy* **2020**, *197*, 105465. [CrossRef]
19. Heydarian, A.; Mousavi, S.M. Application of a mixed culture of adapted acidophilic bacteria in two-step bioleaching of spent lithium-ion laptop batteries. *J. Power Sources* **2018**, *378*, 19–30. [CrossRef]
20. Vest, M. Weiterentwicklung des Pyrometallurgischen IME Recyclingverfahrens für Li-Ionen Batterien von Elektrofahrzeugen. Ph.D. Thesis, RWTH Aachen, Aachen, Germany, 28 January 2016.
21. He, L.-P.; Sun, S.-Y. Recovery of Lithium, Nickel, Cobalt, and Manganese from Spent Lithium-Ion Batteries Using l -Tartaric Acid as a Leachant. *ACS Sustain. Chem. Eng.* **2017**, *5*, 714–721. [CrossRef]
22. Elwert, T.; Frank, F. 2020, Auf dem Weg zu einem geschlossenen Stoffkreislauf für Lithium-Ionen-Batterien. In *Recycling und Sekundärrohstoffe*; Thomé-Kozmiensky, E., Holm, O., Eds.; TK Verlag: Neuruppin, Germany, 2018; pp. 525–530.
23. Huang, B.; Pan, Z. Recycling of lithium-ion batteries: Recent advances and perspectives. *J. Power Sources* **2018**, *399*, 274–286. [CrossRef]
24. Yin, H.; Xing, P. Pyrometallurgical Routes for the Recycling of Spent Lithium-Ion Batteries. In *Recycling of Spent Lithium-Ion Batteries*; Liang, A., Ed.; Springer International Publishing: Cham, Switzerland, 2019; Volume 4, pp. 57–83.
25. Gao, R.; Xu, Z. Pyrolysis and utilization of nonmetal materials in waste printed circuit boards: Debromination pyrolysis, temperature-controlled condensation, and synthesis of oil-based resin. *J. Hazard. Mater.* **2019**, *364*, 1–10. [CrossRef]
26. Beheshti, R.; Tabeshian, A. Lithium-Ion Battery Recycling Through Secondary Aluminum Production. In *Energy Technology*; Zhang, L., Jaroslaw, W., Eds.; Springer International Publishing: Cham, Switzerland, 2017; pp. 267–274.
27. Dorella, G.; Mansur, M.B. A study of the separation of cobalt from spent Li-ion battery residues. *J. Power Sources* **2007**, *170*, 210–215. [CrossRef]
28. Schönberg, A. Mathematische Modellierung metallurgischer Prozesse—Induktive Erwärmung einer Graphitschüttung. Ph.D. Thesis, Montanuniversität Leoben, Leoben, Germany, 11 December 2014.
29. Ponak, C. Carbo-Thermal Reduction of Basic Oxygen Furnace Slags with Simultaneous Removal of Phosphorus via the Gas Phase. Ph.D. Thesis, Montanuniversität Leoben, Leoben, Germany, 2 September 2019.
30. Ponak, C.; Mally, V. Phosphorus Gasification during the Reduction of Basic Oxygen Furnace Slags in a Novel Reactor Concept. *Adv. Mater. Lett.* **2020**, *11*, 20071535. [CrossRef]
31. Mao, J.; Li, J. Coupling reactions and collapsing model in the roasting process of recycling metals from $LiCoO_2$ batteries. *J. Clean. Prod.* **2018**, *205*, 923–929. [CrossRef]
32. Kwon, O.; Sohn, I. Fundamental thermokinetic study of a sustainable lithium-ion battery pyrometallurgical recycling process. *Resour. Conserv. Recycl.* **2020**, *158*, 104809. [CrossRef]

processes

MDPI

Article

Influence of Acids and Alkali as Additives on Hydrothermally Treating Sewage Sludge: Effect on Phosphorus Recovery, Yield, and Energy Value of Hydrochar

Vicky Shettigondahalli Ekanthalu [1,*], Satyanarayana Narra [1,2], Jan Sprafke [1] and Michael Nelles [1,2]

1 Department of Waste and Resource Management, Faculty of Agricultural and Environmental Sciences, University of Rostock, D-18059 Rostock, Germany; satyanarayana.narra@uni-rostock.de (S.N.); jan.sprafke@uni-rostock.de (J.S.); michael.nelles@uni-rostock.de (M.N.)
2 Deutsches Biomasseforschungszentrum GmbH (DBFZ), D-04347 Leipzig, Germany
* Correspondence: vicky.ekanthalu@uni-rostock.de

Abstract: The high moisture content present in sewage sludge hinders the use of sewage sludge in incineration or energy application. This limitation of moisture present in sewage sludge can be obviated by using the hydrothermal carbonization (HTC) process. In sewage sludge management, the HTC process requires less energy compared to other conventional thermo–chemical management processes. The HTC process produces energy-rich hydrochar products and simultaneously enables phosphorus recovery. This study investigates the influence of organic acids, inorganic acid, and alkali as additives on phosphorus transformation, yield, proximate analysis and the heating value of subsequently produced hydrochar. The analysis includes various process temperatures (200 °C, 220 °C, and 240 °C) in the presence of deionized water, acids (0.1 M and 0.25 M; H_2SO_4, HCOOH, CH_3COOH), and alkali (0.1 M and 0.25 M; NaOH) solutions as feed water. The results show that phosphorus leaching into the process-water, hydrochar yield, proximate analysis, and the heating value of produced hydrochar is pH- and temperature-dependent, and particularly significant in the presence of H_2SO_4. In contrast, utilization of H_2SO_4 and NaOH as an additive has a negative influence on the heating value of produced hydrochar.

Keywords: hydrothermal carbonization; sewage sludge; phosphorus recovery; hydrochar; process-water; pH

Citation: Shettigondahalli Ekanthalu, V.; Narra, S.; Sprafke, J.; Nelles, M. Influence of Acids and Alkali as Additives on Hydrothermally Treating Sewage Sludge: Effect on Phosphorus Recovery, Yield, and Energy Value of Hydrochar. *Processes* **2021**, *9*, 618. https://doi.org/10.3390/pr9040618

Academic Editor: Aneta Magdziarz

Received: 1 March 2021
Accepted: 29 March 2021
Published: 31 March 2021

1. Introduction

The management of sewage sludge produced from wastewater treatment plants is an important global issue due to the presence of high moisture content, harmful pathogens, and poor dewaterability. Conventional sewage sludge management involves the direct application on farmland as fertilizer. However, sewage sludge has attracted greater attention as a feedstock for nutrient recovery and renewable biofuels production [1,2]. In the year 2018, about 23% of sewage sludge produced in Germany was managed by applying directly on farmland, and about 65% of the produced sewage sludge was incinerated [3]. Since 2017, new regulation was placed by the German sewage sludge ordinance (AbfKlärV) based on enabling principles of the Circular Economy Act [4] on sewage sludge management. This new regulation is not only making it mandatory to recover phosphorus from sewage sludge in Germany but also prohibits the direct use of sewage sludge on farmland [5]. According to AbfKlärV, sewage sludge must undergo mandatory phosphorus recovery if the phosphorus content is ≥20 g/kg total dry matter (DM) or ≥2% DM. The thermal pretreatment of sewage sludge is still possible; however, the subsequent recovery of phosphorus in the produced incinerated ash or the carbonaceous residue has to be guaranteed. This new obligation applies from January 2029 for the wastewater treatment plants with size >100,000 populations equivalent (PE). The treatment facilities with >50,000 PE must also comply with the new regulation to recover phosphorus from January 2032. After these dates, soil-related disposal of sewage sludge is no longer permitted. Only the smaller wastewater treatment plants ($\leq$50,000 PE) can use their sewage sludge as a soil amender when the phosphorus content in sludge is <20 g/kg total DM.

The option for managing sewage sludge is getting progressively limited as the result of strict environmental legislation placed by Germany over the last decade. The actual problem exists in states of Germany such as Mecklenburg–Western Pomerania, Lower Saxony, and Rhineland–Palatinate where >50% of the produced sewage sludge is managed by direct application for agriculture and landscaping [3]. In this concern, the current number of thermal treatment facilities might not be able to handle the amount of sludge produced, eventually triggering the new problem associated with the storage of untreated sewage sludge.

Incineration is a widely accepted technique to treat sewage sludge in Germany and can significantly reduce the sludge volume and produce reactively hygiene sludge ash residue with high phosphate content [6]. Yang et al. (2019) studied the effect of chlorine-based additives on phosphorus recovery during sewage sludge incineration. The addition of Magnesium chloride ($MgCl_2$) and Calcium chloride ($CaCl_2$) during incineration increases the fixation rate of total phosphorus (TP) to a maximum of 98.5% in sewage sludge treated with 3% (Magnesium) Mg at 900 °C and 97.8% in sewage sludge treated with 5% Calcium (Ca) at 800 °C. Similar to incineration, pyrolysis can also be an effective alternative process to treat and recover phosphorus from sewage sludge. The study by Atienza–Martínez et al. (2014) indicated phosphorus recovered from pyrolysis is temperature-dependent and that more than 90% of phosphorus can be recovered using pyrolysis followed by char combustion in sulfuric acid.

In recent years, hydrothermal carbonization has gained greater attention for treating sewage sludge as it is greatly regarded as an eco-friendly and promising technology. Hydrothermal carbonization is a technology that demonstrates the high potential to treat the moist/wet biomass without it having to be dewatered. HTC uses moisture present in sludge as the reaction medium to process the sewage sludge without pre-drying. During HTC, higher temperature and pressure will aid the moisture present in sewage sludge to serve as a solvent, reactant, and catalyst for converting sewage sludge into hydrochar. The product hydrochar is hygienic, essentially free of pharmaceuticals, easily dewaterable, and likely to be a coal-like product with high energy density [7,8]. Additionally, the HTC process is proved to save up to 53% thermal energy and 69% electrical energy compared to conventional sludge drying methods [9]. Currently, there are several studies on HTC of sewage sludge, particularly in producing hydrochars and subsequent utilization of hydrochar as effective adsorbents [8,10], soil amendment [11], or as feedstock for energy production [1,12]. Further, there are few studies that explain the effects of additives during HTC of lignocellulosic biomass [13–15]. The investigation carried out by Lynam et al. (2012) identifies the benefit of the increased heating value of resulting hydrochar produced using Ca salts (Ca chloride and Ca lactate) as additives during HTC of lignocellulosic biomass.

Previous studies on understanding the effect of feed-water pH during HTC of sewage sludge have mainly focused on investigating phosphorus transformation [6,16,17], risk of heavy metals in hydrochar [18], physicochemical properties of hydrochar [19]. Shi et al. [17] studied the effect of initial pH and HTC reaction temperature on the mobilization of phosphorus. They found that at higher temperatures phosphorus is more likely to be more present in the hydrochar; however, using acid additives in large amounts could shift phosphorus into the liquid phase. Further, the phosphate leached into the process-water can be chemically precipitated by the addition of a coagulant and a mixing of process-water and coagulant. The multivalent metal ions most commonly used are calcium, aluminium, and iron [20,21]. To the authors' knowledge, very few investigations compare the influence of organic acids, inorganic acids, and alkali additives on the HTC of sewage sludge. Wang et al. (2017) investigated the influence of feed water pH (altered by the addition of acetic acid or sodium hydroxide) on phosphorus transformation during HTC of sewage sludge.

The results showed that during the HTC of sewage sludge, metal cations and pH played vital roles in the transformation of phosphorus. The observation made by Ekpo et al. (2016) shows that 94% of the phosphorus in feedstock was recovered into the process-water after hydrothermal treatment of pig manure with a sulfuric acid additive at 170 °C. Reza et al. (2015) [15] studied the influence of using acid and alkali additives on the HTC of wheat straw. However, the behavior of phosphorus transformation greatly differs from the types of biomass, and HTC process conditions and techniques.

The main purpose of this study is to investigate and compare the influence of organic acids (acetic acid and formic acid), inorganic acids (sulfuric acid), and alkali (sodium hydroxide) as additives on the hydrothermal treatment of sewage sludge. Despite the primary objective of this study being to understand the influence of different additives on the P transformation during HTC of sewage sludge, this study also shed light on the effect of additives on dewaterability, yield, and heating value of hydrochar.

2. Material and Methods

2.1. Material

The sewage sludge used in this study was obtained directly from the wastewater treatment plant, Rostock, Germany. The central wastewater treatment plant in Rostock treats both industrial (1/3) and municipal wastewater (2/3) with the capacity to treat wastewater from 320,000 inhabitants [22]. The freshly digested and dewatered sludge was collected in an airtight specimen container and transported immediately to the laboratory. The sewage sludge, after being received in the laboratory, is refrigerated at 4 °C before use. The refrigerated representative samples were directly taken for HTC investigation and respective additives of deionized water, organic acid, inorganic acids, and alkali with known concentration were added and mixed to make a homogeneous slurry. The respective additive solution of 0.1 and 0.25 M concentration was prepared by diluting acetic acid (100%, p.a.), formic acid (≥98.0%, p.a.), sulfuric acid (1 M), and sodium hydroxide (≥97.0% (T), pellets) in the deionized water. The produced additive solution was used on the same day of preparation. The ultimate analysis of the sewage sludge was performed using an organic elemental analyzer by following EN ISO 16948, 2015. Proximate analysis was performed using a LECO Thermogravimetric Analyser (TGA) unit TGA701 to determine moisture content, volatile organic compound, fixed carbon and ash content. The heating value of the sewage sludge and resulting char was determined by Parr 6400 calorimeter (Parr Instruments Inc., Moline, IL, USA) following the method described in EN 14918, 2010. Total phosphate in the obtained sewage sludge was analyzed in an external laboratory following the method described in EN ISO 11885, 2009. All measurements were made in duplicate, and the mean value is reported.

2.2. Hydrothermal Carbonisation Treatment

Hydrothermal carbonization of sewage sludge was carried out in a Parr 4523 reactor (Parr Instrument (Deutschland) GmbH, Zeilweg 15, Frankfurt, Germany) at an autogenic pressure. The processing unit 4523 consists of a reaction vessel of 1-L capacity that can withstand a maximum pressure of 138 bar, a heating jacket equipped with a 2 kW heating coil, a temperature and a pressure sensor, and a stirrer with an attached motor. The reactor temperature and the speed of the stirrer were controlled using a Parr 4848 PID reactor controller. Figure 1 provides an overview of the experimental methodology. The analysis was carried out by charging the reactor with 297.00 g raw sewage sludge (23.5% DM) and it was topped up with 402.00 g of deionized water or additive solution of acetic acid, formic acid, sulfuric acid, or sodium hydroxide in 0.1 M or 0.25 M concentration. The sewage sludge and additives were mixed homogeneously inside the reactor before starting the investigation, and the initial pH of mixed feedstock slurry was noted using WTW pH 3310 m. The defined ratio of sewage sludge to additives was used to produce a homogeneous slurry of 10% DM. The mixture was hydrothermally carbonized at autogenic pressure with a constant heating rate of 4 K/min. The investigation was carried out with the varying temperature of 200 °C, 220 °C, and 240 °C for a retention time of 2 h while keeping the stirrer switched on during the entire process. Later, the reactor was allowed to cool down to room temperature without any additional cooling mechanism.

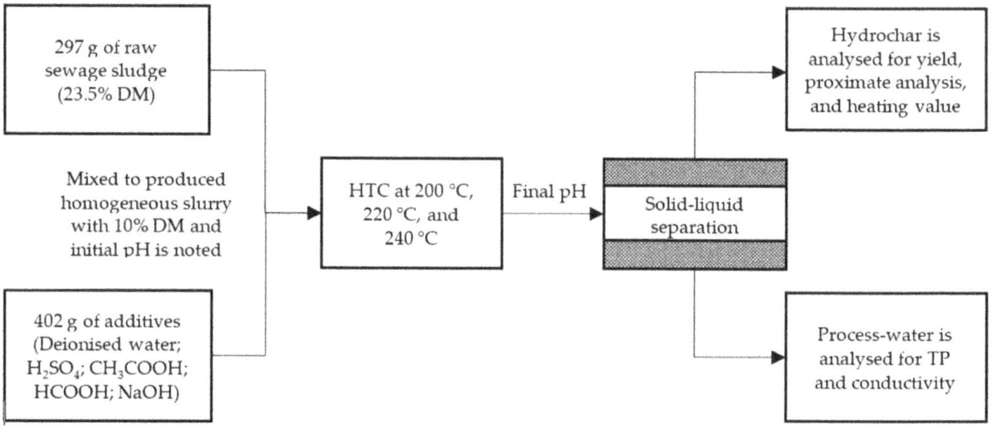

Figure 1. Schematic representation of experimental methodology.

2.3. Product Recovery and Analysis

The final pH of the HTC-slurry obtained after HTC of sewage sludge was noted, and the resulting hydrochar and process-water were separated by using a vacuum filtration apparatus. Vacuum filtration was carried out at the constant process conditions using a top-feeding procedure in a Büchner funnel. The following rules were kept constant for solid-liquid separation of the HTC-slurry and analyzing the dry matter concentration of hydrochar: (1) entire content HTC-slurry after carbonization was poured into the Büchner funnel, (2) the vacuum pump was switched on to generate the vacuum pressure for solid–liquid separation, (3) the solids (hydrochar) thus obtained using vacuum filtration were oven-dried at 105 °C for 24 h and stored in sealed containers for further analysis or usage. Similarly, process-water produced after filtration was collected and stored in a volumetric flask and refrigerated at 4 °C until it was analyzed for total phosphorus (TP) concentration and conductivity.

The yield of the produced hydrochar is calculated as explained in Equation (3). The lower heating value (LHV) of the hydrochar was determined in a similar way to sewage sludge using a Parr 6400 calorimeter following the method described in EN 15170, 2010. TP in the process-water was analyzed spectrophotometrically after acid hydrolysis and oxidation following EN ISO 6878, 2004. The conductivity of the process-water was measured using a Hach HQ 40 d multifunction meter. By determining the conductivity, it was possible to understand the variations of salt content in the process-water produced at different process parameters. Triplicates of all analyzed results were obtained and the mean value was reported.

2.4. The Fraction Phosphorus Recovered on Hydrochar

The TP recovered on the hydrochar was mathematically calculated using the experimental data obtained on TP concentration in sewage sludge and process-water, and the total yield of the hydrochar after HTC. TP recovered from the hydrochar ($TP_{(h)}$) can be mathematically defined as follows:

$$TP_{(h)} = 100 - \left(\frac{Y_{(pw)} \times TP_{(pw)}}{TP_{(fs)}} \times 100 \right) \quad (1)$$

where $TP_{(pw)}$ and $TP_{(fs)}$ is the TP content in process-water and the initial feedstock slurry, respectively, and $Y_{(pw)}$ is the total yield of the process-water after filtration, which is calculated as shown in Equation (2):

$$Y_{(pw)} = \mathrm{m} - \left(\frac{m_o \times Y_{(h)}}{DM_{(h)}} \right) \quad (2)$$

where m is the total weight of the feedstock, m_o is the dry weight of the feedstock, $DM_{(h)}$ is the dry matter percentage of hydrochar after filtration, and $Y_{(h)}$ is the yield (%) of produced hydrochar and was calculated as the applied formula.

$$Y_{(h)}(\%) = \frac{m_h}{m_o} \times 100 \quad (3)$$

where m_h is the total dry weight of produced hydrochar.

3. Results and Discussion

3.1. Characteristic of Sewage Sludge

The results of the proximate and ultimate analysis of sewage sludge are presented in Table 1. The moisture content of sewage sludge was determined to be 76.53%, leaving behind the total solids content of 23.48%. The analysis also demonstrates noticeably lower ash content of 32.83% DM and higher volatile solids (VS) of 61.46% DM, which was inconsistent with the previous investigation ranges [6,23]. The ultimate analysis of the sewage sludge specified the typical C-H-N-S-O content for sewage sludge in Germany [24] with C: 32.5; H: 5.0; N: 4.98; S: 1.50; and O: 21.4 on a dry basis. The dry sewage sludge is known to contain a higher concentration of phosphorus and a relatively higher heating value. The TP content in the feedstock was determined to be 36.1 g/kg, accounting for 3.6% of total dry sludge, and the heating value was observed to be relatively higher with 13.56 MJ/kg (LHV) in comparison with previous studies [6,23,24]. One possible explanation for increased LHV can be the presence of higher volatile solids and lower ash content. Nevertheless, the overall characteristics of the feedstock have the typical composition of sewage sludge in Germany.

Table 1. Proximate and ultimate analysis of sewage sludge.

Parameters	Units	Value
Moisture content	%OS	76.53
Total solids	%OS	23.48
Volatile solids	%DM	61.46
Ash (850 °C)	%DM	32.83
Fixed carbon (FC)	%DM	5.71
LHV	MJ/kg, DM basis	13.56
HHV	MJ/kg, DM basis	14.66
Nitrogen	%DM	4.98
Carbon	%DM	32.5
Sulphur	%DM	1.50
Oxygen	%DM	21.4
Hydrogen	%DM	5.00
Total phosphorus	mg/kg DM	36,100.00

OS = Original Sample, DM = Dry Matter.

3.2. Effect of Additives and Reaction Temperature on Yield of Hydrochar

Figure 2 compares the total yield (%) of hydrochar produced at different temperatures using various additives. An increase in the process temperature from 200 °C to 240 °C has decreased the hydrochar yield on average by about 10%, which agrees with the earlier investigation results demonstrating a decrease in hydrochar yield with an increase in reaction temperature [25,26]. The maximum hydrochar yield was observed with the carbonization method using inorganic acid as additives in comparison with the carbonization method using organic acid, alkali, and deionized water as additives. The maximum hydrochar yield of 69.09% was achieved using a 0.25 M H_2SO_4 additive in feedstock (pH 3.78), with the carbonization temperature of 200 °C and 2 h retention time. In contrast, the same reaction temperature and retention time, using 0.25 M CH_3COOH (pH 5.44), HCOOH (pH 5.38), NaOH (pH 10.68) and deionized water (pH 7.8) additives has resulted in the hydrochar yield of 62.74%, 63.79%, 55.47%, and 59.66%, respectively. Nevertheless, it is interesting to see that at the lower additive concentration (0.1 M), despite having comparatively similar initial pH range (5.8–6.3) of sewage sludge slurry prepared using CH_3COOH (pH 6.3), HCOOH (pH 6.2), and H_2SO_4 (pH 5.8), hydrochar yield was significantly higher with using H_2SO_4 as an additive in comparison with other organic acids.

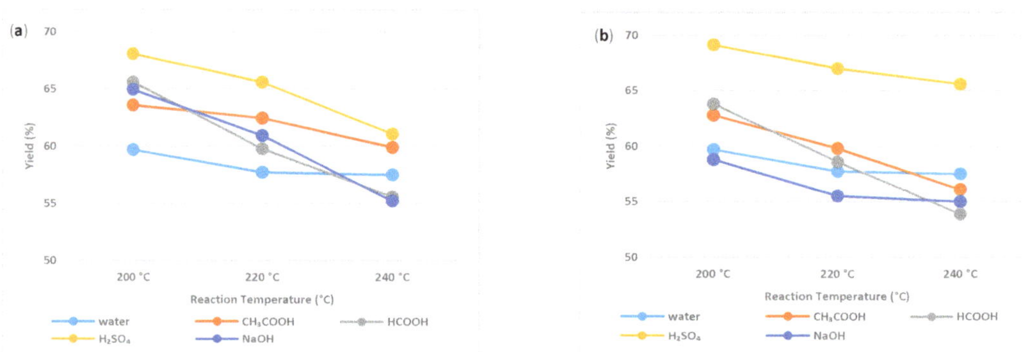

Figure 2. Influence of additives and temperature on yield of hydrochar; (**a**) represents the hydrochar yield produced using additives of 0.1 M concentration; (**b**) represents the hydrochar yield produced using additives of 0.25 M concentration.

The increases in reaction temperature will directly influence eliminating the moisture content in the biomass structure as the effect of hydrolysis reaction and simultaneously foster biomass degradation; this, in turn, decreases hydrochar yield [26,27]. Further, the investigation conducted by Jaruwat et al. (2018) has shown that a longer retention time will increase the yield of the hydrochar as the result of repolymerisation of decomposed biopolymers.

Similar to reaction temperature and retention time, the addition of additives also influences the yield of the hydrochar. Temperature undoubtedly has a greater influence on the mass yield of hydrochar. Nevertheless, similar to temperature, despite having similar pH, retention time, and reaction temperature, using inorganic acid has increased the hydrochar yield in comparison to using organic acids or alkali as additives. An increase in the hydrochar yield can be co-related with the higher molecular mass of H_2SO_4 and changing pH due to strong acid additive utilization in comparison with the utilization of CH_3COOH, HCOOH, NaOH, and deionized water as an additive.

3.3. Effect of Additives and HTC Process Conditions on Solid–Liquid Separation

The dry matter concentration of the various hydrochar residue after filtering the process-water using a vacuum filter at the constant process conditions (top-feeding procedure with a Büchner funnel) is depicted in Figure 3. The HTC treatment was advantageous to sludge dewatering. The dry matter concentration of hydrochar residue after solid–liquid separation increased significantly after the HTC reaction and the use of H_2SO_4 as an additive, significantly favored dewatering. When 0.25 M H_2SO_4 solution was used as an additive, the dry matter of hydrochar residue was 27.68–31.75%, which was significantly higher in comparison with using deionized water as an additive (20.70–24.83%). The influence of H_2SO_4 in enhancing the dewaterability of sewage sludge has also been explained previously [3]. The use of organic acids as an additive did not show any greater difference in the dry matter of hydrochar residue (20.68–26.38%) in comparison with using deionized water as an additive. In contrast, the use of NaOH as an additive had considerably decreased the dry matter of hydrochar residue (1.28–16.51%) at the lower reaction temperature (200 °C); however, at the higher reaction temperatures (220 °C and 220 °C), dry matter of hydrochar residue was higher (27.27–28.82%).

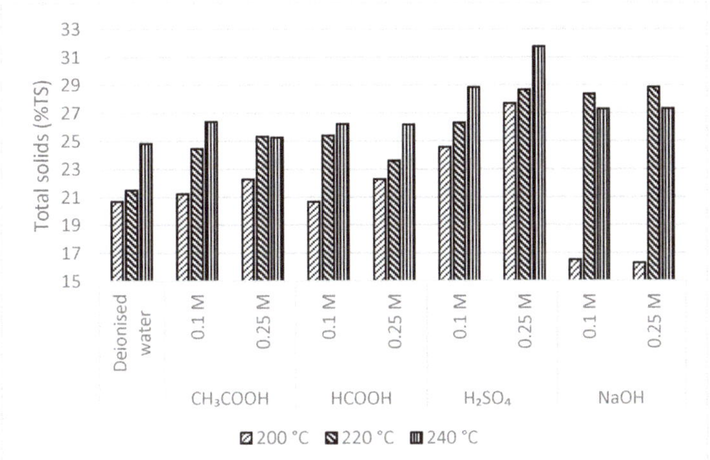

Figure 3. Dry matter concentration of hydrochar residue after filtering the process-water.

The extracellular polymeric materials in the sewage sludge contain viscous protein material that is extremely hydrophilic [28]. The effective way to enhance the sludge dewatering performance is by breaking the cell wall and destroying the sludge flocs to release and hydrolyze the organic matter present in sewage sludge. This phenomenon can be effectively achieved alongside the higher temperature and pressure that occur in the HTC process. The reduction in the binding force of the sludge particles achieved during the HTC process improves the dewatering performance after HTC and is significantly enhanced using H_2SO_4 in the reaction medium. In contrast, the use of NaOH additive at lower reaction temperature (200 °C) was not effective in hydrolyzing the organic matter. This could have influence in retaining of the viscous protein material in sewage sludge, making the HTC-slurry hard to dewater. Nevertheless, using NaOH additive at higher reaction temperatures (220 °C and 240 °C) was effective in hydrolyzing the organic matter present in sewage sludge.

3.4. Effect of Additives on Hydrochar Properties: Proximate Analysis and Heating Value

Following HTC, sewage sludge was carbonized into a brownish-grey solid hydrochar with a nutlike smell. The physical appearance of produced hydrochar implied that hydrochar had a uniform composition and could be readily molded into dense pellets. The proximate analysis and LHV were determined to understand the fuel characteristics of the produced hydrochar. Table 2 represents the results comprising volatile matter, ash content, fixed carbon, and LHV of various hydrochar produced at different process conditions. The hydrochar produced using various additives in this investigation had LHV in the range of 14.24–15.63 MJ/kg, which is similar to the results of earlier studies demonstrating fuel characteristics of hydrochar produced using sewage sludge [1].

The breaking down of biomass at higher temperatures to influence aromatization, polymerization, and condensation to produce hydrochar can be a reason for the increase in fixed carbon content (FC) with increasing reaction temperature [29]. Fixed carbon can be defined as combustible residue present in the char after the volatile matter burned. In general, biomass before carbonization contains high VS content and low FC, but high moisture content [30]. Previously, several studies showed a strong correlation between FC content and calorific value; an increase in the FC content in char can directly increase the heating value of the char [30,31]. The use of H_2SO_4 and NaOH as an additive has negatively influenced the LHV of the produced hydrochar in comparison with the hydrochar produced using organic acids and deionized water as an additive.

Table 2. Proximate analysis and heating value of hydrochar produced using various additives and reaction conditions. (AC-Additive concentration; RT-Reaction temperature; Initial and final pH represents the pH of feedstock slurry before and after HTC. All hydrochars were produced at 2 h retention time.)

Sample Description			pH		Proximate Analysis (wt% Dry Basis)			LHV
Additive	AC	RT (°C)	Initial pH	Final pH	VS	Ash	FC	MJ/kg
Deionized water	-	200	7.8	6.6	49.29	41.89	8.83	15.03
		220	7.9	6.7	44.52	46.55	8.94	15.40
		240	7.8	6.9	47.10	44.31	8.59	15.16
CH_3COOH	0.1 M	200	6.32	6.16	49.77	41.73	8.51	15.02
		220	6.39	6.28	44.97	46.58	8.46	15.41
		240	6.32	6.73	46.24	45.14	8.63	15.20
	0.25 M	200	5.44	5.3	50.45	40.89	8.66	15.50
		220	5.63	5.66	47.25	43.92	8.84	15.56
		240	5.71	5.7	44.74	46.34	8.93	15.75
HCOOH	0.1 M	200	6.23	6.48	49.38	42.94	7.68	15.37
		220	6.22	6.83	44.26	47.57	8.17	15.63
		240	6.14	7.83	46.62	44.90	8.48	14.78
	0.25 M	200	5.38	5.46	52.52	39.76	7.72	15.60
		220	5.45	6.15	48.08	43.49	8.44	15.37
		240	5.28	6.8	44.62	47.35	8.03	14.99
H_2SO_4	0.1 M	200	5.85	6.15	51.59	39.62	8.80	14.70
		220	6.01	6.39	48.90	42.27	8.84	15.19
		240	5.7	6.84	46.84	45.76	7.41	14.83
	0.25 M	200	3.78	4.32	52.37	42.51	5.13	14.24
		220	3.47	4.77	51.09	44.50	4.41	14.32
		240	3.81	5.64	49.16	45.57	5.27	14.68
NaOH	0.1 M	200	9.88	6.92	50.04	43.10	6.87	14.77
		220	9.94	8.4	45.21	48.40	6.40	14.50
		240	9.9	7.02	43.74	47.89	8.37	14.84
	0.25 M	200	10.68	7.5	45.67	48.53	5.81	14.24
		220	10.98	7.7	43.66	49.46	6.88	14.46
		240	10.76	7.9	42.18	50.72	7.11	14.90

The hydrochar produced using organic acids and deionized water as an additive increased the FC content (7.72–8.94%) in comparison to the FC content of the initial feedstock (5.71%). Here, the increase in the FC content in hydrochar might have influenced increasing the LHV (14.99–15.75 MJ/kg). However, the use of H_2SO_4 as an additive at higher concentrations (0.25 M) negatively influenced the FC content of produced hydrochar (4.41–5.27%). Similarly, the use of NaOH as an additive at lower reaction temperatures (200 °C and 220 °C) had no noticeable influence on FC% (5.81–6.87%) in comparison with initial feedstock. On other hand, it was also observed that there was a significant decrease in the VS content and increase in the ash content after HTC of sewage sludge. The decrease in VS content can be attributed to the reaction severity and dissolution of organic material into the liquid phase, and an increase in the ash content can be correlated with the decrease in the mass percentage of VS composition of the hydrochar. However, it is interesting to perceive that the ash content in the hydrochar produced using H_2SO_4 (0.25 M) as an additive is offset more by decreasing FC content than VS content; similar phenomena can also be seen with the hydrochar produced using NaOH as an additive at 200 °C and 220 °C. The decrease in the FC content with the use of H_2SO_4 at higher concentrations and NaOH at lower reaction temperatures (200 °C and 220 °C) as an additive can explain the lower LHV in the respectively produced hydrochars.

3.5. Conductivity of Process-Water

Figure 4 is the graphical representation of the conductivity of the process-water produced using different acids and alkali as additives. The conductivity measurement, in general, provides a reliable means to understand the ion concentration of a solution. The maximum conductivity of 24.10 μS/cm was observed in the process-water produced using alkali additive. Among acid-based additives, the utilization of inorganic acid (H_2SO_4) as an additive had process-water with higher conductivity (17.4–20.12 μS/cm) in comparison with organic acid additives (CH_3COOH and HCOOH).

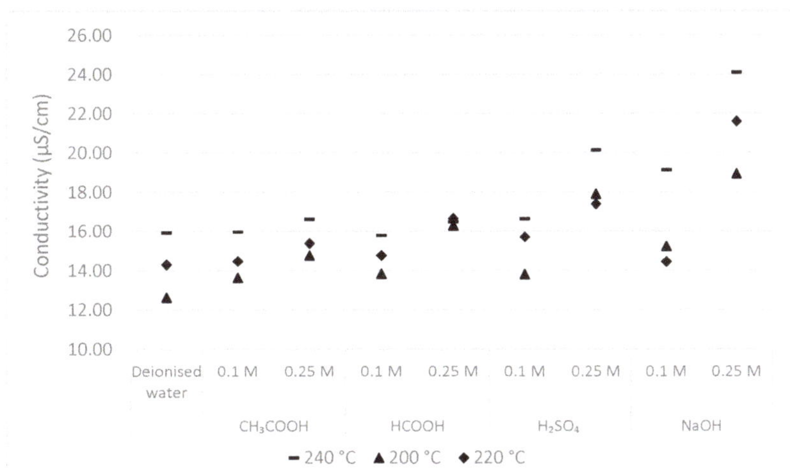

Figure 4. Conductivity of process-water produced using acids and alkali additive at different concentration.

3.6. Effect of HTC Organic Acids, Inorganic Acids, and Alkali Additive on P-Transformation

3.6.1. The pH of Feedstock Slurry, before and after HTC

Table 2 depicts the pH of the feedstock slurry before and after HTC at various temperatures, additives, and additive concentrations. The HTC process comprises hydrolysis, dehydration, decarboxylation, aromatization, and condensation polymerization [32]. During HTC, the pH of the feedstock slurry decreases as the result of the degradation of macromolecular organic matter into an acidic substance (viz., volatile fatty acids) and subsequent dissolution into the liquid phase. Further, the reaction time and temperature also influence the pH of the sludge hydrolysate. The use of organic acids and inorganic acid as additives resulted in the feedstock initial pH between 3.4 and 6.4. In contrast, the use of NaOH additive resulted in the feedstock initial pH ranging from 9.9–11.0. In the baseline condition, the deionized water additive has an initial pH of 7.8. The final pH represents the pH of the feedstock slurry after HTC. The experimental observation demonstrates that, regardless of variations in the initial pH, the final pH value after HTC always tends to move towards neutral. The obtained results were consistent with an idea that the acids formed during hydrolysis were subsequently decomposed or repolymerized at a higher temperature, which influences the pH of feedstock slurry after HTC [28]. Further, it is also possible that the buffering function of the sewage sludge might have a significant effect on the final pH. The obtained results of the shift in final pH towards neutral agree with several earlier HTC studies carried out on sewage sludge [6], swine manure [33], and wheat straw [15].

3.6.2. Effect of Additives on Phosphorus Transformation

Figure 5 shows the concentration of TP in the process-water produced after the HTC of feedstock slurry at various temperatures. For each experiment, TP in the process-water was analyzed spectrophotometrically after acid hydrolysis and oxidation of the process-water sample. Further, the TP in the hydrochar was calculated mathematically using Equation (1). Figure 6 depicts the influence of additives, additive concentration, and pH of the feedstock slurry on the recovering TP from the raw feedstock into hydrochar after HTC at various temperatures. In brief, the results show that even at a similar pH, higher leaching of TP into process-water is achieved by the utilization of inorganic acid (H_2SO_4) as an additive in comparison with organic acids.

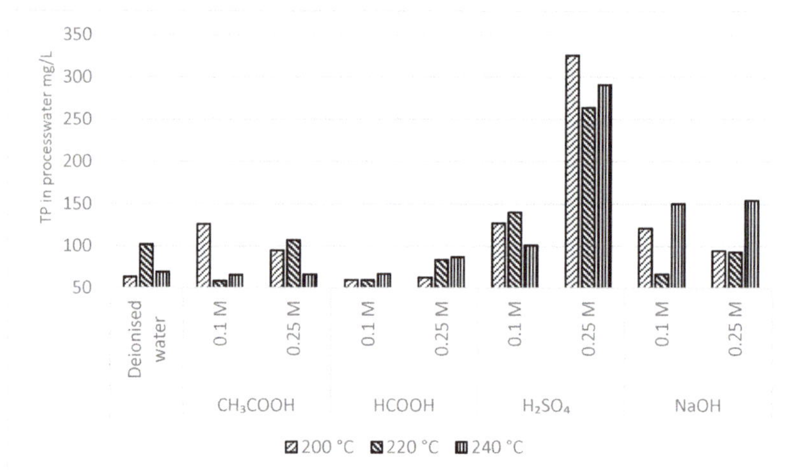

Figure 5. TP concentration of the process-water.

Following the HTC of sewage sludge, the highest TP leaching into the process-water (326 mg/L) was observed by using H_2SO_4 as an additive (pH 3.78), which represents about 93% of TP being recovered from raw feedstock into consequently produced hydrochar. Irrespective of process temperature, using deionised waste as an additive did not have any significant influence on the TP leaching. TP leaching into process-water using deionized water as an additive was observed to be 63–101 mg/L, which represents about 97.7–98.7% of TP being recovered from raw feedstock into consequently produced hydrochar. The TP concentration in the process-water following the treatment at various temperatures and organic acid additives—formic and acetic acid—was observed to be in the range of 58.3–126 mg/L and 59.3–86.6 mg/L, respectively. Likewise, using NaOH as an additive also had comparatively similar TP leaching (66.2–154 mg/L) into the process-water after HTC at various temperatures. The obtained results suggested that organic acids and alkali had a very limited impact on extracting TP from raw feedstock into the process-water, which agrees with the similar results demonstrated by earlier studies [6,33].

During HTC, the extraction of phosphorus into the process-water was generally lower with the utilization of organic acids as additives in comparison with an inorganic acid, regardless of temperature. An increase in H_2SO_4 additive concentration from 0.1 M to 0.25 M increased the TP leaching into process-water by about 3-fold from 101–127 mg/L to 264–326 mg/L, respectively. However, increasing the concentration of organic acid additives from 0.1 M to 0.25 M obviously decreased the pH of the resulting feedstock slurry, but it did not greatly influence TP leaching into process-water.

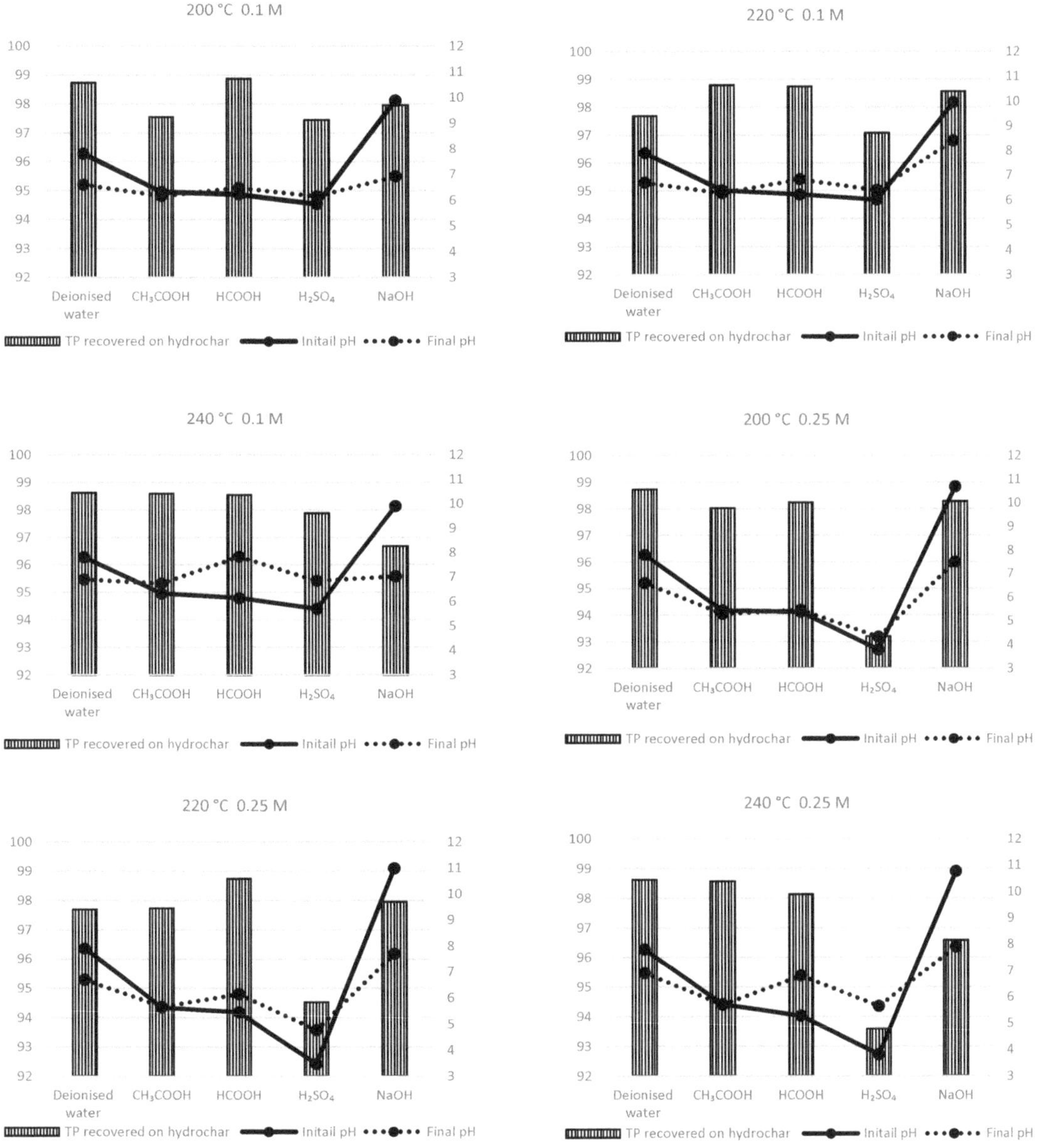

Figure 6. Influence of additives, additive concentration, temperature, and pH on the recovery percentage of TP into the hydrochar.

The initial pH of feedstock slurry produced using CH_3COOH and HCOOH additive was ~6.3 and ~5.6, and ~6.2 and ~5.3, respectively, at 0.1 and 0.25 M concentration. The TP in the process-water was observed to be in the range of 65.5–126 mg/L and 66–105 mg/L when produced using CH_3COOH additive at 0.1 and 0.25 M concentration. Similarly, the

TP in process-water was in a similar range with 59.3–66.9 mg/L and 62–86.3 mg/L when produced using HCOOH additive at 0.1 and 0.25 M concentration.

Factors influencing the TP immobilization during the HTC process include treatment conditions (temperature, reaction time, and additive properties), and the feedstock itself [28]. The formation of phosphorus salts (calcium phosphate, magnesium ammonium phosphate, and magnesium phosphate) are known to immobilize phosphorus into the hydrochar and this immobilization is influenced by the presence of higher inorganic content of the feedstock (such as the level of Ca, Mg, and others), pH, temperature and additives during HTC.

The element composition of the feedstock, particularly the presence of phosphate precipitating metals (viz., Fe, Al, and Ca) has a higher potential in deciding the phosphate retention in the hydrochar product [34]. During HTC of sewage sludge, the presence of a higher concentration of multivalent metal ions such as Al^{3+}, Ca^{2+}, Fe^{3+}, and Mg^{2+} are responsible for forming phosphate with low solubility and in turn enabling the phosphate to be retained in subsequently produced hydrochar. However, the previous studies indicated that the treatment using H_2SO_4 as an additive tends to reduce the level of Ca, Fe, and Mg in hydrochar [33]. Analyzing the conductivity aids in understanding the metal ion concentration in the process-water, and the experimental analysis indicated higher conductivity in the process-water following the use of H_2SO_4 additives in comparison with other organic acids as additives (see Figure 4). The presence of increasing metal ion concentration can explain the higher level of P immobilization into the process-water, particularly with H_2SO_4 additives. Nevertheless, despite having relatively higher conductivity following the use of NaOH as an additive, TP concentration in the process-water was comparatively less. One explanation for increased conductivity following the use of alkali additive can be simultaneously induced ionic salts with NaOH additive utilization. The investigated results suggest that HTC of sewage sludge significantly immobilizes phosphorus into hydrochar in all but mineral acid additives. Results are consistent with another study carried out by Ekpo et al. (2016) demonstrating lower TP leaching into the process-water during HTC of swine manure in the presence of CH_3COOH, HCOOH, and NaOH as additives.

4. Conclusions

The influence of organic acids, an inorganic acid and alkali as additives on phosphorus mobilization, energy value, yield, and dewaterability by hydrothermally carbonizing sewage sludge was analyzed. Phosphorus extraction into the process-water is pH-dependent and particularly significant in the presence of inorganic acid (H_2SO_4). The use of H_2SO_4 and NaOH as additives has decreased the FC content of produced hydrochar, which negatively influences the heating value of the consequently produced hydrochar. A relatively higher reduction in the binding force of the sludge particles was observed during HTC using H_2SO_4 in the reaction medium; this, in turn, improved the hydrochar dewatering performance in comparison with other additives. In conclusion, if the HTC of sewage sludge is designated to leach the phosphorus into the process water, the use of inorganic acid at a higher concentration is favorable; however, compromises will be made in the fuel characteristic of the hydrochar.

Author Contributions: Conceptualization, V.S.E. and S.N.; investigation, V.S.E.; writing—original draft preparation, V.S.E.; writing—review and editing, V.S.E., S.N., J.S., and M.N.; supervision, S.N. All authors have read and agreed to the published version of the manuscript.

Funding: This publication was funded by the German Research Foundation (DFG) and the Open Access Publication Fund of the University of Rostock.

Data Availability Statement: The authors confirm that the data supporting the findings of this study are available within the article.

Conflicts of Interest: The authors declare no conflict of interest.

References

1. He, C.; Giannis, A.; Wang, J.-Y. Conversion of sewage sludge to clean solid fuel using hydrothermal carbonization: Hydrochar fuel characteristics and combustion behavior. *Appl. Energy* **2013**, *111*, 257–266. [CrossRef]
2. Ekanthalu, V.S.; Morscheck, G.; Narra, S.; Nelles, M. Hydrothermal Carbonization—A Sustainable Approach to Deal with the Challenges in Sewage Sludge Management. *Urban Min. Sustain. Waste Manag.* **2020**, 293–302.
3. Statistisches Bundesamt Destatis. Abwasserbehandlung—Klärschlamm. 2020. Available online: https://www.destatis.de (accessed on 15 October 2020).
4. Kreislaufwirtschaftsgesetz—KrWG. *Gesetz zur Förderung der Kreislaufwirtschaft und Sicherung der Umweltverträglichen Bewirtschaftung von Abfällen*; Bundesministeriums der Justiz und für Verbraucherschutz Sowie des Bundesamts für Justiz: Aachen, Germany, 2017.
5. AbfKlär, V. *Verordnung über die Verwertung von Klärschlamm, Klärschlammgemisch und Klärschlammkompost*; Bundesministeriums der Justiz und für Verbraucherschutz sowie des Bundesamts für Justiz: Aachen, Germany, 2017.
6. Wang, T.; Zhai, Y.; Zhu, Y.; Peng, C.; Wang, T.; Xu, B.; Li, C.; Zeng, G. Feedwater pH affects phosphorus transformation during hydrothermal carbonization of sewage sludge. *Bioresour. Technol.* **2017**, *245*, 182–187. [CrossRef] [PubMed]
7. Crocker, M. *Thermo Chemical Conversion of Biomass to Liquid Fuels and Chemicals*; Royal Society of Chemistry: London, UK, 2010.
8. Saetea, P.; Tippayawong, N. *Recovery of Value-Added Products from Hydrothermalc Arbonization of Sewage Sludge*; Hindawi Publishing Corporation: London, UK, 2013.
9. Stucki, M.; Eymann, L.; Gerner, G.; Hartmann, F.; Wanner, R.; Krebs, R. Hydrothermal carbonization of sewage sludge on industrial scale: Energy efficiency, environmental effects and combustion. *J. Energy Chall. Mech.* **2015**, *2*, 38–44.
10. Leng, L.; Yuan, X.; Huang, H.; Shao, J.; Wang, H.; Chen, X.; Zeng, G. Bio-char derived from sewage sludge by liquefaction: Characterization and application for dye adsorption. *Appl. Surf. Sci.* **2015**, *346*, 223–231. [CrossRef]
11. Wang, L.; Chang, Y.; Liu, Q. Fate and distribution of nutrients and heavy metals during hydrothermal carbonization of sewage sludge with implication to land application. *J. Clean. Prod.* **2019**, *225*, 972–983. [CrossRef]
12. Kim, D.; Lee, K.; Park, K.Y. Hydrothermal carbonization of anaerobically digested sludge for solid fuel production and energy recovery. *Fuel* **2014**, *130*, 120–125. [CrossRef]
13. Lynam, J.G.; Reza, M.T.; Vasquez, V.R.; Coronella, C.J. Effect of salt addition on hydrothermal carbonization of lignocellulosic biomass. *Fuel* **2012**, *99*, 271–273. [CrossRef]
14. Lynam, J.G.; Coronella, C.J.; Yan, W.; Reza, M.T.; Vasquez, V.R. Acetic acid and lithium chloride effects on hydrothermal carbonization of lignocellulosic biomass. *Bioresour. Technol.* **2011**, *102*, 6192–6199. [CrossRef]
15. Reza, M.T.; Rottler, E.; Herklotz, L.; Wirth, B. Hydrothermal carbonization (HTC) of wheat straw: Influence of feedwater pH prepared by acetic acid and potassium hydroxide. *Bioresour. Technol.* **2015**, *182*, 336–344. [CrossRef]
16. Ovsyannikova, E.; Arauzo, P.J.; Becker, G.C.; Kruse, A. Experimental and thermodynamic studies of phosphate behavior during the hydrothermal carbonization of sewage sludge. *Sci. Total. Environ.* **2019**, *692*, 147–156. [CrossRef] [PubMed]
17. Shi, Y.; Luo, G.; Rao, Y.; Chen, H.; Zhang, S. Hydrothermal conversion of dewatered sewage sludge: Focusing on the transformation mechanism and recovery of phosphorus. *Chemosphere* **2019**, *228*, 619–628. [CrossRef] [PubMed]
18. Zhai, Y.; Liu, X.; Zhu, Y.; Peng, C.; Wang, T.; Zhu, L.; Li, C.; Zeng, G. Hydrothermal carbonization of sewage sludge: The effect of feed-water pH on fate and risk of heavy metals in hydrochars. *Bioresour. Technol.* **2016**, *218*, 183–188. [CrossRef] [PubMed]
19. Liu, X.; Zhai, Y.; Li, S.; Wang, B.; Wang, T.; Liu, Y.; Qiu, Z.; Li, C. Hydrothermal carbonization of sewage sludge: Effect of feed-water pH on hydrochar's physicochemical properties, organic component and thermal behavior. *J. Hazard. Mater.* **2020**, *388*, 122084. [CrossRef] [PubMed]
20. Lenntech. Phosphorous Removal from Wastewater. Available online: https://www.lenntech.com/phosphorous-removal.htm (accessed on 25 January 2021).
21. TerraNova. The TerraNova®Ultra–Process. Available online: https://terranova-energy.com/en/project/process/ (accessed on 27 January 2021).
22. UBC Sustainable Cities Commission. Enhanced Nutrient Removal with Bio-Filter at Rostock WWTP, Germany. 20 October 2017. Available online: https://www.balticwaterhub.net/solutions/bio-filter-wwtp-rostock (accessed on 20 December 2020).
23. Peng, C.; Zhai, Y.; Zhu, Y.; Xu, B.; Wang, T.; Li, C.; Zeng, G. Production of char from sewage sludge employing hydrothermal carbonization: Char properties, combustion behavior and thermal characteristics. *Fuel* **2016**, *176*, 110–118. [CrossRef]
24. Roskosch, A.; Heidecke, P. *Sewage Sludge Disposal in the Federal Republic of Germany*; German Environment Agency: Dessau-Roßlau, Germany, 2018.
25. Kalderis, D.; Kotti, M.S.; Méndez, A.; Gasco, G. Characterization of hydrochars produced by hydrothermal carbonization of rice husk. *Solid Earth* **2014**, *5*, 477–483. [CrossRef]
26. Jaruwat, D.; Udomsap, P.; Chollacoop, N.; Fuji, M.; Eiad-Ua, A. Effects of hydrothermal temperature and time of hydrochar from Cattail lea-ves. In Proceedings of the International Conference on Science and Technology of Emerging Materials, Pattaya, Thailand, 18–20 July 2018.
27. Sermyagina, E.; Saari, J.; Kaikko, J.; Vakkilainen, E. Hydrothermal carbonization of coniferous biomass: Effect of process parameters on mass and energy yields. *J. Anal. Appl. Pyrolysis* **2015**, *113*, 551–556. [CrossRef]
28. Wang, H.; Yang, Z.; Li, X.; Liu, Y. Distribution and transformation behaviors of heavy metals and phosphorus during hydrothermal carbonization of sewage sludge. *Environ. Sci. Pollut. Res.* **2020**, *27*, 17109–17122. [CrossRef]

29. Mazumder, S.; Saha, P.; Reza, M.T. Co-hydrothermal carbonization of coal waste and food waste: Fuel characteristics. *Biomass Convers. Biorefinery* **2020**, 1–11. [CrossRef]
30. Putra, H.E.; Damanhuri, E.; Dewi, K.; Pasek, A.D. Hydrothermal carbonization of biomass waste under low temperature condition. In Proceedings of the 2nd International Conference on Engineering and Technology for Sustainable Development, Yogyakarta, Indonesia, 18–20 December 2018.
31. Imran, A.M.; Widodo, S.; Irvan, U.R. Correlation of fixed carbon content and calorific value of South Sulawesi Coal, Indonesia. *IOP Conf. Ser. Earth Environ. Sci.* **2020**, *473*, 012106.
32. Funke, A.; Ziegler, F. Hydrothermal carbonization of biomass: A summary and discussion of chemical mechanisms for process engineering. *Biofuels Bioprod. Biorefin.* **2010**, *4*, 160–177. [CrossRef]
33. Ekpo, U.; Ross, A.B.; Camargo-Valero, M.A.; Fletcher, L.A. Influence of pH on hydrothermal treatment of swine manure: Impact on extraction of nitrogen and phosphorus in process water. *Bioresour. Technol.* **2016**, *214*, 617–644. [CrossRef] [PubMed]
34. Huang, R.; Fang, C.; Zhang, B.; Tang, Y. Transformations of phosphorus speciation during (hydro) thermal treatments of animal manures. *Environ. Sci. Technol.* **2018**, *52*, 3016–3026. [CrossRef] [PubMed]

processes

MDPI

Article

Inventory of MSWI Fly Ash in Switzerland: Heavy Metal Recovery Potential and Their Properties for Acid Leaching

Wolfgang Zucha [1,*], Gisela Weibel [1,2], Mirjam Wolffers [1] and Urs Eggenberger [1]

1 Institute of Geological Sciences, University of Bern, 3012 Bern, Switzerland; gisela.weibel@geo.unibe.ch (G.W.); mirjam.wolffers@geo.unibe.ch (M.W.); urs.eggenberger@geo.unibe.ch (U.E.)

2 Zentrum für Nachhaltige Abfall-und Ressourcennutzung ZAR, 8340 Hinwil, Switzerland

* Correspondence: wolfgang.zucha@geo.unibe.ch

Received: 30 November 2020; Accepted: 15 December 2020; Published: 17 December 2020

Abstract: From the year 2021 on, heavy metals from Swiss municipal solid waste incineration (MSWI) fly ash (FA) must be recovered before landfilling. This is predominantly performed by acid leaching. As a basis for the development of defined recovery rates and for the implementation of the recovery process, the authorities and plant operators need information on the geochemical properties of FA. This study provides extended chemical and mineralogical characterization of all FA produced in 29 MSWI plants in Switzerland. Acid neutralizing capacity (ANC) and metallic aluminum (Al^0) were additionally analyzed to estimate the effort for acid leaching. Results show that all FA samples are composed of similar constituents, but their content varies due to differences in waste input and incineration conditions. Based on their geochemical properties, the ashes could be divided into four types describing the leachability: very good (6 FA), good (10 FA), moderate (5 FA), and poor leaching potential (8 FA). Due to the large differences it is suggested that the required recovery rates are adjusted to the leaching potential. The quantity of heavy metals recoverable by acid leaching was estimated to be 2420 t/y Zn, 530 t/y Pb, 66 t/y Cu and 22 t/y Cd.

Keywords: MSWI fly ash; properties of fly ash; acid leaching; heavy metal recovery

1. Introduction

About 4 million tons of waste are incinerated in Switzerland each year in 29 municipal solid waste incineration (MSWI) plants to reduce the mass and volume of waste, destroy organic compounds, and to recover energy. After incineration about 20 wt.% and 2 wt.% of the waste input remains as bottom ash and fly ash (FA). FA precipitates from the flue gas by passing through boiler and electrostatic precipitator. FA has been characterized by numerous studies [1–4]. The major chemical components are Ca, Na, K, Cl, and S. The elevated Cl concentration in FA (often above 10 wt.%) results mostly from the incineration of plastics (PVC). Chlorine forces the volatilization of heavy metals with high vapor pressure by the formation of Cl-complexes [5]. In addition, some heavy metals (e.g., Zn, Pb, Cu, Sb, Sn, and Cd) are chalcophile, and the high S concentration in the waste input additionally supports the transfer into the flue gas. This results in the increased weight percent of several toxic metals in FA. Thus, direct disposal in landfills in Switzerland without previous treatment is prohibited. Furthermore, disposal also means that the metals in FA reach their end of life and are lost as valuable resources. In the current trend towards a circular economy, where urban mining is prominent, FA has become an interesting source for metal recovery. Therefore, the Swiss authorities released the Ordinance of the Avoidance and Disposal of Waste (ADWO), which prescribes the recovery of heavy metals from FA prior to disposal [6]. Currently, FA is either disposed in underground storage of neighboring countries, treated with a neutral leaching, and cemented afterwards, or it is treated by the so-called FLUWA process, which is an acidic leaching process that was established in Switzerland in 1997. The FLUWA process is currently the only feasible state-of-the-art process that achieves the demands of the ADWO. Heavy metals from FA are recovered at varying rates [4,7,8] depending on the type of metal. As a basis for the development of defined recovery rates and for the implementation of the recovery process, the mass flow of metals in FA and their geochemical properties must be known. The metal content in FA depends heavily on the waste input (industrial or household waste), and elemental concentrations differ not only from plant to plant but also on a daily and seasonal basis [9]. The acid arising at the plants' wet flue gas cleaning systems is used as leaching agent. During the FLUWA process, the acid and alkaline scrub water is mixed with FA and reacts in two- to three-stage cascade reactors. After 40–60 min of leaching, vacuum filtration separates the solid metal depleted filter cake from the filtrate (leachate) with dissolved metals. This leachate is used for direct metal recovery [7], or the metals are precipitated as hydroxide sludge by the addition of lime. The sludge is exported, and the metals are recovered by smelting plans. The depleted filter cake is disposed in a Swiss landfill of type D.

The efficiency of the process depends mainly on the pH, Eh, liquid-to-solid ratio (L/S ratio), temperature, and leaching time. The content of heavy metals as well as the mineralogical composition of the FA are important additional factors that influence the efficiency. The FLUWA process is performed at a low pH (3–4) to successfully dissolve the heavy metals from FA. The addition of acid leads to the dissolution of lime (CaO) and calcite ($CaCO_3$) (among other minor phases), which buffer the pH. Lime reacts in a first step with water to form portlandite ($Ca(OH)_2$) before it dissolves, and two hydroxide ions are released (Equations (1) and (2)).

$$CaO + H_2O \rightarrow Ca(OH)_2 \tag{1}$$

$$Ca(OH)_2 \rightarrow Ca^{2+}_{(aq)} + 2OH^- \tag{2}$$

Below a pH of ~7.3, calcite is dissolved by consuming protons and releasing CO_2 in the process (Equation (3)).

$$CaCO_3 + 2H^+ \rightarrow Ca^{2+}_{(aq)} + CO_2 + H_2O \tag{3}$$

If the acid neutralizing capacity of the FA is larger than the amount of acid scrub water, additional acid (e.g., 32% HCl) must be added to achieve low pH conditions, causing additional cost. The oxidation of metallic aluminum (Al^0) in FA forces reducing conditions [10]. Aluminum is usually present in FA as aluminum foil particles, which are entrained with the flue gas. Despite their low content in FA, their presence diminishes the leaching efficiency during the FLUWA process [10]. The oxidation of Al^0 is at the expense of metals such as Pb and Cu (Equation (4)), which are reductively cemented and removed from the leachate [8].

$$2Al^0 + 3Cu^{2+} \rightarrow 2Al^{3+} + 3Cu^0 \downarrow \tag{4}$$

To prevent reductive precipitation, an oxidizing agent (e.g., H_2O_2) is added during acid leaching. It is speculated that other metals such as Fe^0 and Zn^0 in FA may reduce Cu^{2+} and Pb^{2+} during the FLUWA process. This seems, however, to be unlikely as both elements are less reactive than Al^0 [11,12]. The addition of an oxidizing agent at low pH conditions is crucial for Cu and Pb recovery, as it enhances the yield greatly ([13], Table 1).

Regarding the currently limited capacity of only 12 FLUWA facilities and the capacities to be expanded (either by new construction or by external treatment at other plants), an inventory of Swiss FA was made. Knowledge about the FA composition and its properties will help FLUWA operators and authorities in the implementation of the guidelines according to the ADWO. This study therefore presents an overview of all forms of Swiss FA and their chemical and mineralogical composition, acid neutralizing capacity, and content of metallic aluminum. The FA types were divided into exemplary groups (clusters) regarding combined FLUWA processing. In addition, the recovery potential of heavy and valuable metals in FA was calculated.

Table 1. Average metal recovery achieved by acidic leaching with the FLUWA process [13].

	Recovery (%) FLUWA	Recovery (%) Optimized FLUWA (+H_2O_2)
Zn	60–80	60–80
Pb	0–30	50–90
Cu	0–30	40–80
Cd	60–85	85–95

2. Materials and Methods

2.1. Sampling and Sample Processing

The sampling campaign was launched in December 2016 at all 29 MSWI plants in Switzerland. For three weeks, 100–200 g of FA was sampled daily to obtain a representative sample of several kilograms of FA. This timespan was chosen to avoid bias from daily fluctuation [9]. All but three MSWI plants were sampled as requested in January 2017. The others performed the sampling in the following months. One MSWI plant was not included in the study because it was shut down in 2020. The samples were split, stored in sealed bags and aliquots for XRD and XRF, and were dried at 35 and 105 °C, respectively. All FA samples were anonymized.

2.2. Chemical Analysis

Elemental composition was determined by energy dispersive X-ray fluorescence analysis (ED-XRF) using a Xepos spectrometer (SPECTRO, Kleve, Germany); with matrix adjusted calibration. The measurement was performed on pressed powder pellets (32 mm diameter) using 4.0 g of ground sample material and 0.9 g of Hoechst wax C from Merck as binder. Trace elements and rare metals were analyzed by Actlabs (Canada) by the Ultratrace5 program. Actlabs uses INAA and ICP-MS to determine the respective concentrations in the lower ppm and ppb range.

2.3. Mineralogical Analysis

Dried FA samples were mixed with 20 wt.% corundum (internal standard) and milled for 6 min using an XRD-McCrown Mill Retsch GmbH, Haan, Germany). The powder was then filled in a glass capillary (0.3 mm diameter, Hilgenberg Glass no. 10). The measurement was done at the Swiss-Norwegian Beamline (SNBL) (at the European Synchrotron Radiation Facility (ESRF) in Grenoble using a Pilatus 2M (Dectris, Baden, Switzerland) detector measuring from 0.0051 to 34.3751 [°2θ] with a step size of 0.01 [°2θ] and a scan step time of 1 s by radiating the capillary using a focused beam (diameter 100 μ) with a wavelength of 0.69264 Å and 17.9 keV. Phase identification and quantification was done using TOPAS Academics V6 (Coelho Software, Brisbane, Australia) and HighScorePlus 4.6 (Malvern Panalytical, Malvern, UK) using the Rietveld method.

2.4. Acid Neutralizing Capacity

FA (2 g) was added into 20 mL Mili-Q water (L/S ratio of 10). The suspension was then titrated in 40 steps using a 785 DPM Titrino device (Metrohm, Herisau, Switzerland) by adding every 10 min 1 mL 1M HCl under constant stirring. This procedure was determined in a previous study to best describe the behavior of FA for acid leaching conditions [14].

2.5. Metallic Aluminum

Metallic Aluminum was measured by oxidation of Al^0 with H_2O at high pH conditions (Equation (5)) by measuring the amount of produced gas and calculating Al^0 using the ideal gas law.

$$Al^0 + 3H_2O \rightarrow Al(OH)_3 + 1.5H_2 \quad (5)$$

Theoretically, all base metals can from hydrogen in contact with water. A solution of 100 mL 0.5 M NaOH(Merck & Co., Kenilworth, NJ, USA) was filled in a Schott laboratory bottle (1 L) from Smilax and flushed with Ar gas for ~5 min to create an inert atmosphere. FA (25 g) and a magnetic stirring device (300 rpm) were added to the solution, and the bottle was immediately closed and sealed. Temperature and pressure in the thermally isolated bottle were recorded every 10 s by a P/T logger (HOBO U20-001-01, Onset, Cape Cod, MA, USA) for at least 8 h until reaction equilibrium was reached.

2.6. Cluster Analyis

To analyze similarities among FA, a data set with the obtained values of Zn, Al^0, ANC, and the amount of produced FA was compiled. The cluster analysis was done in MATLAB (R2018a, MathWorks, Natick, MA, USA, 2018) using the linkage function to calculate an agglomerative hierarchical cluster tree.

3. Results

3.1. Chemical Composition

The average chemical composition of all FA samples (Table 2) is dominated by the oxides CaO (270,000 mg/kg), SO_3 (13,000 mg/kg), Na_2O (100,000 mg/kg), SiO_2 (80,000 mg/kg), K_2O (70,000 mg/kg), and Cl (130,000 mg/kg) (full details of chemical data can be found in Tables A1 and A2 and in the Supplementary Materials). Further major constituents are Al_2O_3 (35,000 mg/kg), Fe_2O_3 (25,000 mg/kg), P_2O_5 (10,000 mg/kg), MgO (12,000 mg/kg), and TiO_2 (17,000 mg/kg). Of the recoverable elements, Zn (average 36,000 mg/kg) is the most abundant followed by Pb (8000 mg/kg), Cu (2000 mg/kg), and Cd (~200 mg/kg). Precious metals (e.g., Au and Ag) as well as the total content of rare earth elements (REE) show low mg/kg concentration. FA samples have similar constituents, but the content varies heavily due the different waste input (Figure 1). Ca, the dominating contributor to ANC, varies from 150,000 to almost 400,000 mg/kg. S and Cl, which promote the transfer of heavy metals into the flue gas, scatter from 75,000 to 200,000 and 60,000 to 250,000 mg/kg, respectively. Of the total metal content (60,000 to 140,000 mg/kg), the recoverable metals are Zn (15,000–70,000 mg/kg), Pb (2500–16,000 mg/kg), Cu (1000 to 3000 mg/kg), and Cd (100–350 mg/kg). The large concentration range of Zn, Pb, Cu, and Cd indicates again the large differences in the waste input.

3.2. Mineralogical Composition

Phase analyses show that the major solid phases occur in all FA samples. On average, all samples contain an amorphous part of ~41 wt.% including the minor and unidentified phases (Table 3, a complete table can be found in the Supplementary Materials). Crystalline phases are dominated by chlorides, such as halite (NaCl) and sylvite (KCl), which are abundant in all samples (11 wt.% and 4 wt.% on average). K_2ZnCl_4 (5 wt.% on average) occurs in 23 samples and represents the most important phase, which contains easily recoverable Zn. Gehlenite ($Ca_2Al_2SiO_7$), belite (Ca_2SiO_4), and quartz (SiO_2) are the dominant silicate minerals (6.6, 4.0, and 2.4 wt.% on average, respectively). The concentrations of the carbonates calcite ($CaCO_3$) and magnesite ($MgCO_3$) are 4.9 and 2.6 wt.% on average. The dominating oxides are mayenite ($Ca_{12}Al_{14}O_{33}$), perovskite ($CaTiO_3$) (3.3 and 2.8 wt.% on average), and lime (CaO, 1.6 wt.% on average). Rutile (TiO_2) and periclase (MgO) are minor constituents (both 0.7 wt.% on average) but occur in almost all FA samples. Among the sulfates, anhydrite ($CaSO_4$) is the only identifiable phase (8.1 wt.% on average). The broad range of concentrations of the major solid phases in FA is indicated by the large boxes in Figure 2. Beside the main mineralogical constituents, FA samples have many minor phases (<1 wt.%), which cannot be identified or quantified properly.

Table 2. Main mineralogical constituents in FA from 29 municipal solid waste incineration plants.

Phase	Abbr.	Formula	Average Content (wt.%)	Occurrences in FA Samples (max. 29)
Amorphous content (incl. minor)			41	29
Chlorides				
Halite	Ha	$NaCl$	11	29
Sylvite	Sy	KCl	4	28
K_2ZnCl_4	KZn	K_2ZnCl_4	5	23
Sulfate				
Anhydrite	An	$CaSO_4$	8	29
Silicates				
Gehlenite	Ge	$Ca_2Al_2SiO_7$	7	29
Belite	Be	Ca_2SiO_4	4	27
Quartz	Qz	SiO_2	2	29
Carbonate				
Calcite	Cc	$CaCO_3$	5	28
Magnesite	Mc	$MgCO_3$	3	28
Oxides				
Mayenite	My	$Ca_{12}Al_{14}O_{33}$	3	26
Perovskite	Pk	$CaTiO_3$	3	27
Lime		CaO	2	27
Rutile	Rt	TiO_2	1	28
Periclase	Pc	MgO	1	28

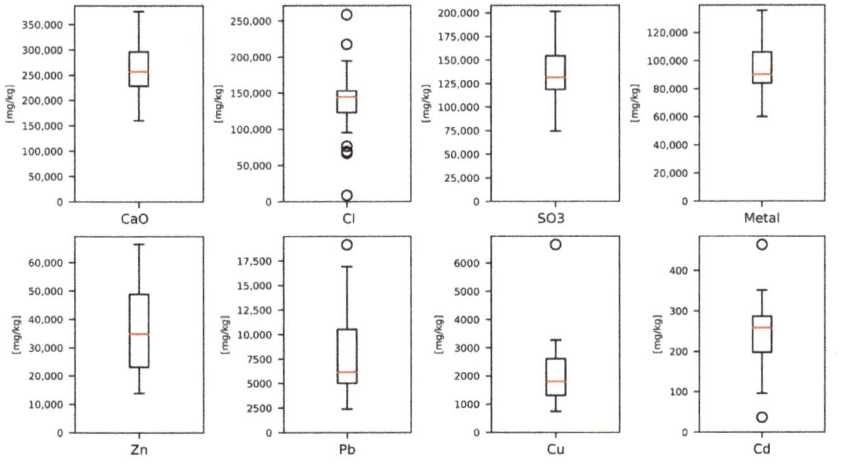

Figure 1. Concentration distribution of selected chemical constituents in the 29 Swiss fly ash (FA) samples.

3.3. Acid Neutralizing Capacity

The titration curves of selected FA samples are shown in Figure 3. Seven FA samples had a starting pH below 8 and reached a pH of 2 before 8 moL H^+ was added. Twelve samples started at pH 10–12 but dropped to a pH below 8 after the addition of a 1 moL H^+. The remaining 10 samples started at pH 12 and required up to 3 moL H^+ before dropping below pH 8, and an additional 5–10 moL H^+ was needed to reach pH 2. There are three major plateaus apparent. The first one is assigned to the dissolution of portlandite ($Ca(OH)_2$). Lime (CaO) reacts immediately with water, forms portlandite, and elevates the pH above 12 (see Equation (2)). The next major step is assigned to the dissolution of calcite ($CaCO_3$, see Equation (4)). The final major step around pH 4 marks the start of the dissolution of Ca-Si phases like belite (Ca_2SiO_4) and (partly) gehlenite ($Ca_2Al_2SiO_7$), and other minor phases start to dissolve. Of all phases, calcite is the dominating phase of the ANC. Ca-Si phases as well as the amorphous part only dissolve partially below pH 4. The required amount of H^+ per kg FA to reach pH 2 with an LS of 10 is shown in Figure 4 as boxplot (a complete table can be found in the Supplementary Materials).

Table 3. Approximate annual flow of metals in Swiss FA.

Annual Flow in Swiss Fly Ash (kg/a) ± Uncertainty (SD)							
Zn	3,052,798	±	78,651	V	1036	±	15
Al	1,392,164	±	24,735	La	994	±	11
Fe	1,340,295	±	24,822	Y	681	±	14
Ti	791,008	±	13,579	Nd	560	±	13
Pb	667,668	±	17,000	Se	473	±	3
Sb	194,269	±	4536	In	276	±	4
Cu	172,269	±	4816	Th	182	±	4
Sn	140,728	±	1090	Sc	178	±	3
Mn	58,243	±	811	Gd	153	±	2
Cr	39,064	±	538	Pr	146	±	2
Sr	28,864	±	278	Tl	105	±	2
Cd	20,735	±	263	Sm	95	±	5
Ba	13,220	±	209	Yb	71	±	1
Ni	10,021	±	182	Dy	64	±	1
Bi	8268	±	125	Be	44	±	2
Rb	7826	±	173	Er	42	±	0.5
Zr	6696	±	94	Ge	42	±	1
As	6293	±	94	Eu	26	±	1.2
Li	4012	±	83	Tb	19	±	0.3
Co	3975	±	43	Nb	16	±	0.5
Ag	3191	±	52	Ho	13	±	0.5
Ce	1761	±	50	Au	12	±	0.2
W	1378	±	22	Te	11	±	0.2
Mo	1090	±	21	Lu	8	±	0.1

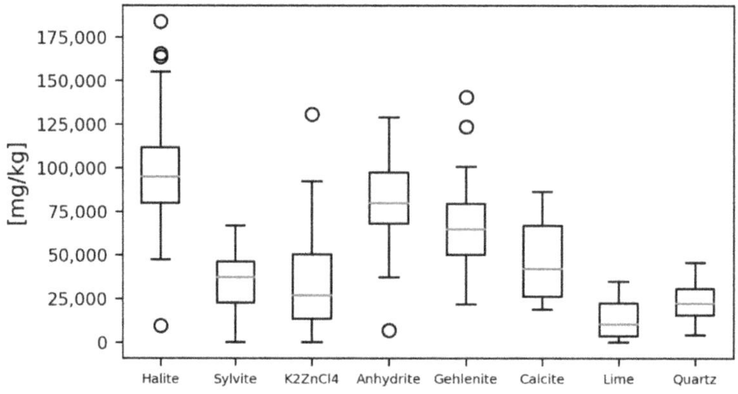

Figure 2. Distribution of the main mineral phases in Swiss FA in wt.%. Outliners are marked as circle.

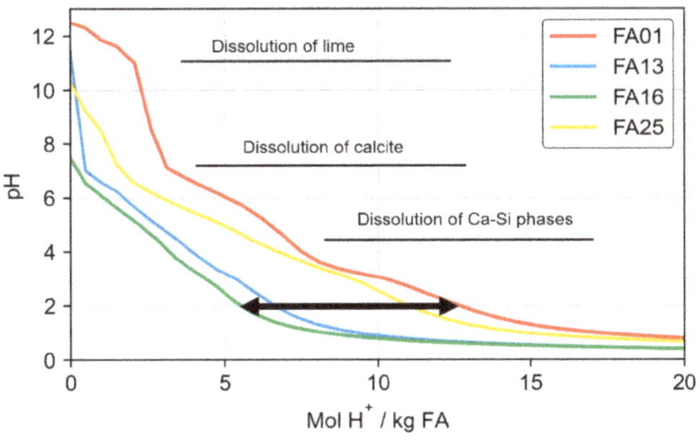

Figure 3. Titration curve of selected FA samples (LS 10). The main plateaus are marked with a black line. The black arrow illustrates the big difference of acid that is required to achieve pH 2.

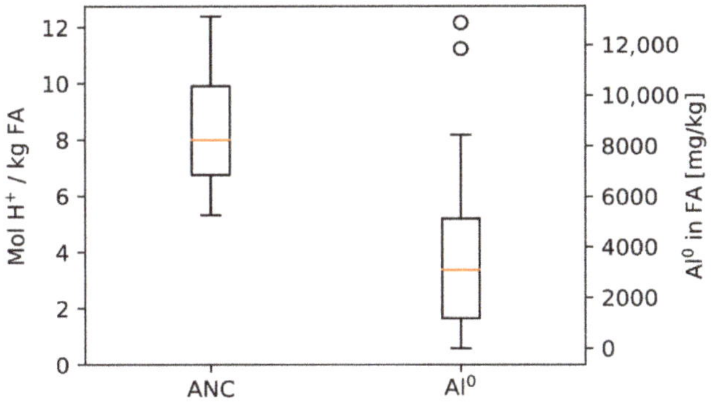

Figure 4. Boxplot of the acid neutralizing capacity (ANC) as amount mol H^+ to reach pH 2 at a L/S ratio of 10 and Al^0 in Swiss FA (n = 29).

3.4. Metallic Aluminum Al^0

The content of Al^0 in FA is shown as a boxplot in Figure 4 (a complete table can be found in the Supplementary Materials). Two samples (FA01, FA04) show content above 1.4 wt.%, five samples (FA02, FA06, FA10, FA14, FA26) contain 0.8–1 wt.%, and most of the samples are between 0.1 and 0.6 wt.%. Two samples contain no Al^0. Overall, the Al^0 content is on average 30% of the total aluminum in FA.

3.5. Cluster Analysis

The dendrogram of the cluster analysis is shown in Figure 5. The cophenetic correlation coefficient, a measure of how faithfully a tree represents the dissimilarities among observation, is 0.71 (the maximum would be 1), which is acceptable. The red cluster on the right has eight samples (FA01, FA04, FA02, FA06, FA14, FA10, FA03, and FA29). On the left side is a branch with a yellow cluster of five samples (FA20, FA21, FA23, FA28, and FA26), a green cluster of six samples (FA16, FA18, FA24, FA22, FA27 and FA25), and finally a blue cluster representing 10 samples (FA13, FA17, FA19, FA11, FA12, FA08, FA11, FA15, FA05, and FA07).

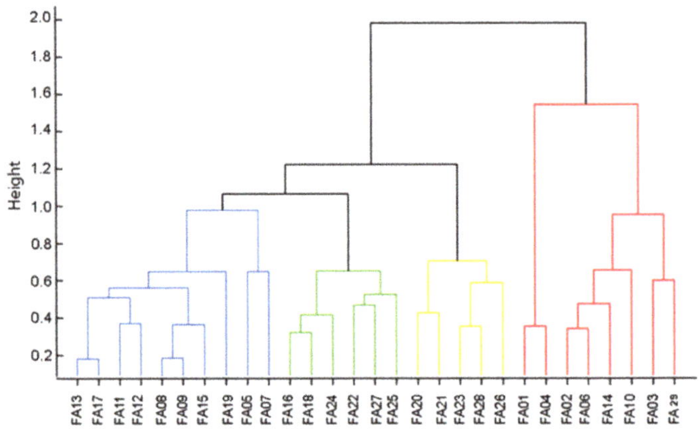

Figure 5. Cluster tree of the FA samples according to their content of Zn and Al^0, the ANC, and the amount of FA produced in 2016. The y-axis shows the distances between the calculated values and hence is a mathematical value that expresses the dissimilarity.

Every cluster represents FA samples with similar properties. The average value of each property and cluster is shown in Figure 6. The green cluster representing FA with very good leaching potential contains FA with the highest Zn concentration of 5.7 wt.% and the lowest content of Al^0 (0.05 wt.%). Each MSWI plant produces almost 4000 tons of FA per year. This cluster is the most interesting for economic metal recovery due to the high Zn recovery with the lowest H_2O_2 consumption and low acid consumption (7.4 mL mol H^+ per kg FA to reach pH 2). The blue cluster representing FA with good leaching potential is the largest cluster with an average Zn concentration of almost 4 wt.% and a content of Al^0 of 0.2 wt.%. These plants produce on average 2200 t of FA per year. The ANC is the lowest because the FA samples required only 7.3 mol H^+ per kg FA to reach a pH of 2 during the titration and a relatively low amount of hydrogen peroxide. The yellow cluster representing FA with moderate leaching potential contains only five MSWI plants. Their FA shows low averaged Zn concentration of 2.6 wt.% but a rather high Al^0 concentration of 0.4 wt.%. Since some MSWI plants are among the largest in Switzerland, the amount of produced FA is 4300 t per year on average. The metal recovery of these ashes requires higher amount of acid (8.6 moL H^+ per kg FA to reach pH 2) and a high amount of H_2O_2. The red cluster containing FA with poor leaching potential shows the lowest Zn concentration of all clusters (2.2 wt.%) but by far the highest concentration of Al^0 (0.8 wt.%). On average, plants in this cluster produces only 1300 t of FA per year. However, this cluster has the highest ANC of 10 mol H^+ per kg FA to reach pH 2.

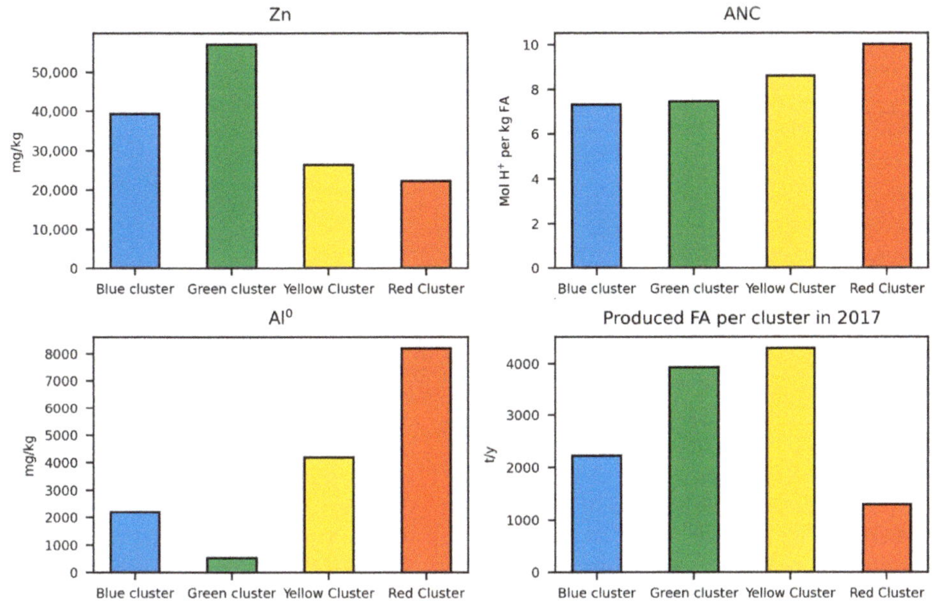

Figure 6. Average Zn and A^{l0} concentration as well as average produced FA per plant and the average ANC for each cluster.

4. Discussion

4.1. Mass Flow of Metals in Swiss FA

The estimated annual mass flow of metals in Swiss FA is illustrated in Table 4. The mass flow was calculated based on concentrations and amount of FA from all plants in 2016 [15]. The total quantity of recoverable metals is: 3052 t/y Zn, 667 t/y Pb, 172 t/y Cu, and 20 t/y Cd. Other base metal contributions are: 1392 t/y Al, 1340 t/y Fe, 791 t/y Ti, and the chalcophile elements Sb 194 t/y and Sn 140 t/y. The annual mass flow for REE (including Sc and Y) is 4.8 t/y, mainly represented by the light-REE Ce (1.8 t/y), La (1 t/y), Y (0.7 t/y), and Nd (0.6 t/y). Other notable metals are Ni (10 t/y), As (6 t/y), Co (4 t/y), Ag (3.1 t/y), and W (1.4 t/y). Gold is a minor constituent in FA, and thus only 12 kg is landfilled each year from FA. The low mass flow of REE and other precious metals (Ag, Au) in FA is due to their low vapor pressure, expressed with low partitioning coefficients < 0.1 [16]. Preferentially chalcophile elements such as Zn, Pb, Cu, Sb, and Sn are expected to be enriched in FA. The origin of the metals in the waste input was not investigated in this study. It can be assumed that abundant metals are mainly from alloys (e.g., Zn, Sn), color pigments (e.g., Ti), or additives in plastic (e.g., Sb). Table 4 shows the annual technically possible acid leaching potential of Zn (2420 t/y), Pb (530 t/y), Cu (66 t/y), and Cd (21.8 t/y) considering the FLUWA process is optimized by using HCl and H_2O_2 as additives [14]. The recovery potential of metals is very high compared to the unwrought metal imported in Switzerland in 2017 [17]. Approximately 30% Zn, 16% Pb, and 1% Cu of the annual import could be replaced by metal recovery from FA. Metal prices fluctuate frequently, and future changes in waste input, e.g., by enhanced metal separation prior to combustion, could drastically change the quantity of heavy metals recoverable by FA and the economic aspects. Ecologically it is, however, beneficial to recover other metals as well, such as Sb and Sn. Heavy metals in landfills are a constant threat to the surrounding environment, especially from a long-term perspective, and primary production (mining, excavation, and extraction) have dramatic impacts on the environment and its inhabitants [18].

Table 4. Total amount of Zn, Pb, Cu, and Cd that is recoverable by the FLUWA process. The data of the raw import in Switzerland in 2017 only refers to the unwrought metal (no. 7901, 7801, 7403, and 8104 of the Swiss Explanatory notes of the Customs Tariff—Tares [19].

Metal	Recovery of the Optimized FLUWA Treatment [10]	Amount (t/y)	Annual Import in Switzerland, 2017 (Unwrought Metal) (t) [17]
Zn	80%	~2420	7338
Pb	85%	~530	3271
Cu	40%	~66	4881
Cd	100%	~21.8	4

4.2. Characterization of FA Regarding the FLUWA Treatment

To estimate implementation planning and effort for FA leaching, the properties of all FA samples were grouped with similar properties into four different clusters. Of all MSWI plants, 16 out of 29 produce FA that shows very high or high leaching potential, and 13 MSWI plants produce FA that shows moderate or low leaching potential. FA with moderate or poor leaching potential require higher effort and has smaller return through the recovery of Zn. Thus, mixing of different FA types could be an expedient means of diminishing the effects of poor leaching potential. For optimized processing, FA composition is suggested to be monitored regularly by simple tests of ANC, Zn (Pb, Cu), and Al^0. The cluster, however, provides preliminary information for general planning and FLUWA design.

4.3. Situation in Switzerland from 2021 on

As of 2021, there will be 29 MSWI plants in Switzerland producing ~80,000 tons of FA per year that must be treated prior to disposal. Although the type of process to recover metals from FA will not be prescribed, all FA is expected to be treated by the FLUWA process, as it represents the state-of-the art process. In 2018, only 12 plants individually conducted acid leaching. The remaining plants produce approximately 43,000 t/y of FA, which has to be treated in external or newly constructed facilities. Considering the costs of 350–450 CHF per ton of FA [4], there is a market potential of CHF 15–20 million per year.

5. Conclusions

This study shows the wide range of chemical and mineralogical differences of the FA in Switzerland and the properties influencing acid leaching. Consequently, the effort to recover heavy metals from FA varies widely is mainly dependent on metal and Al^0 content and on the ANC. It is especially unfavorable for FA with low Zn content but high ANC to recover a fixed rate of heavy metals. This fact was considered in the implementation of the guidelines of the ADWO, which prescribes that the recovery rate of Zn and Pb to achieve is based on the initial concentration in FA [20]. After complete implementation, the FLUWA treatment of all Swiss FA will produce a significant quantity of recovered heavy metals. These quantities are otherwise disposed as pollutants in landfills and removed from the raw metal cycle. However, the remaining filter cake that is deposited still contains a significant load of contaminants, such as Sb and Sn. Further development is therefore needed to increase the recovery of heavy metals and to extend it to less easily recoverable elements.

Supplementary Materials: The following are available online at http://www.mdpi.com/2227-9717/8/12/1668/s1.

Author Contributions: Conceptualization, W.Z. and U.E.; methodology, W.Z., G.W. and U.E.; software, W.Z.; writing—original draft preparation, W.Z.; writing—review and editing, W.Z., U.E., M.W. and G.W.; visualization, W.Z.; supervision, G.W. and U.E.; project administration, U.E.; funding acquisition, U.E. All authors have read and agreed to the published version of the manuscript.

Funding: This research was funded by the Swiss Federal Office for the Environment (FOEN).

Acknowledgments: We thank MSWI plant operators for providing sample material, Kaarina Schenk (FOEN) and Stefan Schlumberger (ZAR) for feedback and support. Analytical support by Christine Lemp, Martin Fisch (both University of Bern) and Ivo Budde is highly acknowledged.

Conflicts of Interest: The authors declare no conflict of interest.

Appendix A

Table A1. Measured elements in Swiss FA.

Name	Unit	Method	Average	Median	Max	Min
Na_2O	mg/kg	ED-XRF	92,728	100,400	144,700	12,000
MgO	mg/kg	ED-XRF	12,733	12,400	27,000	3200
SiO_2	mg/kg	ED-XRF	84,548	81,500	148,100	51,200
P_2O_5	mg/kg	ED-XRF	9798	9192	15,700	6480
SO_3	mg/kg	ED-XRF	136,179	131,400	201,500	74,500
Cl	mg/kg	ED-XRF	137,483	144,200	258,400	8800
K_2O	mg/kg	ED-XRF	65,401	65,800	110,400	12,000
CaO	mg/kg	ED-XRF	268,231	256,800	375,900	160,100
Zn	mg/kg	ED-XRF	36,039	34,830	66,450	13,850
Pb	mg/kg	ED-XRF	7978	6116	19,130	2380
Cu	mg/kg	ED-XRF	2030	1792	6647	737
Cd	mg/kg	ED-XRF	243	258	464	36
Mn	mg/kg	ED-XRF	822	820	1569	534
Fe	mg/kg	ED-XRF	17,641	17,640	31,217	9421
Al	mg/kg	ED-XRF	18,673	17,995	32,390	9632
Ti	mg/kg	ED-XRF	10,582	10,660	16,200	7018
Au	µg/kg	INAA	1829	745	20,600	247
Ag	mg/kg	MULT INAA/TD-ICP-MS	39	37	94	13
Ni	mg/kg	MULT INAA/TD-ICP-MS	123	105	335	72
As	mg/kg	INAA	78	73	163	9
Ba	mg/kg	MULT INAA/TD-ICP-MS	166	174	248	82
Be	mg/kg	TD-MS	0.6	0.5	1.5	0.3
Bi	mg/kg	TD-MS	102	99	264	39
Br	mg/kg	INAA	2071	2130	4290	153
Co	mg/kg	MULT INAA/TD-ICP-MS	51	45	96	22
Cr	mg/kg	INAA	512	463	1460	265
Cs	mg/kg	MULT INAA/TD-ICP-MS	7	8	10	2
Eu	mg/kg	TD-MS	0.3	0.3	0.4	0.2
Hf	mg/kg	INAA	2.1	2.0	4.0	<1
Ge	mg/kg	TD-MS	0.5	0.4	2.0	0.1
In	mg/kg	TD-MS	2.9	1.9	11.4	0.4

Table A2. Measured elements in Swiss FA.

Name	Unit	Method	Average	Median	Max	Min
Li	mg/kg	TD-MS	50	47	151	28
Nb	mg/kg	TD-MS	0.5	0.1	3.6	0.1
Mo	mg/kg	TD-MS	12	8	71	3
Rb	mg/kg	TD-MS	95	96	159	21
Re	mg/kg	TD-MS	0	0	0	0
Sb	mg/kg	INAA	2381	2240	4540	745
Sc	mg/kg	INAA	2.4	2.2	4.9	1.2
Se	mg/kg	MULT INAA/TD-ICP-MS	5.6	4.6	12.5	2.0
Sr	mg/kg	TD-MS	366	363	523	289
Ta	mg/kg	MULT INAA/TD-ICP-MS	<0.1	<0.1	<0.1	<0.1
Te	mg/kg	TD-MS	<0.1	<0.1	<0.1	<0.1
Th	mg/kg	MULT INAA/TD-ICP-MS	2.4	2.3	5.2	1.3
Tl	mg/kg	TD-MS	1.3	1.2	2.1	0.6
U	mg/kg	MULT INAA/TD-ICP-MS	1.2	1.1	3.0	0.7
V	mg/kg	TD-MS	13	14	21	2
W	mg/kg	INAA	30	18	114	<1
Y	mg/kg	TD-MS	9	8	16	5
Zr	mg/kg	TD-MS	90	85	220	39
La	mg/kg	TD-MS	12	12	24	8
Ce	mg/kg	TD-MS	22	20	51	13
Pr	mg/kg	TD-MS	1.9	1.9	2.8	1.2
Nd	mg/kg	TD-MS	7.1	7.2	12.3	4.3
Sm	mg/kg	TD-MS	1.2	1.1	2.4	0.7
Gd	mg/kg	TD-MS	2.0	1.6	7.4	0.9
Dy	mg/kg	TD-MS	0.8	0.8	1.2	0.5
Tb	mg/kg	TD-MS	0.2	0.2	0.9	0.1
Ho	mg/kg	TD-MS	0.2	0.1	1.2	0.1
Hg	μg/kg	TD-MS	528	320	2000	50
Er	mg/kg	TD-MS	0.5	0.5	1.5	0.3
Tm	mg/kg	TD-MS	0.1	0.1	0.1	0.1
Yb	mg/kg	TD-MS	0.8	0.5	8.3	0.3
Lu	mg/kg	TD-MS	0.1	0.1	0.4	<1

References

1. Le Forestier, L.; Libourel, G. Characterization of flue gas residues from municipal solid waste combustors. *Environ. Sci. Technol.* **1998**, *32*, 2250–2256. [CrossRef]
2. Li, M.; Xiang, J.; Hu, S.; Sun, L.-S.; Su, S.; Li, P.-S.; Sun, X.-X. Characterization of solid residues from municipal solid waste incinerator. *Fuel* **2004**, *83*, 1397–1405. [CrossRef]
3. Bayuseno, A.P. Mineral Phases in Raw and Processed Municipal Waste Incineration Residues-Towards a Chemical Stabilisation and Fixation of Heavy Metals. Ph.D. Thesis, Ruhr-Universität Bochum, Bochum, Germany, 2006.
4. Weibel, G. Optimized Metal Recovery from Fly Ash from Municipal Solid Waste Incineration. Ph.D. Thesis, University of Bern, Bern, Switzerland, 2017.
5. Verhulst, D.; Buekens, A.; Spencer, P.J.; Eriksson, G. Thermodynamic Behavior of Metal Chlorides and Sulfates under the Conditions of Incineration Furnaces. *Environ. Sci. Technol.* **1996**, *30*, 50–56. [CrossRef]
6. Swiss Confederation. *Ordinance on the Avoidance and the Disposal of Waste (Waste Ordinance, ADWO)*; Swiss Conferderation: Bern, Switzerland, 2015; Volume 2015, pp. 1–46.
7. Schlumberger, S. *Neue Technologien und Möglichkeiten der Behandlung von Rauchgasreinigungsrückständen im Sinne Eines Nachhaltigen Ressourcenmanagements. New Technologies and Opportunities for the Treatment of Residues from Flue Gas Treatments in a Sustainable Resource Man*; KVA: Sursee, Switzerland, 2010.
8. Bühler, A.; Schlumberger, S. *Schwermetalle aus der Flugasche Zurückgewinnen «Saure Flugaschewäsche—FLUWA-Verfahren» Ein Zukunftsweisendes Verfahren in der Abfallverbrennung*; BSH Umweltservice AG: Sursee, Switzerland, 2010.

9. Budde, I. Charakterisierung der Elementaren Zusammensetzung von Filteraschen und Daraus Entstehender Reaktionsprodukte auf Einer Thermischen Abfallbehandlungsanlage. Aufklärung der Möglichen Blei-, Cadmium- und Kupferextraktion aus Filteraschen und Untersuchung mög. Master's Thesis, Leopold-Franzens-Universität Innsbruck, Innsbruck, Austria, 2015.
10. Weibel, G.; Eggenberger, U.; Schlumberger, S.; Mäder, U.K. Chemical associations and mobilization of heavy metals in fly ash from municipal solid waste incineration. *Waste Manag.* **2017**, *62*, 147–159. [CrossRef] [PubMed]
11. Heuss-Aßbichler, S.; Magel, G.; Fehr, K.T. Abiotic hydrogen production in fresh and altered MSWI-residues: Texture and microstructure investigation. *Waste Manag.* **2010**, *30*, 1871–1880. [CrossRef] [PubMed]
12. Ilyas, A.; Badshan, M.; Praagh, M.V.; Persoon, K.M. Evaluation of H_2 gas Production from MSWI Bottom Ash. In Proceedings of the Proceedings Venice 2010, 3rd International Symposium Energy from Biomass Waste, Venice, Italy, 8–11 November 2010.
13. AWEL. *Stand der Technik für die Aufbereitung von Rauchgasreinigungsrückständen (RGRR) aus Kehrichtverbrennungsanlagen*; AWEL: Zürich, Switzerland, 2013; p. 10.
14. Weibel, G.; Eggenberger, U.; Kulik, D.A.; Hummel, W.; Schlumberger, S.; Klink, W.; Fisch, M.; Mäder, U.K. Extraction of heavy metals from MSWI fly ash using hydrochloric acid and sodium chloride solution. *Waste Manag.* **2018**, *76*, 457–471. [CrossRef] [PubMed]
15. VBSA. In Schweizer KVA Verbrannte Abfallmenge 2016. 2017. Available online: http://vbsa.ch/fakten/abfallverwertung/ (accessed on 26 July 2018).
16. Morf, L.S.; Gloor, R.; Haag, O.; Haupt, M.; Skutan, S.; Di Lorenzo, F.; Böni, D. Precious metals and rare earth elements in municipal solid waste—Sources and fate in a Swiss incineration plant. *Waste Manag.* **2013**, *33*, 634–644. [CrossRef] [PubMed]
17. Eidgenössische Zollverwaltung. Datenbank Swiss Impex. 2.%09Datenbank Swiss Impex, Eidgenössische Zollverwaltung EZV. Available online: https://www.gate.ezv.admin.ch/swissimpex/ (accessed on 20 August 2018).
18. Wäger, P. Scarce metals—Applications, supply risks and need for action. *Not. Polit.* **2011**, *104*, 57–66.
19. Eidgenössische Zollverwaltung. Explanatory Notes of the Customs Tariff—Tares. Available online: https://www.ezv.admin.ch/ezv/de/home/dokumentation/richtlinien/d6-erlaeuterungen-zum-zolltarif.html (accessed on 20 August 2018).
20. Swiss Confederation. *Rückgewinnung von Metallen aus den Filteraschen von Kehrichtverbrennungsanlagen*; Bundesamt für Umwelt BAFU: Bern, Switzerland, 2020; pp. 1–21.

processes

Article

Optimization of Metal Recovery from MSWI Fly Ash by Acid Leaching: Findings from Laboratory- and Industrial-Scale Experiments

Gisela Weibel [1,*], Anna Zappatini [1], Mirjam Wolffers [1] and Stefan Ringmann [2]

[1] Institute of Geological Sciences, University of Bern, Baltzerstrasse 1+3, 3012 Bern, Switzerland; anna.zappatini@geo.unibe.ch (A.Z.); mirjam.wolffers@geo.unibe.ch (M.W.)
[2] KVA Linth, Im Fennen 1A, 8867 Niederurnen, Switzerland; s.ringmann@kva-linth.ch
* Correspondence: gisela.weibel@geo.unibe.ch

Abstract: A major part of Swiss fly ashes (FA) from municipal solid waste incineration (MSWI) are treated with the acid fly ash leaching process (FLUWA) in order to recover heavy metals prior to deposition. The FLUWA process uses scrub water from wet flue gas cleaning to leach heavy metals from FA. The leaching efficiency is strongly dependent on the leaching conditions (e.g., pH, Eh, L/S-ratio). This case study presents the optimization of the FLUWA process at the MSWI plant Linth, Switzerland, through determination of ideal process parameters for optimal metal recovery. By means of laboratory- and industrial-scale experiments, the process was adjusted towards a more efficient leaching of Zn, Pb, Cu, and Cd. This included the use of an oxidizing agent (hydrogen peroxide). Laboratory experiments proved to be a powerful tool for simulating process optimizations at industrial scale. An ideal leaching pH of 3.8 was determined and it was observed that the process stability is significantly influenced by the L/S-ratio applied to the leaching process. In the course of the study, the recovery could be improved to 67% Zn, 66% Pb, 30% Cu, and 91% Cd. It can be concluded that for optimal metal recovery the process has to be individually adjusted to the composition of the processed FA and scrub water of each specific FLUWA process.

Keywords: MSWI fly ash; acid leaching; heavy metal recovery; process optimization

Citation: Weibel, G.; Zappatini, A.; Wolffers, M.; Ringmann, S. Optimization of Metal Recovery from MSWI Fly Ash by Acid Leaching: Findings from Laboratory- and Industrial-Scale Experiments. *Processes* **2021**, *9*, 352. https://doi.org/10.3390/pr9020352

Academic Editors: Daniel Vollprecht and Renato Sarc

Received: 30 November 2020
Accepted: 10 February 2021
Published: 14 February 2021

1. Introduction

In Switzerland, approximately 80,000 tons of fly ash (FA) from municipal solid waste incineration (MSWI) arise annually [1]. Due to the incineration of heavy metal bearing objects (e.g., electronics or batteries) contained in the waste, the FA is highly enriched in heavy metals. Swiss FA contain up to 6.6 wt.% Zn, 1.9 wt.% Pb, 0.6 wt.% Cu, and 0.05 wt.% Cd [2], and such elevated heavy metal concentrations have been observed in ashes from other countries [3]. Thus, FA comprises a large potential for recyclable metals where mainly the elements Zn, Pb, Cu, and Cd are of interest. In recent years, a research focus was put on the optimization of the heavy metal separation and recovery from FA. Acid leaching was shown to be the most effective method to mobilize heavy metals from FA [4–11].

Since 1995, the acid fly ash leaching process (FLUWA process) has been established in Switzerland. Therein, heavy metals are extracted from FA using scrub water [12–14], a liquid residue from wet flue gas cleaning. The heavy metals are accumulated in the leachate which can later be used for heavy metal recovery. The aim of the FLUWA process is to produce a leachate that contains high concentrations of the metals to be recovered. The metal recoveries achieved in the FLUWA process depend on the properties of the FA as well as on the leaching conditions. The main process parameters that influence metal mobilization are pH value and the oxidation-reduction potential [15]. The leaching pH is defined by the alkalinity of the FA and the amount of acid accumulated from the scrub water. In order to reach sufficient acidic process conditions for mobilizing the heavy metals, the addition of additional acid is often necessary. The pH value defines whether the mobilized metals reprecipitate during extraction, making them unavailable for recovery. In order to create ideal conditions for the extraction of redox-sensitive metals, an oxidant (e.g., hydrogen peroxide) is necessary during the FLUWA process. Oxidizing conditions during metal extraction are a prerequisite to suppress the reductive precipitation of elements such as Cu, Pb, and to a minor extent, Cd [16–18]. Other factors influencing FA leaching are temperature and reaction time. A temperature of ca. 60 °C arises from the mixing of FA and scrub water during the FLUWA process. The reaction time is dependent on the number of extraction reactors used. Three reactors connected in series are prevalent in Switzerland. An extraction time of one hour, cumulated from 20 min per reactor, has commonly been established. Filtration of the ash slurry by vacuum-belt filtration yields a metal-depleted filter cake and a metal-enriched leachate. The filter cake has less impact on the environment and can be deposited together with bottom ash on a landfill type D according to the Swiss Waste Ordinance [19]. The metalliferous leachate is precipitated to a hydroxide sludge. The hydroxide sludge can so far only be processed abroad as hazardous waste. With the foundation of SwissZinc AG, a central treatment facility for hydroxide sludge is pursued. Implementation is expected by the end of 2023 [20]. This plant will apply a system based on the FLUREC process to produce high-purity zinc and as byproducts Pb, Cu and Cd in a recoverable form [12]. The prescription of recovering heavy metals from FA and recovery of metals is prescribed from 2026 onwards in Switzerland [19]. Therefore, the investigation of the FLUWA process is of increasing interest to improve heavy metal separation and to estimate the limiting factors of metal depletion.

With this study, the optimization of the FLUWA process at the Swiss MSWI plant Linth is shown. The goal of the study was to determine the process parameters for an optimal recovery for Zn, Pb, Cu and Cd. This was achieved by performing experiments both on laboratory and industrial scale. The laboratory-scale experiments allow the preliminary examination of the influence of different parameters on the FA leaching process. As the operation on the industrial process relies on many more variables which cannot all be varied or excluded, the laboratory-scale experiments render it possible to better isolate and examine the influence of selected parameters. These initial experiments also deliver an idea of metal recoveries attainable for the specific FLUWA process. In the course of the study, three industrial-scale experiments were carried out. Such experiments represent the reality of a large-scale operation which runs continuously over hours and days where the input materials may also vary over time. Parallel to each industrial-scale experiment, laboratory-scale experiments were carried out with the same process parameters. This indicates to what extent laboratory-scale experiments can model the industrial FLUWA process. In a first experiment, the "current state" of the FLUWA process was sampled in order to make statements about the operational state and the efficiency of the metal recovery. The successive two industrial-scale experiments were performed to implement and verify the optimized operating parameters determined in the laboratory.

2. Materials and Methods

2.1. FLUWA Process at MSWI Plant Linth: Process Conditions and Composition of FA

A total of 6500 t/y of FA are acid leached in the FLUWA process at MSWI plant Linth. In addition to FA from the plant's own waste incineration, external FA from three other Swiss MSWI plants are treated with the FLUWA process. The FLUWA of the MSWI plant Linth has three reactors operating in series, each with a volume of about 3 m^3 with an average residence time of about 45 min (Figure 1).

Approximately 9000 m^3 of acid scrub water with a hydrochloric acid load of 1600 t HCl (32%) and 5500 m^3 of alkaline scrub water with a sulfur load of 275 t from the wet flue gas cleaning process are available annually for the extraction of the FA. Depending on the experiment, an amount of ca. 2–3 t of FA was leached with 6 m^3 of acid and 3 m^3 of alkaline scrub water per hour of FLUWA operation. When necessary, HCl 32% was added for pH control. For the optimization tests, a temporary hydrogen peroxide (H_2O_2, 30%) feed line was built into reactor 1.

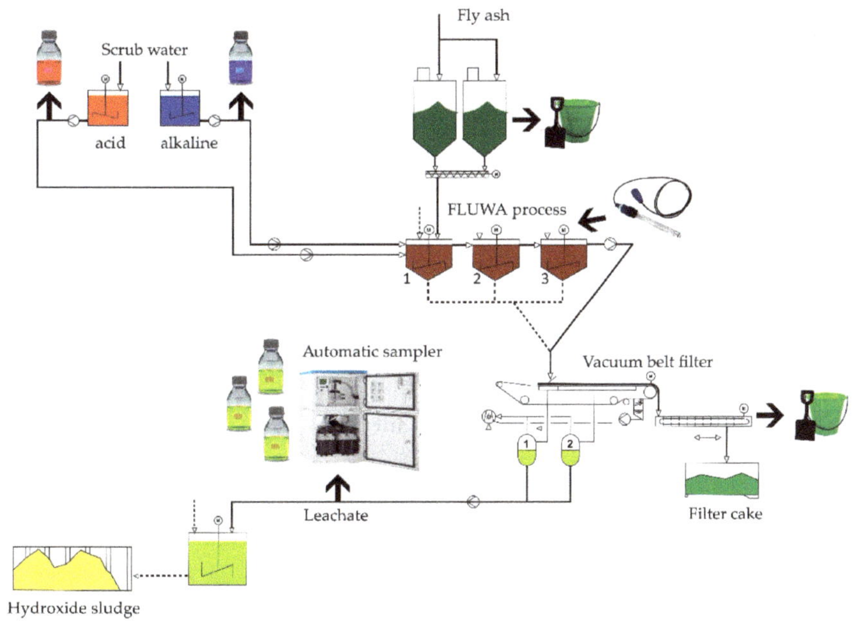

Figure 1. Scheme of the acid fly ash leaching (FLUWA) process and sampling locations during the industrial-scale experiments.

Table 1 summarizes the mean value of the FA composition over the entire project duration and indicates the standard deviation for each element concentration. Compared to the average heavy metal concentrations of Swiss fly ash [2], the FA treated at MSWI plant Linth shows elevated Zn (52,593 mg/kg), Pb (9378 mg/kg), Cu (2671 mg/kg) and Cd (332 mg/kg) concentrations.

Table 1. Mean elemental composition in mg/kg and standard deviation of the fly ash (FA) treated at municipal solid waste incineration (MSWI) plant Linth.

	Mean Concentration of FA Treated at MSWI Plant Linth		
Al	30,278	±	2268
Ba	2152	±	208
Br	3255	±	174
Ca	144,928	±	8696
Cd	332	±	34
Cl	97,707	±	2892
Cr	684	±	120
Cu	2671	±	291
Fe	19,274	±	1271
K	52,755	±	3590
Mn	881	±	67
Ni	132	±	16
P	3851	±	117
Pb	9378	±	973
S	61,619	±	9450
Sb	2621	±	273
Si	77,689	±	15,562
Sn	1447	±	139
Ti	9783	±	887
Zn	52,593	±	7371

2.2. Chemical Analysis

All solids were dried at 105 °C until constant weight and then ground with a ball mill. Then, 4.0 g of the ground sample was mixed with 0.9 g binder (Fluxana Cereox®) and pressed into a powder pill. The analysis of solids was carried out using X-ray fluorescence analysis on a Spectro Xepos ED-XRF. The elements Al, Ca, Cd, Cu, Fe, K, Mn, Pb, S, Sb, Si, Ti and Zn were determined with matrix-adapted calibration for FA or filter cake. The elements Ba, Br, Cl, Cr, Ni, Pb and Sn were determined with the Turboquant method.

The liquid samples (leachates) were diluted with 1% HNO_3 (HNO_3, 1%, VWR Chemicals, AnalaR Normapur) and analyzed with a Spectroblue SOP ICP-OES. The calibration was performed using Merck Multielement Standard IV and the analyses were checked with a check standard (Merck VIII Multielement Standard). The analytical error amounts to ±5% for all elements except Na, K, Ca, Sb and S that showed ±10% error based on multiple measurements of certified standard solutions.

2.3. Laboratory-Scale Experiments

The preliminary laboratory-scale experiments were carried out in order to determine the influence of different conditions and parameters on the heavy metal recovery. It is already known from preliminary studies that mainly pH value, H_2O_2 dosage and L/S-ratio need to be investigated to optimize a specific FLUWA process [15]. Therefore, six different experiments were performed at laboratory scale (Table 2): at extraction pH of 3.8 and 4.5, with a concentration of 0, 20, 40 and 60 L H_2O_2/t ash and with two different L/S ratios of 2 and 3. The experiments were performed twice to ensure reproducibility and the mean value used as result. In addition to the three industrial-scale experiments, laboratory-scale experiments were performed under conditions simulating the specific process conditions as good as possible.

Table 2. Experimental setups for laboratory-scale experiments.

Parameter	pH	Redox (L/t H_2O_2)	L/S (m^3/t)
Experiment 1	3.8	40	2
Experiment 2	4.5	40	2
Experiment 3	3.8	0	2
Experiment 4	3.8	20	2
Experiment 5	3.8	60	2
Experiment 6	3.8	40	3

For all laboratory tests, 75 g FA was added to the corresponding amount of acid scrub water required for the given L/S-ratio. For each experiment, 70% acid scrub water and 30% alkaline scrub water (HCl = 1.26–1.42 mol/L) were used. Upon mixing the FA with scrub water, the neutralization heat almost instantly led to a rise in temperature in the reactor to approximately 60–65 °C. A heating stirrer including a thermostat was used to maintain a temperature of 60 °C while stirring the mixture at ca. 400 rpm. The neutralization occurring upon mixing also led to a pronounced increase in pH value. The pH was corrected to the desired value (pH 3.8 or 4.5) by the dropwise addition of HCl, 32% (VWR Chemicals, AnalaR Normapur) to the mixture. The dosing of H_2O_2 (30%, Merck, stabilized, for synthesis) was carried out in several small portions within the first 20 min of leaching to simulate the continuous peroxide addition to reactor 1 on the industrial scale. The total leaching time applied was 60 min. The pH and temperature of the suspension were documented as well as the quantities of each reagent added. At the end of the leaching time the mixture was hot filtered. The filter cake was washed twice with deionized water (50 mL each, 18.2 MΩ·cm) and the combined leachate was rapidly conserved by diluting it 1:100 (vol/vol) with nitric acid (HNO_3, 1%, VWR Chemicals, AnalaR Normapur).

2.4. Industrial-Scale Experiments

A main goal of the industrial-scale experiments was to collect time-dependent data on the industrial operation. This in turn allowed to identify and address the problems related to the continuous large-scale industrial process. The experimental setups of the three industrial-scale experiments are shown in Table 3. All material flows were sampled at defined time intervals adapted to the operation (mostly in 30-min intervals). The sampling included FA, scrub water (separated acid and alkaline), filter cake from vacuum belt filter, and leachate (Figure 1). The leachate samples were taken at defined time intervals (sampling in 7 min intervals and mixed to samples of 28 min) by means of an automatic sampler (CSF 48, Endress+Hauser). The sample preparation for the leachates was carried out as soon as possible on site. This includes the uptake of the mass flow (present volume of leachate per unit of time), the filtration and the subsequent stabilization of the sample with HNO_3 (30%, VWR Chemicals, AnalaR Normapur) prior to analysis. In addition, pH value, redox potential, and oxygen content were measured in all three reactors at defined time intervals (usually every 15 min).

Table 3. Experimental setups for industrial-scale experiments.

Parameter.	pH	Redox (L/t H_2O_2)	L/S (m^3/t)
Experiment A	6	0	3
Experiment B	4.5	40	3
Experiment C	4	40	4

2.5. Calculation of the Leaching Recovery

When performing laboratory-scale experiments, mass balance is easily determinable because all volumes, concentrations and weights can be determined. Therefore, the calculation of the metal recovery was performed as follows (Formulas (1) and (2)):

$$recovery\ (\%) = 100\cdot\left(1 - \frac{c(filter\ cake)\left[mg\cdot kg^{-1}\right]\cdot m_{filter\ cake}[kg]}{c(fly\ ash)\left[mg\cdot kg^{-1}\right]\cdot m_{fly\ ash}[kg] + c(scrub\ water)\left[mg\cdot L^{-1}\right]\cdot V_{scrub\ water}[L]}\right) \quad (1)$$

where m is the mass of filter cake or FA. For the laboratory tests, the recovery can also be calculated via the concentration in the leachate:

$$recovery\ (\%) = 100\cdot\frac{c(leachate)\left[mg\cdot L^{-1}\right]\cdot V_{leachate}[L]}{c(fly\ ash)\left[mg\cdot kg^{-1}\right]\cdot m_{fly\ ash}[kmg] + c(scrub\ water)\left[mg\cdot L^{-1}\right]\cdot V_{scrub\ water}[L]} \quad (2)$$

where $V_{leachate}$ is the volume of the leachate at the end of the experiment. The mass balance is then calculated using Formula (3):

$$mass\ balance\ (\%) = 100\cdot\frac{\left(c(leachate)\left[mg\cdot L^{-1}\right]\cdot V_{leachate}[L]\right) + \left(c(filter\ cake)\left[mg\cdot kg^{-1}\right]\cdot m_{filter\ cake}[kg]\right)}{c(fly\ ash)\left[mg\cdot kg^{-1}\right]\cdot m_{fly\ ash}[kg] + c(scrub\ water)\left[mg\cdot L^{-1}\right]\cdot V_{scrub\ water}[L]} \quad (3)$$

As the FLUWA process at MSWI plant Linth is conducted in a continuous process, this weight difference is in most cases not known. The mobilization of the acid-soluble components of the FA (e.g., salts) leads to a reduction of mass of the filter cake. Therefore, the mass loss on the industrial scale is often calculated indirectly by measuring and balancing the "inert elements" (e.g., Al, Si, P, Ca, Ti, Cr, Fe, Ni, Sb and Ba). These elements are hardly mobilized under the respective extraction conditions and therefore accumulate in the filter cake relative to its total mass. Assuming that the absolute amounts of these elements in FA and filter cake are equal and only their concentrations differ, the average enrichment factor of the inert elements can be used as a proxy for the mass loss (Formulas (4) and (5)).

$$enrichment\ factor = \frac{\sum\left(\frac{c_{inert}(filter\ cake)\left[mg\cdot kg^{-1}\right]}{c_{inert}(fly\ ash)\left[mg\cdot kg^{-1}\right]}\right)}{number\ of\ inert\ elements\ used} \quad (4)$$

where *c* is concentration of the respective element.

$$recovery\ (\%) = 100\cdot\left(1 - \frac{c(filter\ cake)\left[mg\cdot kg^{-1}\right]}{c(fly\ ash)\left[mg\cdot kg^{-1}\right]\cdot enrichment\ factor}\right) \quad (5)$$

3. Results

3.1. Laboratory-Scale Experiments

Figure 2 shows the results obtained in the laboratory-scale experiments. The results of these experiments allowed the identification of optimal parameters for the industrial-scale FLUWA operation at MSWI plant Linth.

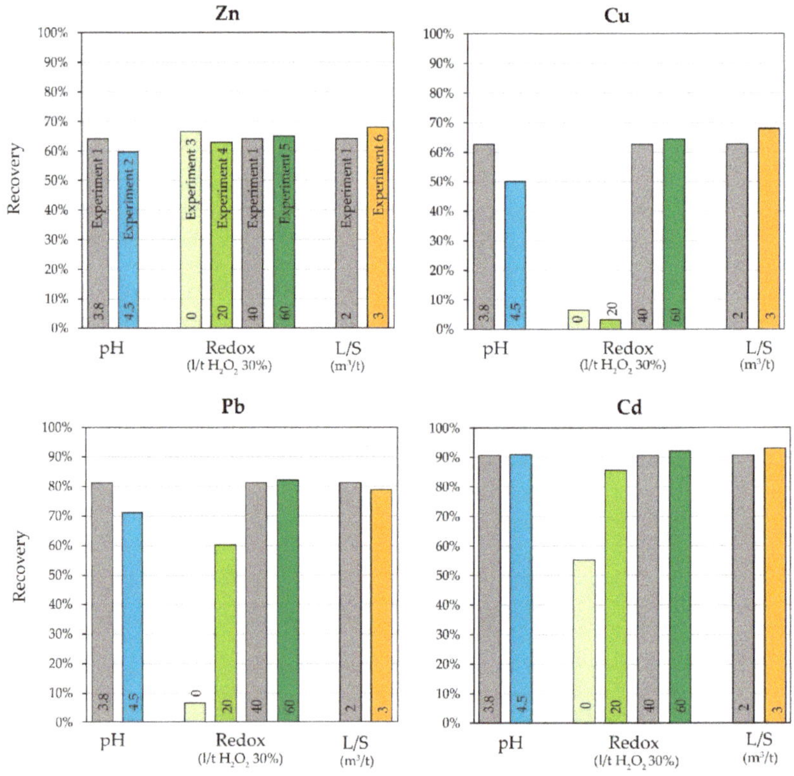

Figure 2. Recoveries of laboratory-scale experiments for Zn, Cu, Pb and Cd to simulate the FLUWA process under different leaching conditions.

The variation of the three parameters pH, redox and L/S ratio throughout the experiments affected the metal recoveries in different ways. A leachate pH value of 4.0 and above at the end of the experiment resulted in diminished recoveries of all four elements studied (Experiment 2). This effect was most pronounced for Cu and Pb. If the pH is adjusted to a value of 3.8 at the end of the leaching process, the recoveries of Cu and Pb are 10% higher (Experiment 1). The oxidizing conditions when adding H_2O_2 to the ash slurry mainly affect the elements Cu and Pb and to a minor extent Cd. Without the addition of an oxidizing agent, 67% Zn and 55% Cd are recovered from FA whereas the recoveries of Pb and Cd are extremely low (<10%) (Experiment 3). The addition of 20 L H_2O_2/t FA increases the recoveries of Pb (60%) and Cd (86%) (Experiment 4). To achieve significant recoveries for Cu, 40 L H_2O_2/t FA are required (63%) (Experiment 1). Table 4 lists the analytical results of laboratory experiment 1 as well as the mass balance calculated using Formula (3). An addition of more than 40 L H_2O_2 to the ash slurry did not significantly improve Cu recovery (Experiment 5). In addition, the L/S-ratio did not have a large effect on the recoveries measured for the laboratory-scale experiments, at least not in the applied range between L/S 2 and 3 (Experiment 6).

Table 4. Mass balance of laboratory-scale experiment 1.

	Amount of material	Elemental concentrations (ppm)			
		Cd	Zn	Pb	Cu
Fly ash (FA)	75 g	364 ± 10	47,877 ± 2200	10,080 ± 275	2942 ± 115
Scrub water	150 mL	<0.1	<0.1	<0.1	<0.1
Filter cake †	164 ± 1 g	48 ± 2	24,715 ± 841	2740 ± 23	1571 ± 16
Leachate	180 ± 3 mL ††	128 ± 1	11,519 ± 112	3242 ± 66	879 ± 7
Mass balance		95%	93%	97%	94%

† dry (105 °C). †† comprising 60 mL of deionized water used for rinsing the filter cake.

The optimal leaching conditions determined in the laboratory for the FLUWA process at MSWI plant Linth are as follows: Extraction pH of 3.8, 40 L H_2O_2/t FA and a L/S of 2 (Experiment 1). These parameters served as a basis for the optimization of the industrial FLUWA process.

3.2. Industrial-Scale Experiments

The industrial-scale experiments at the MSWI plant Linth provided time-resolved data over one day of the FLUWA process (Figure 3). The time-resolved sampling of input-FA, leachate and filter cake allowed to determine the metal recoveries (Formulas (4) and (5)). The simultaneous pH monitoring in the last reactor of the FLUWA process helped to identify the potential and limitations of the recovery for the metals Zn, Cu, Pb and Cd.

Experiment A depicted and investigated the "current state" of the FLUWA process before any optimizations. No H_2O_2 was added during this experiment and the pH value and L/S ratio were not adjusted regularly. The concentration of Cu (2648 to 2927 mg/kg) and Cd (250 to 320 mg/kg) in the input-FA were rather constant within the entire day (Figure 3). On the other hand, the concentration of Zn (46,830 to 80,300 mg/kg) and Pb (8743 to 10,680 mg/kg) that enters the FLUWA process with the input-FA was increasing at the beginning of the process and decreasing again with increasing process duration. Depending on the different composition of the FA added to the FLUWA process, the pH value and metal recovery show strong fluctuations over time (Figure 3). The pH value of the ash slurry in reactor 3 started low (pH 2) and gradually increased over time, leveling off at ca. pH 6 at the end of the process. Under these conditions, little Cu and Pb could be recovered throughout the day. Less pH and redox sensitive elements such as Zn and Cu could be recovered better under the prevailing conditions. Especially Zn showed a constant recovery between 58 and 68%, whereas the recovery of Cd fluctuated in the range of 34 to 69%.

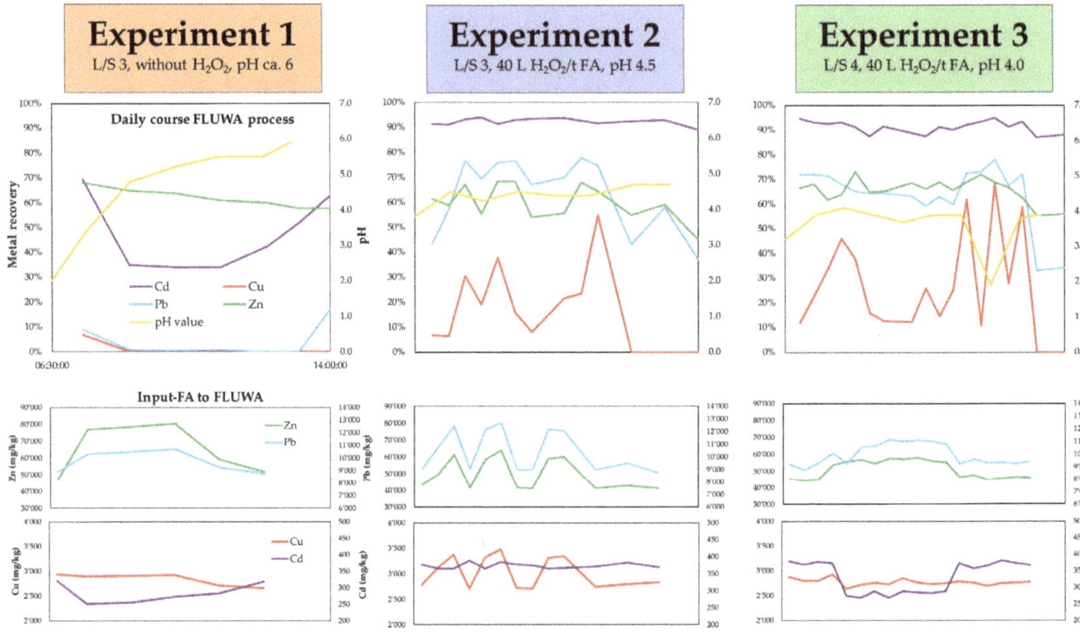

Figure 3. Results of the daily course of the three industrial-scale experiments of the FLUWA process at MSWI plant Linth. The concentrations in the input-FA (mg/kg) and the recoveries (%) of the metals Zn, Pb, Cu and Cd as well as the pH value at the end of the process are shown for each experimental day.

In Experiment B, the initial plan was to be able to better control the pH by a lower FA dosing (increased L/S) and to introduce oxidizing leaching conditions. However, the input-FA dosage could be reduced only slightly due to the large quantities of FA present and the limited capacities in the storage silos. Thus, the L/S-ratio was on a daily average at the same value as in Experiment A (L/S 3). In addition, large concentration fluctuations of Zn (41,080 to 63,860 mg/kg), Pb (9498 to 12,670 mg/kg) and Cu (2699 to 3474 mg/kg) in the input-FA affected the extraction recovery during this experiment. Despite these difficult conditions, it was possible to keep the pH value at a lower level than in Experiment A by the additional dosage of HCl 32%. The pH-value at the end of the leaching process in reactor 3 was maintained at pH 4.5. In addition to the more acidic leaching conditions, oxidizing conditions were established with the addition of 40 L/t H_2O_2 to the FLUWA process. The recoveries of Pb, Cu and Cd were significantly better due to the oxidizing, more acidic leaching conditions compared to Experiment A. The daily curves of the leaching recoveries of Zn, Pb and Cu show a clear correlation with the concentrations of the added input-FA. The higher the concentrations of FA, the higher the metal recovery. For Cd, no such correlation is apparent, and the recoveries are very stable at a high level.

Experiment C was again performed under oxidizing conditions using 40 L/t H_2O_2. However, the input-FA was homogenized compared to experiment B by mixing the FA of the different storage silos before dosing (Zn (44,520 to 57,990 mg/kg), Pb (9222 to 11,100 mg/kg), Cu (2645 to 2929 mg/kg), Cd (269 to 381 mg/kg)). In addition, the FA dosage could be reduced leading to a leaching L/S-ratio of 4. The higher L/S led to the pH value being kept more stable at pH 4.0 during the entire experimental day. Due to the more homogenic FA-input and the more stable process pH-value, the recoveries for Zn and Pb were much more stable during the day. Despite the stable input concentrations, however, there were still exceptionally large fluctuations in the Cu recovery during Experiment C.

In addition to the daily course shown in the mean recoveries for Zn, Pb, Cu and Cd over the entire day of the industrial-scale experiments are listed in Table 5. It can be summarized that the recovery of Zn remains at a very stable level (60–70%), irrespective of the adjustments made, whereas for Cd, the recovery could be increased from 52% to >90%. For the more pH- and redox-sensitive elements Pb and Cu, increased recoveries were also achieved with the optimization experiments, but at a lower level (Pb 11% to 66%, Cu 0% to 30%).

Table 5. Mean recoveries of Zn, Pb, Cu and Cd of the industrial-scale experiments of the FLUWA process within one day at MSWI plant Linth.

Experiment	Zn [%]	Pb [%]	Cu [%]	Cd [%]
A	66	11	0	52
B	60	65	21	92
C	67	66	30	91

The results of the industrial-scale experiments have shown that it is difficult to implement the ideal leaching conditions determined in the laboratory on a large-scale. For this reason, experiments were carried out in the laboratory with the materials (FA and scrub water) and conditions (pH, Eh and L/S) obtained at the large scale-experiments. This allows a direct comparison of laboratory- and industrial-scale experiments (Figure 4). The results show that the recoveries for Cd, Pb and Zn in the laboratory were comparable to those of the industrial-scale experiments. However, despite the detailed reconstruction of the operating conditions, large deviations could be observed in the recovery of Cu. Laboratory-scale experiments show higher leaching recoveries for Cu (7%, 50% and 68%) compared to the corresponding industrial-scale experiments (0%, 22% and 30%).

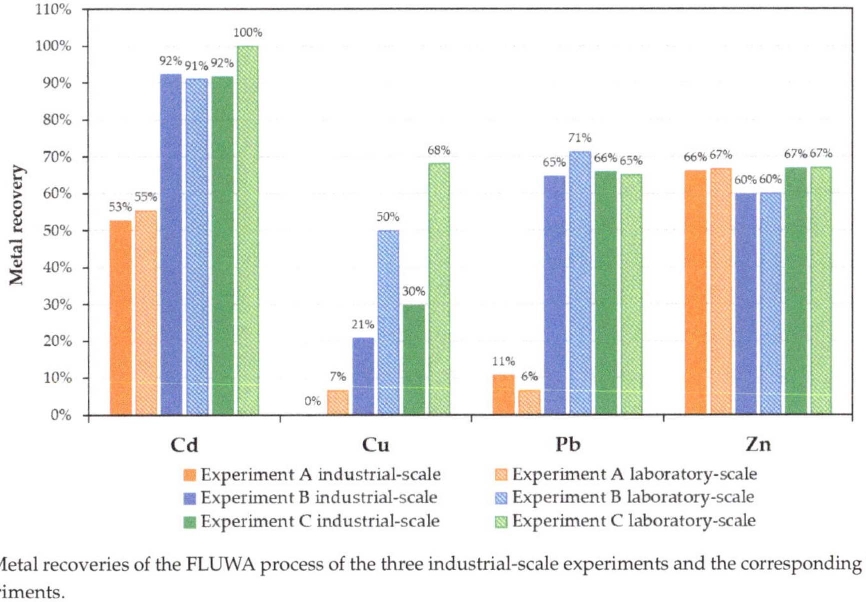

Figure 4. Metal recoveries of the FLUWA process of the three industrial-scale experiments and the corresponding laboratory-scale experiments.

4. Discussion

4.1. *Process Stability and Leaching Behaviour of Zn, Cu, Pb, Cd*

The recovery of Zn for both laboratory and industrial-scale experiments were between 60% and 70% regardless of the conditions (Figure 4). The use of H_2O_2 had no influence at all, whereas an increased L/S-ratio and lower pH value slightly but not significantly increased the recovery (Figure 2). Zn recovery thus seems to mainly depend on its input concentration and binding forms. The more Zn is present in FA, the more can usually be mobilized [15].

The redox conditions during extraction are of central importance for the recovery of redox-sensitive elements such as Cu (E°: +0.35 V), Pb (E°: −0.31 V) and Cd (E°: −0.40 V). Without the addition of H_2O_2, the reducing conditions during the FLUWA process led to the cementation reaction of dissolved Cu^{2+}, Pb^{2+} and Cd^{2+} on the surface of less noble metals present in the FA, such as Al^0 (E°: −1.66 V), Zn^0 (E°: −0.76 V) or Fe^0 (E°: −0.41 V) (Reaction Equation (6)).

$$3Cu^{2+} + 2Al^0 \rightarrow 3Cu^0\downarrow + 2Al^{3+} \tag{6}$$

The addition of H_2O_2 to the FLUWA process oxidizes most of the metallic components in FA, resulting in more noble metals such as Cu and Pb being retained in solution and thus recovered (Reaction Equation (7)).

$$2Al^0 + 3H_2O_2 + 6H^+ \rightarrow 2Al^{3+} + 6H_2O \tag{7}$$

The recoveries for Cu and Pb are further strongly dependent on the pH value during the leaching process. The time-resolved monitoring of the process conditions in reactor 3 have shown that in a continuous process, the L/S-ratio significantly influences process stability (Figure 3). This is by enhancing or diminishing the effects of naturally occurring fluctuations in both FA and acid scrub water compositions [2]. The FA as well as the scrub water composition depend strongly on the waste input as well as the type and condition of the flue gas cleaning system or dust removal system. They show large variations over time, even in one MSWI plant. In addition to varying heavy metal concentrations, FA also differ in terms of their alkalinity. The carbonate present in the FA thereby buffers the pH of the suspension to >5, which promotes hydroxide precipitation of the metals dissolved from the FA [21]. In order to keep the extracted metals in solution, the pH in the leaching reactors usually has to be lowered with additional HCl (32%). The amount of acid required to reach a certain pH during the FLUWA process varies depending on the FA extracted. The lower the L/S-ratio, the more difficult it is to maintain a constant pH value. With a pH value of ≤3.8 at the end of the FLUWA process, the hydroxide precipitation of most metals can be kept at a satisfactorily low level. The hydroxide precipitation can be well demonstrated by laboratory-scale experiments (Figure 2). An increase in pH value of 0.7 units between a value of 3.8 and 4.5 led to lower recoveries of Pb and Cu. Cu begins to precipitate as copper hydroxide (or basic copper salts) at a pH of 4.0–4.5 and thus accumulates in the leached filter cake (Reaction Equation (8)) [22].

$$Cu^{2+} + 2OH^- \rightarrow Cu(OH)_2\downarrow \tag{8}$$

In addition to hydroxide precipitation, Pb can also form poorly soluble sulfates and hence be removed from the leachate by lead-sulfate precipitation [22]. The sulfate supplied by the scrub water does mainly precipitate as gypsum. If the sulfate load in the scrub water exceeds the Ca available from the FA, Pb can form lead sulfate and hence precipitates. This problem can be somewhat alleviated as lead has the ability to form chloride complexes, which in turn are relatively soluble [23]. Alternatively, the scrub water could be treated with a Ca containing reagent in order to reduce the sulfate load before the leaching. As the pH value strongly affects whether the extracted metals re-precipitate and are thus lost, it is important that the pH value of the leachate is closely monitored.

4.2. Comparability Between Laboratory-and Industrial-Scale Experiments

An important aspect of the present project was to verify the transferability of the FLUWA process parameters determined by laboratory-scale experiments to operational practice. Generally, there is a good agreement between the laboratory-scale and the corresponding industrial-scale experiments (Figure 4). Therefore, laboratory-scale experiments are appropriate to simulate the large-scale FLUWA process. Operating parameters determined in the laboratory for optimal recoveries of heavy metals could further be implemented on a large-scale. In the laboratory, the variation of the L/S-ratio had only a small influence on heavy metal recoveries because there, the complication of temporal change of FA composition present at the industrial scale is omitted. Cu, reacting most sensitive to slight pH value and redox condition changes, is however more challenging to reliably extract on the industrial-scale in comparison to the laboratory-scale leaching. This can be explained by the fact that the pH value in the laboratory reactor can be controlled more precisely than during continuous, large-scale operation.

5. Conclusions

This work has confirmed the importance of a stable process control of the FLUWA in order to achieve a high metal recovery. This is particularly important for the recovery of pH- and redox-sensitive elements such as Cu and Pb. Important parameters for a high metal recovery are the use of a suitable oxidizing agent (e.g., hydrogen peroxide) and precise pH control to assure maintenance of a final pH ≤ 3.8 to avoid metal precipitation. Stable process control can be achieved through good mixing of FA as well as elevated L/S-ratio. The optimal leaching parameters have to be determined and adjusted periodically. FLUWA process parameters for one MSWI plant cannot directly be applied to another plant and must be evaluated in laboratory-scale experiments for each specific FLUWA process. With this study it could be successfully demonstrated that this transferability of laboratory-scale experiments into the operational practice is possible.

Author Contributions: Conceptualization, G.W., A.Z. and U.E.; methodology, G.W., A.Z., U.E. and S.S.; formal analysis, G.W., A.Z.; investigation, G.W., A.Z.; resources, G.W.; data curation, G.W., A.Z.; writing—original draft preparation, G.W., A.Z.; writing—review and editing, M.W., S.R., U.E.; visualization, G.W., A.Z.; supervision, U.E., S.S.; project administration, G.W.; funding acquisition, G.W., U.E. All authors have read and agreed to the published version of the manuscript.

Funding: This research was partially funded by MSWI plant Linth, where the industrial-scale experiments were also performed. Further funding was provided by the Swiss Federal Office for the Environment (FOEN) and the association of operators of swiss waste treatment industrials (VBSA).

Acknowledgments: Many thanks to Markus Stäger, Walther Gasser and Thomas Heckendorn (KVA Linth) for their technical support during sampling and industrial-scale experiments. Special thanks to Urs Eggenberger (U.E., University of Bern), Stefan Schlumberger (S.S., ZAR) for project feedback and discussion. Analytical support and assistance with experiments by Andreas Glauser and Christine Lemp (University of Bern) is highly acknowledged.

Conflicts of Interest: The authors declare no conflict of interest.

References

1. Federal Office for the Environment (FOEN). Indikator Abfall. 2016. Available online: https://www.bafu.admin.ch (accessed on 30 October 2020).
2. Zucha, W.; Weibel, G.; Wolffers, M.; Eggenberger, U. Inventory of MSWI Fly Ash in Switzerland: Heavy Metal Recovery Potential and Their Properties for Acid Leaching. *Processes* **2020**, *8*, 1668. [CrossRef]
3. Bayuseno, A.P.; Schmahl, W.W. Characterization of MSWI fly ash through mineralogy and water extraction. *Resour. Conserv. Recycl.* **2011**, *55*, 524–534. [CrossRef]
4. Katsuura, H.; Inoue, T.; Hiraoka, M.; Sakai, S. Full-scale plant study on fly ash treatment by the acid extraction process. *Waste Manag.* **1996**, *16*, 491–499. [CrossRef]
5. Van der Bruggen, B.; Vogels, G.; Van Herck, P.; Vandecasteele, C. Simulation of acid washing of municipal solid waste incineration fly ashes in order to remove heavy metals. *J. Hazard. Mater.* **1998**, *57*, 127–144. [CrossRef]
6. Ludwig, B.; Khanna, P.; Prenzel, J.; Beese, F. Heavy metal release from different ashes during serial batch tests using water and acid. *Waste Manag.* **2005**, *25*, 1055–1066. [CrossRef] [PubMed]
7. Chiang, K.-Y.; Jih, J.-C.; Chien, M.-D. The acid extraction of metals from municipal solid waste incinerator products. *Hydrometallurgy* **2008**, *93*, 16–22. [CrossRef]
8. Yang, R.; Liao, W.-P.; Lin, C.-Y. Feasibility of lead and copper recovery from MSWI fly ash by combining acid leaching and electrodeposition treatment. *Environ. Prog. Sustain. Energy* **2012**, *32*, 1074–1081. [CrossRef]

9. Tang, J.; Steenari, B.M. Solvent extraction separation of copper and zinc from MSWI fly ash leachates. *Waste Manag.* **2015**, *44*, 147–154. [CrossRef] [PubMed]
10. Quina, M.J.; Bontempi, E.; Bogush, A.; Schlumberger, S.; Weibel, G.; Braga, R.; Funari, V.; Hyks, J.; Rasmussen, E.; Lederer, J. Technologies for the management of MSW incineration ashes from gas cleaning: New perspectives on recovery of secondary raw materials and circular economy. *Sci. Total Environ.* **2016**, *635*, 526–542. [CrossRef] [PubMed]
11. Ferraro, A.; Farina, I.; Race, M.; Colangelo, F.; Cioffi, R.; Fabbricino, M. Pre-treatments of MSWI fly-ashes: A comprehensive review to determine optimal conditions for their reuse and/or environmentally sustainable disposal. *Rev. Environ. Sci. Biotechnol.* **2019**, *18*, 453–471. [CrossRef]
12. Schlumberger, S.; Schuster, M.; Ringmann, S.; Koralewska, R. Recovery of high purity zinc from filter ash produced during the thermal treatment of waste and inerting of residual materials. *Waste Manag. Res.* **2007**, *25*, 547–555. [CrossRef] [PubMed]
13. Bühler, A.; Schlumberger, S. Recovering heavy metals from fly ash "Acid fly ash leaching–FLUWA process" a forward-looking process in waste in-cineration. MSWI residues in Switzerland–The raw material with added value. *Fed. Off. Environ.* **2010**, *1*, 185–192.
14. Karlfeldt Fedje, K.; Andersson, S. Zinc recovery from Waste-to-Energy fly ash—A pilot test study. *Waste Manag.* **2020**, *118*, 90–98. [CrossRef] [PubMed]
15. Weibel, G.; Eggenberger, U.; Schlumberger, S.; Mäder, U.K. Chemical associations and mobilization of heavy metals in fly ash from municipal solid waste incineration. *Waste Manag.* **2017**, *62*, 147–159. [CrossRef] [PubMed]
16. Free, M. *Hydrometallurgy Fundamentals and Applications*; John Wiley & Sons, Inc.: Hoboken, NJ, USA, 2013; pp. 1–427.
17. Karavasteva, M. Kinetics and deposit morphology of copper cementation onto zinc, iron and aluminium. *Hydrometallurgy* **2005**, *76*, 149–152. [CrossRef]
18. Demirkıran, N.; Ekmekyapar, A.; Künkül, A.; Baysar, A. A kinetic study of copper cementation with zinc in aqueous solutions. *Int. J. Miner. Process.* **2007**, *82*, 80–85.
19. Swiss Confederation. Verordnung über die Vermeidung und die Entsorgung von Abfällen (VVEA). Available online: https://www.fedlex.admin.ch/eli/cc/2015/891/de (accessed on 15 November 2020).
20. Stiftung Zentrum für Abfall- und nachhaltige Ressourcennutzung (ZAR). SwissZinc—National Plant for the Recycling of MSWI Hydroxide Sludge (Fact Sheet no. 2). 2016. Available online: https://zar-ch.ch/fileadmin/user_upload/Contentdokumente/Oeffentliche_Dokumente/Projektblatt_SwissZinc_Nr2.pdf (accessed on 30 November 2020).
21. Hartinger, L. *Handbook of Wastewater and Recycling Technology*; Fachbuchverlag Leipzig: Munich, Germany, 1991; Volume 2, ISBN 978-3-446-43170-6.
22. Hollemann, A.F.; Wiberg, E. *Lehrbuch der Anorganischen Chemie*; Walter de Gruyter: Berlin, Deutschland, 2007; Volume 2, ISBN 978-3-11-017770-1.
23. Weibel, G.; Eggenberger, U.; Kulik, D.A.; Hummel, W.; Schlumberger, S.; Klink, W.; Fisch, M.; Mäder, U.K. Extraction of heavy metals from MSWI fly ash using hydrochloric acid and sodium chloride solution. *Waste Manag.* **2018**, *76*, 457–471. [CrossRef] [PubMed]

processes

MDPI

Article

Waste Wood Fly Ash Treatment in Switzerland: Effects of Co-Processing with Fly Ash from Municipal Solid Waste on Cr(VI) Reduction and Heavy Metal Recovery

Mirjam Wolffers [1,*], Gisela Weibel [1,2] and Urs Eggenberger [1]

1 Institute of Geological Sciences, University of Bern, Baltzerstasse 1 + 3, 3012 Bern, Switzerland; gisela.weibel@geo.unibe.ch (G.W.); urs.eggenberger@geo.unibe.ch (U.E.)
2 Zentrum für nachhaltige Abfall- und Ressourcennutzung ZAR, Wildbachstrasse 2, 8340 Hinwil, Switzerland
* Correspondence: mirjam.wolffers@geo.unibe.ch

Abstract: In Switzerland, waste wood fly ash (WWFA) must be treated before deposition on landfills due to its high pollutant load (Cr(VI) and heavy metals). Acid fly ash leaching, the process used for heavy metal recovery from municipal solid waste incineration fly ash (MSWIFA), represents a possible treatment for heavy metal depletion and Cr(VI) reduction in WWFA. The co-processing of WWFA with MSWIFA during acid fly ash leaching was investigated in laboratory- and industrial-scale experiments with different setups. Of interest were the effects on heavy metal recovery efficiency, the successful outcome of Cr(VI) reduction and consumption of neutralizing chemicals (HCl, H_2O_2). Detailed chemical and mineralogical characterization of two WWFA types and MSWIFA showed that MSWIFA has higher concentrations in potentially harmful elements than WWFA. However, both WWFA types showed high concentrations in Pb and Cr(VI), and therefore need treatment prior to deposition. Depending on the waste wood proportion and quality, WWFA showed chemical and mineralogical differences that affect leaching behavior. In all experimental setups, successful Cr(VI) reduction was achieved. However, WWFA showed higher consumption of HCl and H_2O_2, the latter resulting in a particularly negative effect on the recovery of Pb and Cu. Thus, co-processing of smaller WWFA portions could be expedient in order to diminish the negative effects of Pb and Cu recovery.

Keywords: wood ash treatment; MSWI fly ash; heavy metal recovery; acid leaching; chromate reduction; hot alkaline extraction

Citation: Wolffers, M.; Weibel, G.; Eggenberger, U. Waste Wood Fly Ash Treatment in Switzerland: Effects of Co-Processing with Fly Ash from Municipal Solid Waste on Cr(VI) Reduction and Heavy Metal Recovery. *Processes* **2021**, *9*, 146. https://doi.org/10.3390/pr9010146

Received: 30 November 2020
Accepted: 11 January 2021
Published: 13 January 2021

1. Introduction

The demand for renewable heat and energy production using the CO_2 neutral energy source wood has been growing enormously in Switzerland over the last decades—leading to strongly increasing amounts of wood ashes. In Switzerland, an annual load of 60,000 t wood ash arises from automatic firings through the energetic use of natural wood (e.g., forest) and from the thermal utilization of waste wood (e.g., coated, painted wood) [1]. A quarter thereof represents wood ash from waste wood enriched in heavy metals and Cr(VI). Depending on the incinerator plant and furnace, wood ash can be divided into up to three different fractions: grate ash, cyclone ash and filter ash [2]. The coarse-grained grate ash arises directly from the grate and is equivalent to bottom ash residing from municipal solid waste incineration (MSWI). This is the biggest fraction with roughly 60–90 wt % of the thermal residue [2]. The cyclone ash and the filter ash arise at the flue gas cleaning system and are often collected together and referred to as wood fly ash. Compared to grate ash, wood fly ash is enriched in volatile elements (e.g., Cl, heavy metals) since their low boiling point makes them evaporate during combustion and later precipitate at the flue gas cleaning system [3]. The chemical composition of wood ash is mainly dependent on wood quality and incineration conditions [3,4]. Factors affecting wood quality are wood type, compartment, growing environment and possible treatments (e.g., impregnation) prior to combustion [3,5]. While wood ash from natural wood may be enriched in organic pollutants (e.g., PAH, PCDD-/F [6]) and can contain naturally incorporated heavy metals in elevated concentrations, the contaminated waste wood ash is mostly characterized by highly elevated heavy metal concentrations (e.g., Zn, Pb, Cu, Cr) remaining from paints, coatings or impregnation [7]. Due to the oxidative conditions during combustion, Cr(III) compounds are mainly oxidized to very toxic and highly mobile Cr(VI) [8]. Thermal residues from waste wood which have been impregnated with Cr-bearing compounds are often severely enriched in Cr(VI). In water extraction tests, studies report Cr(VI) concentrations in waste wood ash that exceed the threshold value for landfilling [8,9].

Because of the possibly high contaminant load, wood ash is considered as waste and must be dumped on landfills, although opportunities for recycling are being sought (e.g., in concrete production [10]). The less polluted grate ash and fly ash from natural wood can be deposited without further treatment on landfill type D and E, according to the Swiss Waste Ordinance [11]. Waste wood fly ash (WWFA), however, must be treated before deposition due to the elevated concentrations in Cr(VI) and possibly environmentally harmful heavy metals. As there is currently not enough capacity available in Cr(VI) reducing facilities in Switzerland to treat the entire quantity of WWFA before landfilling, waste wood fly ashes can be deposited temporarily without a prior treatment on landfill type D or E (depending on their total organic carbon (TOC) content (<2 or <5 wt %, respectively) until 2023. From 2023 on, WWFA must be treated in order to reduce Cr(VI) and recover the heavy metals. Acid fly ash leaching with the FLUWA process [12,13] represents a promising method for treating WWFA prior to deposition. The FLUWA process represents the state-of-the-art process in Switzerland for recovering heavy metals (mainly Zn, Cd, Pb, Cu) from the similarly generated MSWI fly ash (MSWIFA). The ash is thereby leached with acid scrub water, acid from the flue gas cleaning system (~5% HCl). Filtration of the ash slurry yields a heavy metal enriched leachate that is then precipitated to a hydroxide sludge for subsequent heavy metal recovery and a filter cake depleted in heavy metals that is deposited on landfills. Because WWFA can yield heavy metal concentrations in the same range as MSWIFA, but occur in smaller quantities, a cotreatment of both ashes could be expedient. Beginning 2021, all Swiss MSWIFA must be treated before deposition [11] and depending on the heavy metal recovery guideline in revision, the use of an oxidant (e.g., hydrogen peroxide (H_2O_2)) will be necessary during the FLUWA process. Oxidizing conditions during metal extraction are a prerequisite to suppress the reductive precipitation of the redox-sensitive elements Cu, Pb, and to a minor extent, Cd. The co-processing of WWFA in the FLUWA process is already carried out at this study's investigation site Energiezentrale Bern, where it is economically favorable to co-process the arising WWFA40, a WWFA with 40% waste wood content. However, the heavy metal extraction efficiency of cotreating WWFA40 and the completeness of Cr(VI) reduction during the FLUWA process have not been investigated in detail.

The aim of the study was, therefore, to investigate the heavy metal recovery, as well as the Cr(VI) reduction efficiency of the FLUWA process when WWFA40 is co-processed. The actual state was investigated (reducing conditions) as well as the future state (oxidizing conditions) when oxidizing conditions become state-of-the-art for the FLUWA process. The industrial process was first simulated in laboratory-scale experiments in order to evaluate the leaching behavior of the different ash types (MSWIFA, WWFA40 and MFA (a mix of the two ashes to simulate the co-processing)) and to quantify the leaching efficiency in terms of heavy metal recovery and amount of neutralizing chemicals. The same experimental setups were later implemented at an industrial scale. Of special interest of the experiments was the heavy metal recovery efficiency, successful outcome of Cr(VI) reduction under reducing and oxidizing processing conditions and consumption of neutralizing chemicals (HCl, H_2O_2). The ashes used in the experiment were characterized with respect to their chemical and mineralogical composition, and their acid-neutralizing capacity (ANC) was determined. For comparison, the sample WWFA100 with 100% waste wood content was analyzed. To assess both the hazard potential of WWFA and the completeness of Cr(VI) reduction, the water-soluble and total content of Cr(VI) were determined for WWFA and filter cakes.

2. Materials and Methods

2.1. Sample Origin, Sampling and Sample Preparation

MSWIFA and WWFA40 samples were collected in 2017 at the waste and wood power plant Energiezentrale Bern. An annual amount of 135,000 t municipal solid waste and 65,000 t of wood (25% water content, fluidized-bed combustion) was incinerated in separate incinerators in 2017. MSW and wood are combusted separately but treated together prior to landfilling with the FLUWA process in two consecutive extraction reactors (1 m^3 each). The co-processing of WWFA40 in the FLUWA is economically favorable since excess acidity of their scrub water is consumed by the alkalinity of the WWFA40, and the use of lime milk is minimized. At present, the FLUWA process at the incineration plant is performed under reducing conditions (without the addition of H_2O_2) with MSWIFA and WWFA40 proportions in the ratio as they are produced. Adjustments of the ash ratio are made in the current process such that a favorable extraction pH of 3.8 is achieved. To perform industrial experiments at oxidizing conditions, a pumping system was installed for continuous dosing of H_2O_2.

In order to understand the geochemical differences between MSWIFA and WWFA40, three representative composite samples of each ash type were investigated in terms of chemical and mineralogical composition. The sampling duration varied between one and three weeks. Samples were taken twice a day and mixed into composite samples. Additionally, three samples of WWFA40 (weekly composite samples) and their corresponding filter cakes were made available for Cr(VI) analyses. For comparison, the sample WWFA100 (100% waste wood, monthly composite sample) from a Swiss biomass power plant was investigated.

Approximately 10 kg of ash was collected in each sampling campaign. The ashes were homogenized and split into 1 kg working batches and dried at 105 °C until constant weight for chemical analysis and at 40 °C for mineralogical analysis.

2.2. Chemical Analysis

The elemental composition of the ashes was obtained through energy-dispersive X-ray fluorescence (ED-XRF) analysis performed on pressed powder pellets (4.0 g ash, 0.9 g wax as a binder) using a Spectron Xepos (SPECTRO, Kleve, Germany) spectrometer with matrix adjusted calibration. For quality control, the samples were analyzed in duplicates. The accuracy of the method was previously verified by the authors [14] through multiple determinations of similar ash samples and the analysis of the standard reference material BCR 176R [15]. The ED-XRF measurements showed good reproducibility within <2% for the elements Cu, Zn, Cd, Sb, Pb, Br, Sn, Ba, within 5% for Al, Si, S, Cl, Ca, Ti, Mn, Fe, Cr, Sr and within <10% for K, Na, Mg. Extract solutions were analyzed by inductively coupled plasma optical emission spectroscopy (ICP-OES) on a Thermo Scientific iCAP 700 Series (Waltham, MA, USA) after dilution with HNO_3 1% and calibrated with the multielement standards CertiPUR IV and X (Merck, Kenilworth, NJ, USA). The analytical error accounts for ±5% for all elements except Na, K, Ca, Sb and S that showed ±10% error based on multiple measurements of certified standard solutions.

2.3. Mineralogical Analysis

The mineralogical composition was obtained throughout X-ray powder diffraction (XRD) analysis using a Panalytical X'Pert Pro diffractometer (CuKα-radiation) (Malvern Panalytical, Almelo, Netherlands). 4 g of material was mixed with 1 g of internal standard (corundum) and ground dry for 6 min at 55 Hz in an XRD McCrone mill from Retsch (Haan, Germany). Measurement was performed on disoriented samples from 5 to 75° 2Theta at 40 kV acceleration voltage and at an electron generating current of 40 mA. An automatic divergence slit was used. For quantification, TOPAS-Academic software (V6, Coelho Software, Brisbane, Australia) was used (Rietveld refinement). The extended uncertainty is attributed to 50% for concentrations <1%, 20% for concentrations <5% and 10% to concentrations >5%. The structural data (*.cif files) of the inorganic crystal structure database (ICSD) was used, and phase concentrations and amorphous part were calculated based on the internal standard.

2.4. Acid-Neutralizing Capacity

Acid-neutralizing capacity (ANC) titration was performed using a 785 DPM Titrino device from Metrohm (Herisau, Switzerland). with 2 g of ash 20 mL ultrapure water. Every 10 min, 1 mL of HCl 1 M was added under continuous stirring until the ash slurry reached a pH of 2.

2.5. Water-Extractable Cr(VI)

The content of water-soluble Cr(VI) was determined in WWFA samples in order to make statements about the hazard potential of WWFA. Furthermore, the content of water-soluble Cr(VI) in the filter cakes was determined to examine the completeness of the Cr(VI) reduction during the FLUWA process. Cr(VI) analyses were performed on the eluates obtained with the standard eluate test F-22 [16]. The ash and filter cake samples were thereby eluted with ultrapure water for 24 h at a liquid to solid (L/S) ratio of 10 (4 g ash, 40 mL ultrapure water). For quality control, as well as to investigate transformations in the redox state during the eluate test, Cr(III) and Cr(VI) spikes were added, and the eluate test was performed two times per sample, each spiked and unspiked. Selected samples were further eluted as duplicates. For the spiking, the concentration of 50 mg/kg Cr(III) in the form of $CrNO_3{\cdot}9H_2O$ and 10 mg/L $CrO_4{}^{2-}$ for the Cr(VI) spike were added for the eluate test. The comparison of the Cr(VI) concentrations of the two eluates per sample allowed calculating the recovery of the spike. The water-extractable Cr(VI) content in the eluates was determined spectrophotometrically with a Merck Spectroquant Pharo 100 after complexation with diphenylcarbazide (DPC) at an absorption maximum of 540 nm. For complexation, a Spectroquant chromate test set (Merck, No 1.14758, Darmstadt, Germany) was used. No determination of Cr(VI) in the FLUWA leachate was performed as the Cr concentration in the leachates were below the detection limit in all samples.

2.6. Total Cr(VI)

For a more comprehensive assessment of the hazard potential of Cr(VI), the total content of Cr(VI) was determined for two WWFA samples (WWFA40_2 and WWFA100). Total Cr(VI) extraction on WWFA was performed on duplicates by hot alkaline extraction (method 3060A [17]). For quality control, the standard reference material NIST 2701 [18] was analyzed for total Cr(VI) content, and the determined value was within the given uncertainty. The accuracy of the method on similar materials was previously verified [19] by the authors, and reproducibility within 10% was attributed to the method based on spike recoveries and multiple measurements of the standard reference material [20]. Analysis of Cr(VI) concentration in the extract solutions was performed with ICP-OES after the use of CHROMAFIX PS-H+ cation exchange cartridges (Macherey–Nagel, Düren, Germany) in order to retain Cr(III). The effectiveness of the cartridges on similar extract solutions was tested in previous studies [19]. In order to investigate Cr redox transitions during extraction, a parallel extraction with Cr(VI) spiking was performed for each sample by adding a concentration of 100 mg/kg $PbCrO_4$.

2.7. Laboratory Experiments

In order to investigate the effects of co-processing WWFA40 in the FLUWA process, the process was simulated on a laboratory-scale at conditions feasible on an industrial-scale. At a laboratory-scale, mass balances can be quantified precisely, and process conditions (pH, Eh) can more easily be controlled. Three different ash types were used for the experiment:

MSWIFA (sample MSWIFA_2), WWFA40 (sample WWFA40_2) and MFA (the ash mix of WWFA40_2 and MSWIFA_2, mixed at a ratio of 1:2, to simulate a cotreatment). Two different experimental setups were performed at a laboratory-scale (Table 1): without the use of H_2O_2 (30%) and with a concentration of 40 L H_2O_2/t ash, which represents standard plant conditions. All experiments were performed two times in order to assure reproducibility.

Table 1. Experimental setups for laboratory- and industrial-scale experiments.

		Laboratory-Scale			Industrial-Scale		
	Ash Type	**MSWIFA**	**MFA**	**WWFA40**	**MSWIFA**	**MFA**	**WWFA40**
Without H_2O_2	Extraction pH	3.8	3.8	3.8	3.7–4.1	3.8–4.2	3.7–4.3
	Leachate pH	4.8	4.8	5.4	3.8–4.4	4.0–4.5	4.2–4.8
	Experiment duration (h)	1	1	1	8	4	5
40 L H_2O_2/t ash	Extraction pH	2.5	2.5	2.5	3.3–3.8	3.5–4.2	-
	Leachate pH	4.5	4.5	4.2	3.9–4.4	3.8–4.4	-
	Experiment duration (h)	1	1	1	21	18	-
60 L H_2O_2/t ash	Extraction pH	-	-	-	3.5–3.8	3.7–4.3	-
	Leachate pH	-	-	-	3.8–4.3	3.8–4.5	-
	Experiment duration (h)	-	-	-	24	6	-

150 mL of artificial acid scrub water (HCl 5% with 25 g/L $NaSO_4$) was heated in a beaker glass to 40 °C before adding the ash (50 g, liquid to solid ratio L/S = 3). Due to the exothermic reaction between the scrub water and the ash, the temperature rose immediately to 60 °C. Under continuous stirring, the ash slurry reacted for 60 min at 55–60 °C. The pH and Eh values (Ag/AgCl) were recorded temperature-compensated. The pH was controlled by adding HCl (32%). Before filtration, the pH was adjusted to a value of 3.8 with NaOH (65%). The slurry was filtered with a vacuum filter device. 100 mL of deionized water was used to wash the filter cake. For the experiments performed with H_2O_2, a concentration of 40 L/t H_2O_2 was added by 10 consecutive portions (at a 3 min interval) to the ash slurry in the first half of the experiment. The leachates were diluted with HNO_3 1% and further analyzed by ICP-OES. The filter cake was weighed and dried at 105 °C until a constant mass was reached. The heavy metal recovery was determined from mass balance calculations.

2.8. Industrial-Scale Experiments

Both the current state (reducing conditions) as well as the future state (oxidizing conditions) were investigated at an industrial scale (Table 1). Each ash type (MSWIFA, WWFA40 and MFA (ratio WWFA40 to MSWIFA of 1:2.5)) was treated separately without the addition of H_2O_2. Additionally, experiments with 40 and 60 L/t H_2O_2 (30%) (Merck, Darmstadt, Germany) were performed for MSWIFA and MFA. At laboratory-scale experiments, it was shown that WWFA40 required higher H_2O_2 dosages for oxidizing conditions to persist. Therefore, industrial-scale experiments were also conducted with 60 L/t. The efficiency of the cotreatment of WWFA40 and MSWIFA was compared to the scenarios where only WWFA40 or MSWIFA was treated solely. It was found that the treatment of wood ash solely is difficult at an industrial scale due to difficulties with filtration. Therefore, the industrial-experiments with WWFA40 solely were not carried out with H_2O_2 and the focus was put on the cotreatment with MSWIFA.

The different experimental setups were performed at Energiezentrale Bern on different days for time periods between 4 and 24 h. A calculated amount of H_2O_2 was added continuously to the first extraction reactor. During each experiment, multiple samples (in 30- or 60-min intervals, depending on the experiment duration) were taken from the input ash and from the filter cake and combined into composite samples. The pH and Eh were monitored. An average L/S-ratio of 15 was calculated on annual mass flux balances since flow measurement of the leachate is not implemented. The ash and filter cake samples were dried at 105 °C and analyzed for their elemental composition with ED-XRF in order to calculate the recovery.

3. Results

3.1. Chemical Composition

The main constituents of MSWIFA are Ca, Cl, Si, S, Na (>75,000 mg/kg, Table 2). Zn, K, Al (>35,000 mg/kg) and Fe (>10,000 mg/kg) are subsidiary constituents. WWFA differ in chemical composition from MSWIFA, and there are also chemical differences between WWFA40 and WWFA100. WWFA generally shows higher Ca and Si concentrations (>100,000 mg/kg) than MSWIFA, followed by K, Al, Cl, S, Fe (>20,000 mg/kg). Notable is the higher concentration in S and heavy metals (mainly Pb, Zn, Cu) in WWFA100 compared to WWFA40, the latter showing higher concentrations in matrix elements (Ca, Si, K, Mg) in contrast. The concentrations of the main- and subsidiary constituents in the three MSWIFA samples vary within 10–20%, but Cu shows variations by more than 30%. The variations in chemical composition are smaller for WWFA40 than for MSWIFA. The element concentrations vary mostly within 10% between the three WWFA40 samples (except for about 20% for Cr, Mn and Pb). MSWIFA shows strongly elevated concentrations in the potentially harmful heavy metals Pb, Cu, Sb, Cd, some exceeding the threshold for landfilling by multiple times (Pb) or orders of magnitude (Sb, Cd). WWFA shows considerably lower Cd and Sb concentrations (lower by almost two and three orders of magnitude, respectively) but shows strongly elevated Pb concentrations. In two samples of WWFA40, Pb concentration is only 10% lower than in MSWIFA. In WWFA100, Pb concentrations exceed those in MSWIFA by more than double. Zn and Cu concentrations in WWFA40 are about one-third of the concentration in MSWIFA, and in WWFA100, about half. In contrast, WWFA shows higher concentrations in Ba, Cr, Fe, Mn and Ti concentrations than MSWIFA. WWFA100 further shows a very high TOC content of 74,400 mg/kg—which exceeds the threshold value for landfilling.

3.2. Mineralogical Composition

The difference in chemical composition between MSWIFA and WWFA is also represented by a different mineralogical composition (Table 3). As a result of the high Cl content, Cl salts such as halite, sylvite and the Zn bearing K_2ZnCl_4 are important phases in MSWIFA. Anhydrite represents another main phase, together with several silicates (e.g., gehlenite) and calcite. The main mineralogical differences between WWFA40 and WWFA100 are in calcite and anhydrite content. WWFA40 show very high calcite concentrations (17–28 wt %) and high concentration in quartz (7–10 wt %). Sylvite, periclase and Ca-, Al-, Na- silicates form minor components in WWFA40. WWFA100 shows calcite, gehlenite, anhydrite and magnesite as main phases (>7 wt %), followed by minor concentrations in quartz and Ca-, Al-, Na- silicates. The presence of amorphous phases is clearly visible in all spectra by a bump in the background between 20 and 40° 2Theta and was calculated to make 35–50% of the total content.

Table 2. Elemental composition of the analyzed municipal solid waste fly ash (MSWIFA) and waste wood fly ash types WWFA40 and WWFA100, determined by energy-dispersive X-ray fluorescence (ED-XRF).

	MSWIFA				WWFA40		WWFA100
	mg/kg				mg/kg		mg/kg
	MSWIFA_1	MSWIFA_2	MSWIFA_3	WWFA40_1	WWFA40_2	WWFA40_3	WWFA100
Al	39,315	45,558	37,210	46,912	44,985	42,520	32,200
Ba	1928	1996	1679	5091	4681	5107	5370
Ca	165,984	170,617	163,200	288,884	285,400	307,600	235,650
Cd	277	248	311	26	23	28	71
Cl	82,120	85,100	114,500	31,140	32,800	32,530	36,050
Cr	360	350	323	582	597	443	1221
Cu	2131	3411	1771	602	683	581	1131
Fe	18,930	20,123	12,980	22,562	22,350	20,890	26,265
K	52,045	46,290	60,030	61,974	71,182	70,050	41,475
Mg	2529	3024	901	20,685	20,218	20,390	11,075
Mn	820	811	876	4274	5327	4296	4856
Na	75,190	76,720	91,410	6905	7590	7940	8485
Ni	94	92	74	95	91	96	76
P	4774	4531	4485	9383	8730	9107	3798
Pb	8143	8688	8204	7876	7590	5152	21,015
S	80,762	70,840	72,500	25,845	25,042	29,250	65,215
Sb	3150	3533	2988	<3	<3	<3	22
Si	101,232	107,883	86,750	144,334	148,950	126,300	129,000
Ti	11,712	11,472	8606	n.a.	n.a.	n.a.	16,535
Zn	44,607	39,810	39,570	12,569	13,047	13,330	21,550
TOC	5190	5750	6010	10,150	17,043	12,030	74,400

Table 3. Mineralogical composition of the analyzed MSWIFA, WWFA40 and WWFA100, determined by XRD.

		MSWIFA				WWFA40		WWFA100
		wt %				wt %		wt %
Phases	Formula	MSWIFA_1	MSWIFA_2	MSWIFA_3	WWFA40_1	WWFA40_2	WWFA40_3	WWFA100
Chlorides								
Halite	NaCl	9	-	14	<1	<1	1	1
Sylvite	KCl	3	3	5	4	4	5	-
K_2ZnCl_4	K_2ZnCl_4	4	3	3	-	-	-	-
Sulfates								
Anhydrite	$CaSO_4$	10	10	11	-	-	-	8
Silicates								
Quartz	SiO_2	3	2	2	10	10	7	5
Gehlenite	$Ca_2Al(AlSi)O_7$	7	5	3	4	2	4	9
Alpha belite	Ca_2SiO_4	4	3	-	5	5	3	3
Albite	$NaAlSi_3O_8$	4	3	3	5	3	5	2
Sanidine	$KAlSi_3O_8$	-	-	-	-	-	-	5
Microcline	$KAlSi_3O_8$	3	4	2	-	-	-	-
Carbonates								
Calcite	$CaCO_3$	5	6	6	17	28	20	8
Magnesite	$MgCO_3$	<1	2	1	-	-	-	7
Ankerite	$CaFeCO_3$	-	-	-	-	1	-	-
Oxides								
Lime	CaO	-	-	-	2	-	4	-
Hematite	Fe_2O_3	<1	1	-	1	<1	<1	1
Rutile	TiO_2	-	-	-	2	1	3	1
Periclase	MgO	1	-	-	3	4	3	-
Mayenite	$Ca_{12}Al_{14}O_{33}$	2	2	3	-	-	-	-
Perovskite	$CaTiO_3$	2	2	2	-	-	-	-
Phosphates								
Monetite	$CaHPO_4$	5	4	3	5	6	5	2
Amorphous		37	41	43	42	35	39	49

3.3. Acid-Neutralizing Capacity

Acid-neutralizing capacity (ANC) is similar for ashes of the same type but shows differences between the ash types (Figure 1). The amount mol H^+ needed to reach an optimal extraction pH of 3 varies within 10% and 5%, respectively, for MSWIFA and WWFA40. On average, MSWIFA consumed 4.7 mol H^+, WWFA100 6.2 mol H^+, and WWFA40 with 8.9 mol, almost double the amount of H^+ to reach a pH of 3. The titration curve of WWFA40 is characterized by a small plateau at pH 12.7 and a big plateau at pH 7. Although WWFA100 shows a similarly high initial pH as WWFA40, the ANC titration curve resembles more that of MSWIFA than that of WWFA40, as the plateau starting at pH 7 is less pronounced. The pH starts dropping rapidly from an initial value of 12.7 to pH 7, where calcite buffering starts. For MSWIFA, the initial pH of the titration curve is lower (pH 11.5) and drops rapidly towards pH 5—where an almost linear decrease in pH initiates.

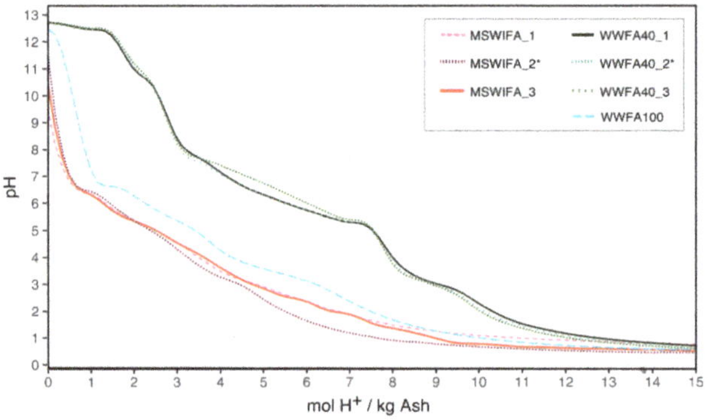

Figure 1. Titration curves of acid-neutralizing capacity (ANC) for MSWIFA, WWFA40 and WWFA100. The samples used for the leaching experiments are indicated with *.

3.4. Water-Extractable and Total Cr(VI)

Multiple determination of the water-soluble Cr(VI) concentration of the samples revealed reproducibility within 10%. The Cr(III) spike was fully retained during all eluate tests, which proves that no oxidation of Cr(III) occurred. The Cr(VI) spike was fully retained during the majority of the eluate tests. In the experiments where the Cr(VI) spike was not fully retained, redox transformations leading to the reduction of Cr(VI) occurred (indicated with * in Table 4).

Table 4. Water-extractable Cr(VI) and total Cr(VI) concentrations of WWFA and the filter cakes of WWFA40. Cr(VI) concentrations indicated with * are from extractions with poor Cr(VI) spike recoveries (<2%), implying erroneously low values. Where not indicated, the Cr(VI) spike was fully retained.

	WWFA40						WWFA100
	mg/kg						mg/kg
	WWFA40_1	WWFA40_2	WWFA40_3	WWFA40_4	WWFA40_5	WWFA40_6	WWFA100
Water-extractable Cr(VI)	58	117	95	83	96	110	1 *
Water-extractable Cr(VI) of filter cake	-	<0.05	-	<0.05	<0.05	0.22	-
Total Cr(VI)	-	1 *	-	-	-	-	87

All eluates of the 6 analyzed WWFA40 samples showed water-extractable Cr(VI) concentrations that exceed the threshold limit for landfilling (0.5 mg/kg) by more than two orders of magnitude. The water-extractable Cr(VI) content made up for 10–20%

of the Cr concentration in WWFA40. For WWFA100, the determined water-extractable Cr(VI) concentration was as low as 1 mg/kg. Since the Cr(VI) spike recovery was only 2% for WWFA100, this implies that a major part of the sample's native Cr(VI) also has been reduced. All six analyzed filter cakes from WWFA40, including all filter cakes from the experiments performed in this study, showed Cr(VI) concentrations below the given threshold for landfilling, independent of the applied scale (laboratory or industrial) and the redox conditions.

For the hot alkaline extraction of sample WWFA40_2, none of the Cr(VI) spike was recovered, in neither of the duplicates. The total Cr(VI) concentration was expected to be similar or higher to the measured water-extractable Cr(VI) concentration. Instead, a concentration of 1 mg/kg was measured. For WWFA100, Cr(VI) spike recovery was 80% in the hot alkaline extraction and the measured total Cr(VI) concentration 87 mg/kg. The results of the double determination agreed within 3%. It is assumed that matrix interferences occurred during the hot alkaline extraction of the sample WWFA40_2 (and to a minor extent in sample WWFA100), leading to a strong diminution in Cr(VI) concentration. This might have been favored by the strongly reducing conditions during the hot alkaline extraction with WWFA40_2.

3.5. Laboratory-Scale Leaching Experiments

The heavy metal recovery achieved for the two different experimental setups is shown in Figure 2. The reproducibility of the experiments performed in duplicates was very good (within 5–10%). Only for Cu, the reproducibility was within 20% since the solubility of Cu is strongly pH-dependent, and a small increase in the filtrate pH value can enhance precipitation of Cu hydroxides [21]. Given the attributed uncertainty of 10%, Zn recovery can be considered equal for MSWIFA and MFA, whereas it was lower by 30% for WWFA40. This lower yield is associated with a high leachate pH of 5.4, which assumedly led to Zn precipitation. For Cd, recovery is 40% lower for MFA compared to MSWIFA, whereas no Cd was recovered from WWFA40. No Pb and Cu were mobilized for any of the ash types without the use of H_2O_2.

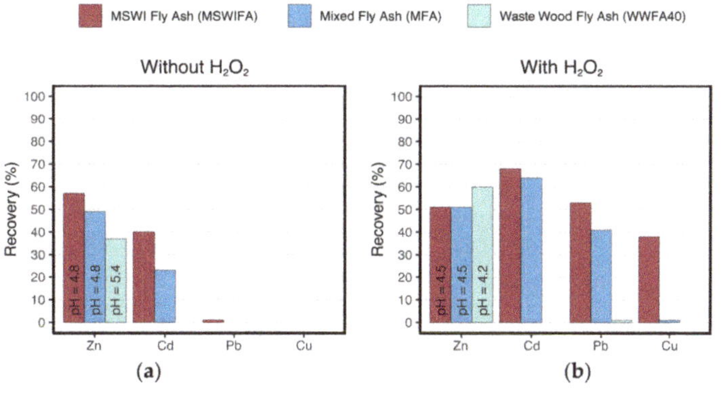

Figure 2. Heavy metal recovery in % for the target heavy metals Zn, Cd, Pb, Cu for the 3 different ash types in laboratory experiments. The pH of the leachate is indicated in the bar for Zn recovery. (**a**) recovery without the use of H_2O_2 (**b**) recovery for experiments with 40 L/t H_2O_2.

When a quantity of 40 L/t H_2O_2 was used, Zn recovery did not change. The recovery for Cd could almost be doubled from MSWIFA, and it was achieved to recover Pb (53%) and Cu (38%). This significantly higher recovery for Pb and Cu (and to a minor extent Cd) when using H_2O_2 was observed in previous studies [14]. The recovery for MFA is equal (within the uncertainty) to that of MSWIFA for Zn and Cd. For Pb, recovery is lower by 25%, and almost no Cu was mobilized during the experiment. For WWFA40, the same recovery for Zn was achieved as for the other ash types, but any of the other heavy metals could be recovered. In the experiment with H_2O_2, the amount of HCl 32% needed to keep extraction pH at a level of 2.5 was twice as high for MFA compared to the experiments with MSWIFA (17 vs. 9 mL, respectively), and the amount needed for WWFA40 was 33 mL.

The ashes showed a strong redox buffer, visible by the subsequent drop in redox potential after each H_2O_2 dosage (Figure 3). For MSWIFA, the redox potential dropped to strongly negative values shortly after the H_2O_2 dosage, whereas for WWFA40, Eh was still positive before the next H_2O_2 dosage. Thus, H_2O_2 consumption seemed to be slower. However, the amount of H_2O_2 added was not enough to maintain oxidative conditions over the entire extraction time for MFA and WWFA40. Only for MSWIFA, it was possible to maintain a stable positive redox potential over the entire experiment with 40 L/t H_2O_2.

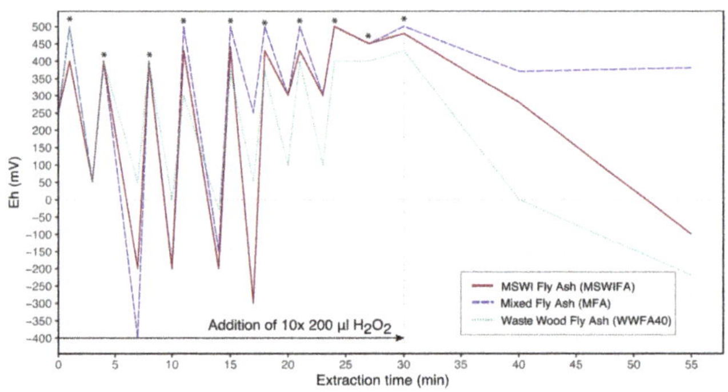

Figure 3. Evolution of redox potential (Eh) during laboratory experiment with H_2O_2. **Left**: Eh before and after each dose of H_2O_2 (indicated with *). **Right**: evolution of Eh after the last H_2O_2 dosage.

3.6. Industrial-Scale Leaching Experiments

Taking into account the attributed uncertainty of 10%, the recovery for Zn was the same for the three ash types independent of the amount of H_2O_2 added. The recovery for Cd reflected the trends observed from the laboratory experiments: a lower Cd recovery by one-third for MFA and a negligible Cd recovery for WWFA40. As already observed in the laboratory experiments, the recovery for Pb and Cu was negligible without H_2O_2 (Figure 4a). With 40 L/t H_2O_2, 55% of Pb and 16% of Cu could be mobilized from MSWIFA, but only 12% Pb and 3% Cu from MFA (Figure 4b). The recovery of both Pb and Cu was thus significantly lower for MFA compared to MSWIFA.

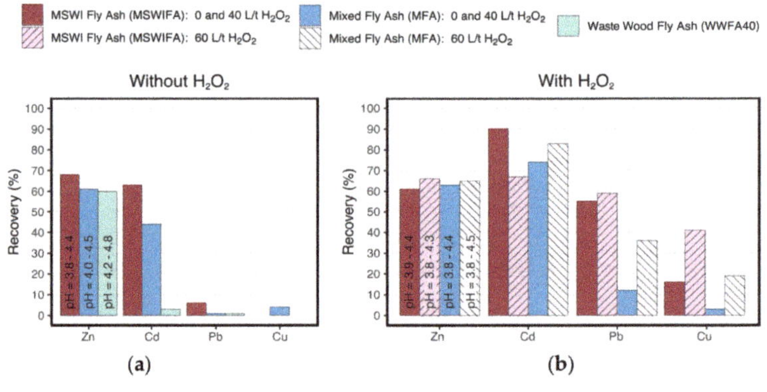

Figure 4. Recovery for the industrial-scale experiments (**a**) without and (**b**) with H_2O_2 (40 and 60 L/t H_2O_2). The pH of the leachate is indicated in the bar for Zn recovery. The industrial-scale experiments with H_2O_2 were not performed for WWFA40.

With the higher dosage of 60 L/t H_2O_2, a higher Cd fluctuation in recovery could be observed, but considering the attributed uncertainty of 10%, the recoveries were comparable for the two ash types. For Pb recovery, a strong increase by a factor of three could be observed for MFA, whereas Pb recovery did not increase with the higher dosage for MSWIFA. A clear increase in Cu mobilization could be observed for both ash types, which resulted in recoveries 2.5 and 6 times higher than with 40 L, respectively. Table 5 lists the heavy metal concentrations of the ashes investigated in the industrial-scale leaching experiments. Considerable differences in heavy metal concentrations could be observed, which must be taken into account when comparing the recoveries.

Table 5. Heavy metal concentrations of the different ashes investigated in industrial-scale leaching experiments. The concentrations for the experiments with 40 and 60 L/t H_2O_2 are indicated with * and **, respectively.

<table>
<tr><th></th><th colspan="4">Experiment without H_2O_2</th><th colspan="5">Experiments with H_2O_2</th></tr>
<tr><th></th><th colspan="4">mg/kg</th><th colspan="5">mg/kg</th></tr>
<tr><th></th><th>Zn</th><th>Cd</th><th>Pb</th><th>Cu</th><th>Zn</th><th>Cd</th><th>Pb</th><th>Cu</th><th></th></tr>
<tr><td rowspan="2">MSWIFA</td><td rowspan="2">44,710</td><td rowspan="2">220</td><td rowspan="2">6070</td><td rowspan="2">1360</td><td>43,700</td><td>260</td><td>10,700</td><td>2500</td><td>*</td></tr>
<tr><td>57,350</td><td>320</td><td>12,910</td><td>3090</td><td>**</td></tr>
<tr><td rowspan="2">MFA</td><td rowspan="2">31,950</td><td rowspan="2">170</td><td rowspan="2">6760</td><td rowspan="2">2010</td><td>31,330</td><td>190</td><td>6460</td><td>1970</td><td>*</td></tr>
<tr><td>31,830</td><td>160</td><td>5710</td><td>1770</td><td>**</td></tr>
<tr><td>WWFA40</td><td>13,890</td><td>30</td><td>5520</td><td>630</td><td>-</td><td>-</td><td>-</td><td>-</td><td></td></tr>
</table>

4. Discussion

4.1. Chemical and Mineralogical Differences and its Effects on Acid-Neutralizing Capacity

The chemical and mineralogical characterization showed considerable differences between the ash types. As the difference in matrix elements, WWFA showed significantly higher Ca and Si contents and significantly lower concentrations in Cl than MSWIFA. Within WWFA, chemical differences were also observed, depending on the waste wood content. The S content, for example, was as high as in MSWIFA for WWFA100, but less than half the concentration in WWFA40. The fluctuations in the elemental composition of MSWIFA are attributed to the compositional differences in waste input. Remarkable is the constant chemical and mineralogical composition between the different WWFA40 samples, as waste wood is also a very heterogeneous feedstock [22]. This could be an effect of the low waste wood content in WWFA40. The significantly higher Cl concentration in MSWIFA is associated with the combustion of plastics (PVC). It is known that elevated Cl concentrations in the flue gas favor the evaporation and transport of heavy metals (e.g., Cd, Cu, Pb, Zn, Sb [23–25]), and elevated heavy metal concentrations in MSWIFA have been observed by many authors (e.g., [15,26]). The high concentrations of these hazardous metals in MSWIFA are associated with the combustion of, e.g., batteries, paints, alloys, plastics [27]. The high Sb concentrations in MSWIFA are problematic due to their high toxicity (especially of the trivalent species [28]). Antimony is not soluble at the low pH conditions prevailing in the FLUWA process and accumulates in the filter cake. To fully assess the hazard potential of Sb and to evaluate Sb mobilization during the FLUWA process, additional studies are in progress. The high Pb and Cr(VI) concentrations in WWFA are problematic and justify the need for treatment prior to deposition. WWFA100, the WWFA with higher waste wood content, showed significantly higher heavy metal concentrations, which reflects findings made in previous studies on WWFA (e.g., [22]). The elevated Pb and Zn concentration in WWFA is probably due to pigments (e.g., $PbCO_3$, ZnO) from paints and coatings, whereas the high Cu content in WWFA could arise from the combustion of pickled wood products [29].

During combustion, alkali- and alkaline earth metals in the wood transform to oxides and are subject to successive hydrogenation and carbonation during cooling [30], which could explain the high calcite content in WWFA40. The high Si content stands in relation to the usage of quartz sand as bed material during fluidized bed combustion [10]. In the presence of SO_2 and O_2, CaO often forms sulfate compounds (e.g., anhydrite) [31], as present in WWFA100. WWFA100 further showed higher TOC content (associated with incomplete combustion [32]) and different matrix composition (e.g., less calcite, more sulfates) than WWFA40, the latter affecting ANC. The high ANC of WWFA40 can be explained by the very high calcite content and dissolution of Ca-silicates (e.g., gehlenite, belite). The high initial pH of all the three ash types suggests the presence of non- or microcrystalline CaO or $Ca(OH)_2$ since no CaO or lime was identified in all ashes. WWFA100 and MSWIFA both showed poorly acid buffering sulfates and chlorides as main constituents, which explains the lower amount of H^+ needed to reach the acid conditions required for the FLUWA process.

Thus, although WWFA showed different chemical and mineralogical characteristics than MSWIFA, there are also different geochemical properties within WWFA, depending on their waste wood content and certainly also depending on differences in the waste wood composition.

4.2. Water-Extractable and Total Cr(VI) in WWFA and Filter Cakes

The measured average water-extractable Cr(VI) concentration of 93 mg/kg in WWFA40 justifies the strong need for Cr(VI) reduction prior to landfilling. The treatment of WWFA40 with the FLUWA process successfully reduced water-soluble Cr(VI). All analyzed filter cakes showed Cr(VI) concentrations below the threshold limit for landfilling, even when the experiments were performed under oxidizing conditions with H_2O_2. It was shown that acidic conditions are sufficient to dominate the reduction of water-extractable Cr(VI), as also observed in other studies [33]. It was also reported that H_2O_2 could act as a reductant in acidic solutions [34]. It is assumed that the water-soluble Cr(VI) is being reduced to Cr(III) during the FLUWA process, followed by precipitation as Cr(III) phase, probably as hydroxide. This is supported by the absence of dissolved Cr in the filtrate. Unfortunately, no Cr phases could be identified with XRD as the concentrations are very low and the precipitated phases possibly amorphous. It was further shown that the high Cr(VI) concentrations in the eluates were not caused by the oxidation of Cr(III) during the eluate test. However, Cr(VI) reduction occurred during the eluate test for WWFA100 due to its highly reductive character—leading to erroneously low water-extractable Cr(VI) concentrations. Similar matrix interferences occurred during the hot alkaline extraction for WWFA40. These observations show that spiking during Cr(VI) extraction tests with reactive material such as WWFA is indispensable. Poor Cr(VI) spike recovery is not indicative of method failure, but rather an indication for the potential of the sample to reduce the spiked Cr(VI) and not sustain its native Cr(VI) [35]. They further report that the presence of high TOC contents, as well as considerable S^{2-} or Fe^{2+} concentrations in the sample, are most likely the reason for low Cr(VI) spike recovery or reduction of native Cr(VI). Although oxidative conditions prevail during combustion, it cannot be excluded that locally reducing conditions occur, where Fe^{2+} and S^{2-} persist. While no mineral phases containing Fe^{2+} and S^{2-} were observed, their presence in minor concentrations cannot be excluded. For further interpretation, a more detailed investigation on possible reductants other than C_{org} in WWFA must be performed, with special focus on the content of Fe^{2+} and S^{2-}.

4.3. Leaching Experiments: Heavy Metal Recovery and Consumption of Neutralizing Chemicals

The laboratory-scale experiments were able to predict well the recovery trends of the industrial-scale. Differences in the recoveries between laboratory- and industrial-scale are primarily attributed to differences in pH and to element contents of the ash, as well as to a larger L/S in the industrial scale, which will increase the recovery. For WWFA40, a higher H_2O_2 dosage was needed to achieve oxidizing conditions during extraction. This could be caused due to the high content in organic matter or the presence of metals in their metallic form—leading to rapid consumption or even catalytic destruction of the added

H_2O_2. The high ANC of WWFA40 led to higher acid consumption in the FLUWA compared to MSWIFA (3× higher). Thus, WWFA40 can represent a heavy metal-rich replacement for the often-used pH neutralizing agent lime milk. The high acid consumption is even more pronounced when H_2O_2 is used, since the oxidation of, e.g., metallic compounds consumes H^+.

For the element Zn, recovery was equally high for the different ash types independent of the H_2O_2 dosage, as Zn mobility is independent of the redox conditions during extraction but controlled by pH and binding form [36]. The achieved Zn recovery was lower by 15% in the laboratory-scale experiments compared to the industrial-scale experiment. It is assumed that this is due to the higher L/S used in the industrial-scale experiment (L/S of 15 compared to L/S of 3). The low Zn recovery of WWFA40 (laboratory-scale) is associated with enhanced precipitation of Zn due to the high leachate pH. Remarkable is the fact that the Zn recovery seems not strongly affected by the Zn concentration in the ash. As observed in other studies [36], the Zn yield stagnates at about 70% (in this study at about 65% in the industrial-scale-experiments). It is assumed that the majority of the Zn in the ashes is readily available for dissolution (e.g., as Cl- or S-salts). The remaining 30–35% of the Zn seems, however, to be present in the insoluble form under these conditions (e.g., as glassy particles, as Ca replacement in gehlenite or associated with iron [36]). The recovery of Cd showed the same trends for laboratory- and industrial-scale, but with lower recoveries by 15–50% in the laboratory-scale. The higher recoveries at the industrial-scale are again attributed to the higher L/S. The lower Cd yields in the laboratory-scale experiment without H_2O_2 are associated with the higher leachate pH. Since Cd concentration is about one order of magnitude lower than the other elements, it is subject to larger fluctuations as inhomogeneities in Cd concentration in the sample are more pronounced. The low Cd recovery from WWFA40 is attributed to the very low Cd concentration in the ash. An increase in Cd recovery can be observed for each ash type when the experiments are performed under oxidizing conditions, as observed in other studies [36]. As WWFA40 showed a higher redox buffer than MSWIFA, it is assumed that the lower Cd recoveries in the experiments without H_2O_2 are a result of reductive precipitation of Cd since Cd recoveries are comparable to those of MSWIFA when using H_2O_2. The mobility of Pb and Cu is highly dependent on the redox conditions, as well as on the pH (especially for Cu). The recoveries for Pb and Cu for 40 L H_2O_2 are higher on the laboratory-scale than on the industrial-scale. Besides differences in elemental concentrations in the ashes, the pH and redox conditions are more easily controllable on the laboratory-scale than on the industrial-scale. Additionally, the industrial-scale experiment runs over longer timespans and is subject to fluctuations of ash input and neutralizing chemicals. The recoveries for Pb and Cu were significantly lower for MFA compared to MSWIFA, which is attributed to the higher redox buffer of WWFA40. However, further data are required for quantifying the negative effects.

The FLUWA process nevertheless represents a valuable option for treating WWFA as the heavy metal concentrations in WWFA are in the same range as for MSWIFA, and WWFA shows comparable heavy metal recoveries for Zn and Cd. The negative effects of the higher consumption of H_2O_2 affecting the Pb and Cu recovery may be diminished by a cotreatment of lower WWFA ratios.

5. Conclusions

Heat and energy production in Switzerland using waste wood incineration is growing, and new treatment pathways must be implemented to recover heavy metals from the ashes and to reduce Cr(VI) content. Acid leaching, already established for MSWIFA, was found to be a valuable option for the treatment of WWFA. Laboratory-scale experiments were found to be suitable when evaluating the co-processing of MSWIFA and WWFA before implementing at the industrial scale.

Comparison of the chemical and mineralogical composition of WWFA with MSWIFA showed that WWFA could contain heavy metals (especially Pb) in elevated concentrations similar to that of MSWIFA. The investigated WWFA samples showed Cr(VI) concentrations more than two orders of magnitude above the threshold value for landfilling. It was found that the concentrations in heavy metals, Cr(VI) and matrix minerals differed within the two WWFA types, depending on waste wood content. The elevated heavy metal and Cr(VI) concentrations in WWFA justify the need for treatment prior to deposition. The treatment with the FLUWA process allowed to successfully reduce the Cr(VI) in the filter cake until below the threshold value for landfilling, even when the process was performed under oxidizing conditions. The co-processing of WWFA required higher acid dosages due to its high ANC, but the Zn and Cd recovery were not negatively affected by the co-processing. Nevertheless, the co-processing of WWFA had a particularly negative effect on the recovery of the redox-sensitive elements Pb and Cu, as WWFA showed a strong redox buffer and thus a higher consumption of the oxidant H_2O_2. Therefore, higher dosages of H_2O_2 are needed to maintain oxidizing conditions during the process required for Pb and Cu mobilization. The use of a stronger oxidizing agent (e.g., permanganate) could be expedient and should be further tested with regard to successful Cr(VI) reduction. Alternatively, smaller percentages of WWFA could be co-processed in existing FLUWA plants in order to diminish the negative effects due to the higher demand for neutralization chemicals.

Within the next years, the implementation of co-processing the two ash types could contribute significantly to the growing demand for treatment capacities in Switzerland.

Author Contributions: Conceptualization, U.E., G.W. and M.W.; methodology, M.W., U.E. and G.W.; formal analysis, M.W.; investigation, M.W.; resources, M.W.; data curation, M.W.; writing—original draft preparation, M.W.; writing—review and editing, U.E. and G.W.; visualization, M.W.; supervision, G.W. and U.E.; project administration, M.W.; funding acquisition, U.E. All authors have read and agreed to the published version of the manuscript.

Funding: This research was partially funded by the Swiss Federal Office for the Environment (FEN) and Office for Water and Waste (AWA, Canton of Berne).

Data Availability Statement: The data presented in this study are available on request from the corresponding author.

Acknowledgments: Special thanks to Thomas Bücherer, Thomas Andres and Roland Furrer from Energiezentrale Bern for enabling the project, providing sample material and process details, as well as for their technical support during sampling and industrial-scale experiments. Many thanks to Stefan Schlumberger (ZAR) and Kaarina Schenk (FEN) for project feedback and discussion. Analytical support and/or assistance with industrial-scale experiments by Anna Zappatini, Wolfgang Zucha, Christine Lemp (University of Bern) and Stephan Fromm (ZAR) is highly acknowledged.

Conflicts of Interest: The authors declare no conflict of interest.

References

1. Schweizerische Eidgenossenschaft. *Schweizerische Holzenergiestatistik—Erhebung für das Jahr 2019*; Bundesamt für Energie: Bern, Switzerland, 2020.
2. Obernberger, I. Nutzung fester Biomasse in Verbrennungsanlagen unter besonderer Berücksichtigung des Verhaltens aschebildender Elemente. In *Schriftenreihe Thermische Biomassenutzung*; dbv-Verlag: Graz, Austria, 1997.
3. Zimmermann, S.; Hässig, J.; Landolt, W. *Literaturreview Holzasche—Wald: Nährstoffentzug durch Holzernte, Ascherückführung in den Wald, abiotische und biotische Wirkungen*; Swiss Federal Office of Environment: Birmensdorf, Schweiz, 2010.
4. Tarelho, L.A.C.; Teixeira, E.R.; Silva, D.F.R.; Modolo, R.C.E.; Silva, J.J.F. Characteristics, management and applications of ashes from thermochemical conversion of biomass to energy. In Proceedings of the Conference Exhibition on Biomass for Energy, Jonkopin, Sweden, 29–31 May 2012.
5. Etiégni, L.; Campbell, A.G. Physical and chemical characteristics of wood ash. *Bioresour. Technol.* **1991**, *37*, 173–178. [CrossRef]
6. Wunderli, S.; Zennegg, M.; Dolezal, I.S.; Gujer, E.; Moser, U.; Wolfensberger, M.; Hasler, P.; Noger, D.; Studer, C.; Karlaganis, G. Determination of polychlorinated dibenzo-p- dioxins and dibenzo-furans in solid residues from wood combustion by HRGC/HRMS. *Chemosphere* **2000**, *40*, 641–649. [CrossRef]
7. Huron, M.; Oukala, S.; Lardière, J.; Giraud, N.; Dupont, C. An extensive characterization of various treated waste wood for assessment of suitability with combustion process. *Fuel* **2017**, *202*, 118–128. [CrossRef]
8. Pohlandt-Schwandt, K. Treatment of wood ash containing soluble chromate. *Biomass Bioenergy* **1999**, *16*, 447–462. [CrossRef]

9. Pohlandt-Schwandt, K.; Salthammer, T.; Marutzky, R. Reduction of soluble chromate in wood ash by formaldehyde. *Biomass Bioenergy* **2002**, *22*, 139–143. [CrossRef]
10. Teixeira, E.R.; Camões, A.; Branco, F.G. Valorization of wood fly ash on concrete. *Resour. Conserv. Recycl.* **2019**, *145*, 292–310. [CrossRef]
11. Swiss Federal Council. *Ordinance on the Avoidance and the Disposal of Waste (Waste Ordinance, ADWO)*; Bundesamt für Umwelt: Bern, Switzerland, 2015.
12. Schlumberger, S.; Schuster, M.; Ringmann, S.; Koralewska, R. Recovery of high purity zinc from filter ash produced during the thermal treatment of waste and inerting of residual materials. *Waste Manag. Res.* **2007**, *25*, 547–555. [CrossRef] [PubMed]
13. Bühler, A.; Schlumberger, S. Schwermetalle aus der Flugasche zurückgewinnen «Saure Flugaschewäsche—FLUWA-Verfahren» ein zukunftsweisendes Verfahren in der Abfallverbrennung. KVA-Rückstände in der Schweiz—Der Rohstoff mit Mehrwert. *Bundesamt für Umwelt BAFU* **2010**, *1*, 185–192.
14. Weibel, G.; Eggenberger, U.; Schlumberger, S. and Mäder, U.K. Chemical associations and mobilization of heavy metals in fly ash from municipal solid waste incineration. *Waste Manag.* **2017**, *62*, 147–159. [CrossRef] [PubMed]
15. Held, A.; Kramer, G.N.; Robouch, P.; Wätjen, U. *The Certification of the Mass Fractions of As, Cd, Co, Cr, Cu, Fe, Hg, Mn, Ni, Pb, Sb, Se, Tl, V and Zn in Fly Ash BCR-176R*; European Commission Directorate-General Joint Research Centre Institute for Reference Materials and Measurements: Luxembourg, 2007.
16. Swiss Federal Office of Environment. *Messmethoden im Abfall- und Altlastenbereich*; Bundesamt für Umwelt: Bern, Switzerland, 2017.
17. US Environment Protection Agency. *EPA Method 3060A. Alkaline Digestion for Hexavalent Chromium*; EPA: Washington, DC, USA, 1996.
18. National Institute of Standards & Technology. *Certificate of Analysis Standard Reference Material 2701. Certificate of Analysis Standard Reference Material 2701*; National Institute of Standards & Technology: Gaithersburg, MD, USA, 2009.
19. Weibel, G.; Waber, H.N.; Eggenberger, U.; Mäder, U.K. Influence of sample matrix on the alkaline extraction of Cr(VI) in soils and industrial materials. *Environ. Earth Sci.* **2016**, 75. [CrossRef]
20. Weibel, G. Optimized Metal Recovery from Fly Ash from Municipal Solid Waste Incineration. Ph.D. Thesis, University of Bern, Bern, Switzerland, 2017.
21. Hartinger, L. *Handbuch der Abwasser- und Recyclingtechnik*, 2nd ed.; Fachbuchverlag Leipzig: Deutschland, Germany, 1991.
22. Gehrmann, H.J.; Mätzing, H.; Nowak, P.; Baris, D.; Seifert, H.; Dupont, C.; Defoort, F.; Peyrot, M.; Castagno, F. Waste wood characterization and combustion behaviour in pilot lab scale. *J. Energy Inst.* **2020**, *93*, 1634–1641. [CrossRef]
23. Belevi, H.; Moench, H. Factors determining the element behavior in municipal solid waste incinerators. 1. Field studies. *Environ. Sci. Technol.* **2000**, *34*, 2501–2506. [CrossRef]
24. Morf, L.S.; Brunner, P.H.; Spaun, S. Effect of operating conditions and input variations on the partitioning of metals in a municipal solid waste incinerator. *Waste Manag. Res.* **2000**, *18*, 4–15. [CrossRef]
25. Luan, J.; Li, R.; Zhang, Z.; Li, Y.; Zhao, Y. Influence of chlorine, sulfur and phosphorus on the volatilization behavior of heavy metals during sewage sludge thermal treatment. *Waste Manag. Res.* **2013**, *31*, 1012–1018. [CrossRef] [PubMed]
26. Bayuseno, A.P.; Schmahl, W.W. Characterization of MSWI fly ash through mineralogy and water extraction. *Resour. Conserv. Recycl.* **2011**, *55*, 524–534. [CrossRef]
27. European Commission DG ENV E3. *Heavy Metals in Waste—Final Report*; Project ENV.E.3/ETU/2000/0058; COWI: Copenhagen, Denmark, 2002.
28. Gad, S.C. Antimony. In *Encyclopedia of Toxicology*; Gad Consulting Services: Cary, NC, USA, 2014; Volume 1. [CrossRef]
29. Gras, B. *Untersuchung von Holzaschen aus Kleinfeuerungsanlagen: Erkennen von Brennstoffmissbrauch*; Institut für Hygiene und Umwelt: Hamburg, Germany, 2006.
30. Holmberg, S.L.; Claesson, T. Mineralogy of granulated wood ash from a heating plant in Kalmar, Sweden. *Environ. Geol.* **2001**, *40*, 820–828. [CrossRef]
31. Anthony, E.J.; Jia, L.; Laursen, K. Strength development due to long term sulfation and carbonation/sulfation phenomena. *Can. J. Chem. Eng.* **2001**, *79*, 356–366. [CrossRef]
32. Park, J.H.; Eom, J.H.; Lee, S.L.; Hwang, S.W.; Kim, S.H.; Kang, S.W.; Yun, J.J.; Cho, J.S.; Lee, Y.H.; Seo, D.H. Exploration of the potential capacity of fly ash and bottom ash derived from wood pellet-based thermal power plant for heavy metal removal. *Sci. Total Environ.* **2020**, *740*, 140205. [CrossRef] [PubMed]
33. Wittbrodt, P.R.; Palmer, C.D. Reduction of Cr(VI) in the Presence of Excess Soil Fulvic Acid. *Environ. Sci. Technol.* **1995**, *29*, 255–263. [CrossRef] [PubMed]
34. Pettine, M.; Campanella, L.; Millero, F.J. Reduction of Hexavalent Chromium by H_2O_2 in Acidic Solutions. *Environ. Sci. Technol.* **2002**, *36*, 901–907. [CrossRef]
35. Vitale, R.J.; Mussoline, G.R.; Petura, J.C.; James, B.R. Hexavalent Chromium Extractions from Soils: Evaluation of an Alkaline Digestion Method. *J. Environ. Qual.* **1994**, *23*, 1249–1256. [CrossRef]
36. Weibel, G.; Eggenberger, U.; Kulik, D.A.; Hummel, W.; Schlumberger, S.; Klink, W.; Fisch, M.; Mäder, U.K. Extraction of heavy metals from MSWI fly ash using hydrochloric acid and sodium chloride solution. *Waste Manag.* **2018**, *76*, 457–471. [CrossRef] [PubMed]

processes

MDPI

Article

Municipal Solid Waste as Secondary Resource: Selectively Separating Cu(II) from Highly Saline Fly Ash Extracts by Polymer-Assisted Ultrafiltration

Christine Hettenkofer *,†, Stephan Fromm † and Michael Schuster

Division of Analytical Chemistry, Department of Chemistry, Technical University of Munich, Lichtenbergstraße 4, 85748 Garching, Germany; stephan.fromm.sw@t-online.de (S.F.); michael.schuster@tum.de (M.S.)

* Correspondence: hettenkofer.christine@gmail.com

† These authors contributed equally to this work.

Received: 29 November 2020; Accepted: 14 December 2020; Published: 16 December 2020

Abstract: Urban mining from fly ash resulting from municipal solid waste incineration (MSWI) is becoming more and more important due to the increasing scarcity of supply-critical metals. Metal extraction from acid fly ash leaching has already been established. In this context selective Cu recovery is still a challenge. Therefore, our purpose was the separation of Cu(II) from MSWI fly ash extracts by polymer-assisted ultrafiltration (PAUF). We investigated three polyethyleneimines (PEIs) with regard to metal retention, Cu(II) selectivity, Cu(II) loading capacity, and the viscosity of the PEI containing solutions. A demanding challenge was the highly complex matrix of the fly ash extracts, which contain up to 16 interfering metal ions in high concentrations and a chloride content of 60 g L^{-1}. Overcoming that, Cu(II) was selectively enriched and separated from real fly ash extract at pH 3.0. At pH 1.0, a PEI-free Cu(II) concentrate was obtained and PEIs could be regenerated for reuse in further separation cycles. The PAUF conditions developed at laboratory scale were successfully transferred to pilot scale, and hyperbranched PEI (HB-PEI) was found to be the most suitable reagent for PAUF in a technical scale. Moreover, HB-PEI enables photometric control of the Cu(II) enrichment.

Keywords: selective Cu(II) separation; sustainable waste treatment; municipal solid waste; polymer-assisted ultrafiltration; real fly ash extracts; urban mining; pilot installation

1. Introduction

For decades, prosperity has increased worldwide, which is reflected in a strong rise in resource consumption. To cover the resulting resource demand, the use of secondary raw materials is becoming more and more important. Processing of residues derived from municipal solid waste incineration (MSWI) contributes to a sustainable circular economy. Beneath the widely used bottom ash [1], a mixture of the finer boiler and filter ash, called fly ash, represents an attractive source for raw materials in urban mining. The chemical composition and resource potential of MSWI fly ash was intensively investigated [2–4], with Zn, Cu, and Pb identified as being particularly suitable due to the high concentration of these metals in MSWI fly ash and their good extractability [5].

Acid washing is an established method for extracting heavy metals from MSWI fly ash. The fly ash is treated with a hydrochloric acid solution, preferably stemming from wet flue gas cleaning in MSWI plants. Under slightly acidic conditions (pH 3.0–4.0), heavy metals are extracted with an efficiency > 90% for Cd and Pb and 70–80% for Zn and Cu [2,3,6–8]. Vacuum belt filtration is used for dewatering, resulting in a chloride-rich fly ash extract containing the extracted metals and a heavy metal-poor filter cake. Acid fly ash washing is widely established in Swiss MSWI plants, but currently only two plants outside of Switzerland are treating fly ash this way (MVA Ingolstadt, Germany, and Termizo Liberec, Czech Republic).

For the recovery of heavy metals from fly ash extract, the so-called FLUREC process is a proven method [9]. Pb, Cu, and Cd are precipitated by cementation with Zn powder and collected as metal sludge. Subsequently, solvent extraction is used for selective extraction and enrichment of Zn. From a sulfuric acid concentrate, high-grade Zn is then recovered by electrowinning. This process is performed in the Zuchwil MSWI plant (Switzerland).

In most MSWI plants, including MVA Ingolstadt, the extract from acid fly ash washing is processed by hydroxide precipitation during wastewater treatment. The drained sludge contains, among other components, hydroxides of Zn, Pb, Cd, and Cu. For metal recovery, the sludge may be treated outside of the MSWI plant by applying the Waelz process [10], a pyrometallurgical technique primarily used for steelwork dusts. Here, Zn and Pb are recovered as Waelz oxide, used as secondary raw material in hydrometallurgical Zn production, while Cu is slagged and therefore lost for recovery.

Cu, however, became one of the most valuable metals through the 20th century, ranking after iron and aluminum in importance for infrastructure and technology [11]. It is commonly used in a broad range of industrial applications, becoming even more important hand-in-hand with the electrification of transport technologies. According to Elshkaki et al. [12], the Cu demand could rise from 275 to 350% by the year 2050 and exceed the projected Cu mineral resources. Consequently, preventing resource scarcity by improving the efficiency of the Cu cycle and enhancing Cu recycling rates is crucial.

In the so-called SESAM project funded by the German Federal Ministry of Education and Research (Bundesministerium für Bildung und Forschung), the resource potential of German MSWI fly ash and the recovery of valuable metals from MSWI fly ash was investigated [4,5,13]. Special attention was given to the acid fly ash leaching process as applied in the MSWI plant in Ingolstadt (Germany). In particular, the separation of valuable metals from the fly ash extract prior to hydroxide precipitation was targeted. In order to extend the spectrum of available techniques, the application of polymer-assisted ultrafiltration (PAUF) for selective separation of Cu from fly ash extract was studied.

This ultrafiltration technique is also called polymer-enhanced ultrafiltration (PEUF), polymer-supported ultrafiltration (PSU), liquid-phase polymer-based retention (LPR), and polymer filtration (PF) [14]. It is based on the pH-dependent, reversible binding of metal ions to functionalized water-soluble polymers. In a membrane ultrafiltration step, the polymer–metal complexes are retained by the membrane and remain in the retentate solution, whereas non-bound ions pass the membrane and are found in the permeate solution. Polymer-assisted ultrafiltration is a single-phase process examined for the complexation, enrichment, and separation of metal ions from wastewater.

So far, many different polymers have been studied regarding PAUF applications [15]. An outstanding and extensively investigated water-soluble polymer class are polyethyleneimines (PEIs). Their amino groups are able to form ammine complexes with various (heavy) metal ions, such as Zn(II), Ni(II), and Cu(II). In recent years, Cu(II) separation has been performed using, inter alia, branched PEI and partially ethoxylated PEI; selective retention of Cu(II) using PEI was carried out at pH $\geq$ 3.0 [16,17], pH $\geq$ 5.0 [18], and pH $\geq$ 6.0 [19,20]. By decreasing the pH value, the respective metal ion can be released from the polymer.

A challenge in PAUF is the technical implementation of the filtration step in continuous operation. The polymer held back in the ultrafiltration retentate forms a layer on the membrane surface, severely decreasing the permeate flux. In order to control the thickness of this so-called gel layer, cross- or tangential flow filtration is applied. The polymer containing retentate is passed over the membrane surface tangentially to the permeate flow at high velocity. This causes turbulence in the retentate and reduces the gel layer formation. When assessing the technical application of water-soluble polymers, determining their hydrodynamic behavior in tangential flow filtration is an important factor [21–23].

Yet despite a substantial body of research polymer-assisted ultrafiltration still has not reached wide industrial application. Up to now, only a few PAUF studies have been carried out with operation in continuous mode with a pilot installation [22,24–27]. Schulte-Bockholt et al. [28] investigated the processing of industrial phosphating rinsing baths using a PAUF pilot plant. Apart from a few other applications like metal removal from chlorine free pulp and paper industry wastewater [29] and from mine drainage water [30] with metal concentrations in the mg L^{-1} range, PAUF has primarily been investigated using synthetic metal ion solutions without testing the applicability with real wastewater. To the best of our knowledge, PAUF has not yet been applied for highly complex matrices, such as extracts of municipal solid waste or other applications in terms of urban mining.

In our study, the selective enrichment and separation of Cu(II) from MSWI fly ash extracts by PAUF was investigated. The high salinity of fly ash extracts with competing ions in concentrations up to the g L^{-1} range, is extremely challenging not only for PAUF applications. Three commercially available PEIs were examined in laboratory-scale experiments with fly ash extracts from two MSWI plants. Based on these results, the hydrodynamic behavior of hyperbranched PEI was studied in the operation of a membrane ultrafiltration pilot plant. The technical scale retention, enrichment, purification, and regeneration of Cu(II) was thoroughly investigated and is discussed in the following sections.

2. Materials and Methods

2.1. Elemental Analysis of Fly Ash Extracts

Waste incineration plants of Ingolstadt, Germany (MVA Ingolstadt), and Zuchwil, Switzerland (KEBAG Zuchwil), kindly provided fly ash extract samples. The samples were filtered using syringe filter holders (0.45 µm, polyether sulfone; Sartorius, Göttingen, Germany). For each sample, 60 elements were analyzed by matrix-adapted inductively coupled plasma optical emission spectrometry (ICP-OES; iCAP 7600 Duo, Thermo Fisher Scientific, Waltham, MA, USA). The corresponding results were validated by inductively coupled plasma mass spectrometry (ICP-MS; Agilent 7900, Santa Clara, CA, USA). Chloride concentrations of the fly ash extracts were determined by potentiometric titration with silver nitrate solution (Titrando 904, Metrohm, Herisau, Switzerland).

2.2. Laboratory-Scale PAUF Experiments

2.2.1. Reagents

Hyperbranched polyethyleneimine (HB-PEI; MW ~25 kDa, 56 wt.% in H_2O; Lupasol HF) and modified polyethyleneimine (MOD-PEI; MW ~2000 kDa, 24 wt.% in H_2O; Lupasol SK) were kindly supplied by BASF SE (Ludwigshafen, Germany). Partially (80%) ethoxylated polyethyleneimine (PE-PEI; MW ~70 kDa, 35–40 wt.% in H_2O) was purchased from Sigma Aldrich (St. Louis, MO, USA). Sodium hydroxide (for analysis; Merck, Darmstadt, Germany) and nitric acid (65%, for analysis; Merck) were used to adjust the working pH of the feed solution.

2.2.2. Ultrafiltration Setup and Analytical Methods

The ultrafiltration system consisted of a peristaltic pump (Ismatec, Zürich, Switzerland), an ultrafiltration module (Sartorius, Göttingen, Germany), a pH electrode (SI Analytics, Mainz, Germany), and a pH meter (Schott, Mainz, Germany) (Figure S1). The feed solution was magnetically stirred. Vivaflow® 50 filtration units (Sartorius) consisting of a flat membrane made of polyether sulfone (molecular weight cut-off (MWCO) = 5 or 10 kDa, A = 50 cm^2) were used as ultrafiltration modules. During the ultrafiltration process, different material flows appear, while the feed solution is piped through the ultrafiltration membrane. Permeate is defined as the flow passing the membrane, whereas retentate is the flow retained by the membrane. Sampling of feed (1, Figure S1) and permeate (2, Figure S1) was performed 15 min after each pH adjustment. All permeate and feed samples were analyzed by ICP-OES. The polymer concentration in permeate and feed was determined by total organic carbon (TOC) measurement using a TOC analyzer (Shimadzu, Kyoto, Japan). Calibration of the latter was performed by determining the TOC of various concentrations of the respective polymer in ultrapure water. The resulting metal and polymer retentions were calculated using Equation (1):

$$R_i\ [\%] = (1 - (c_{p,i}/c_{f,i})) \times 100 \tag{1}$$

where i is the metal species or polymer, R_i is retention of i, $c_{p,i}$ is concentration of i in the permeate, and $c_{f,i}$ is concentration of i in the feed.

2.2.3. Pretreatment of PEIs

Polydisperse substances such as (modified) PEIs show a broad molar mass distribution. Low molecular weight fractions have to be removed to ensure that the polymers are retained by the membrane in the subsequent PAUF experiments. Thus, the polymers were pretreated in an ultrafiltration step, to remove small polymer molecules that can pass the membrane. For this purpose, ultrafiltration membranes (MWCO = 10 kDa) were used and the permeate was led into a separate vessel. During precleaning, polymer concentration in permeate and feed was continuously determined by TOC measurement and the polymer retention was calculated according to Equation (1). The precleaning of the polymer was terminated after reaching a polymer retention of more than 98%.

2.2.4. Determination of Metal Retention Behavior of PEIs

Prior to their use in ultrafiltration experiments, the fly ash extracts were filtered using syringe filter holders. The feed solutions of the PAUF experiments consisted of diluted fly ash extract (1:2 in ultrapure water) from MVA Ingolstadt and KEBAG Zuchwil and a final concentration (7.9–10.0 g L^{-1}) of HB-PEI, PE-PEI, or MOD-PEI. Control experiments without the addition of polymer were also carried out. The ultrafiltration experiments using membranes with a MWCO of 10 kDa were carried out in total recirculation mode, which means permeate and retentate flows were returned to the feed. At first, the pH value of the feed solutions was decreased to pH 0.7–1.0 by the addition of nitric acid (65%). Afterwards, the pH was successively increased up to 5.0–5.8 by the addition of sodium hydroxide solution (6*M*). Finally, to investigate the metal release from the PEIs, the pH was decreased to 1.0. The permeate and feed samples that were taken after each pH adjustment were analyzed by ICP-OES. Additionally, TOC measurements were carried out to control the polymer retention.

In addition, all feed samples were centrifuged for 10 min at 4427 *g* (Z 206A, Hermle, Wehingen, Germany) and the resulting supernatants were analyzed by ICP-OES.

2.2.5. Determination of Cu(II) Loading Capacity of PEIs

To investigate the maximum Cu(II) loading capacity of each PEI, further ultrafiltration experiments were carried out. The feed solutions consisted of fly ash extract from KEBAG Zuchwil and HB-PEI at a final concentration of 4.6 g L^{-1}, PE-PEI at 4.7 g L^{-1}, and MOD-PEI at 3.7 g L^{-1}. Membranes with a MWCO of 5 kDa were used. During the whole experiment, the pH was kept constant at 4.0 (HB-PEI, MOD-PEI) or 4.3 (PE-PEI) and the ultrafiltration system was operated in total recirculation mode (see Section 2.2.4). After the pH was adjusted to 4.0 or 4.3, the Cu(II) concentration of the feed solutions was stepwise increased by adding a Cu(II) nitrate trihydrate solution (for analysis; Merck) diluted with ultrapure water. Finally, pH was adjusted to 1.0 by adding nitric acid (65%) to examine the Cu(II) release from the PEIs. Fifteen minutes after each Cu(II) addition and after adjusting the final pH to 1.0, samples of permeate and feed were taken and analyzed by ICP-OES and TOC.

2.3. *Viscosity Determination of Aqueous Polymer Solutions*

The viscosity of aqueous solutions containing HB-PEI, PE-PEI, and MOD-PEI was determined using a DV-III+ rheometer (Brookfield, Middleboro, MA, USA) with cone/plate setup (CPE40). Temperature was controlled at 30 °C by a circulation thermostat (LAUDA RE204, Koenigshofen, Germany). In order to remove membrane-permeable size fractions of the polymers, pretreatment was performed according to Section 2.2.3.

2.4. Pilot-Scale PAUF Experiments

2.4.1. PAUF Pilot Plant

For pilot-scale PAUF experiments, a tangential flow ultrafiltration plant according to Figure 1 was used (Andreas Junghans GmbH, Frankenberg/Sachsen, Germany). Three ceramic multichannel tubular membranes (CA-TiO_2/Al_2O_3; Inopor, Veilsdorf, Germany) with a mean pore width of 10 nm were mounted in the membrane module.

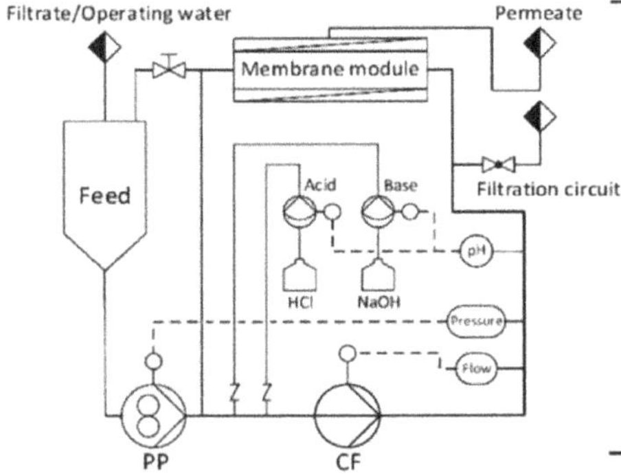

ceramic membrane:	multi-channel tube TiO_2/Al_2O_3 (active/support)
number of membranes:	3
mean pore wide:	10 nm
membrane area:	0.75 m²
filtration circuit volume:	11 l
operating pressure:	5 bar
operating temperature:	40 °C
operating crossflow:	600 – 3 500 l/h
operating pH value:	0–10

Figure 1. Scheme and operating data of tangential flow ultrafiltration pilot scale plant used for Cu(II) enrichment and separation (Andreas Junghans GmbH, Frankenberg/Sachsen, Germany). PP, pressure pump; CF, cross-flow pump.

The pilot plant was operated with HB-PEI (Lupasol HF, MW ~25 kDa, 56 wt.% in H_2O). For startup, the polymer solution was added into the feed tank and fed to the filtration circuit. Polymer solutions were pretreated with water in order to flush out the membrane-permeant size fraction until polymer retention > 99.8% was reached. In operation, the polymer was trapped in the filtration circuit (volume 11 L) and the feed reservoir was polymer-free. Polymer solutions were reused in various experiments after regeneration at pH 1.

The parts of the pilot plant that were in contact with polymer containing media were made of PVC. A gear pump was used to feed solutions from the feed reservoir into the filtration circuit and to release the buildup of filtration pressure. Cross-flow was generated by a centrifugal pump and regulated by automatic flow measurement. The pH value in the filtration circuit was continuously measured and controlled by automatic dosing of 6*M* hydrochloric acid and 10*M* sodium hydroxide solution directly into the filtration circuit. For tempering of the pilot plant, a heat exchanger was optionally mounted in the filtration circuit but removed during the experiments in order to avoid additional flow resistance. Heat generated by the cross-flow pump was sufficient to keep the temperature in the filtration circuit at 40 °C. The pilot plant provided a human–machine interface based on Siemens SIMATIC S7.

2.4.2. Fly Ash Extracts Used for Pilot-Scale PAUF Experiments

For investigations with the pilot plant, fly ash extract from the acid washing at the Ingolstadt MSWI plant was used. Suspended matter was allowed to settle and the fly ash extract was then filtered through a bag filter (50 μm). The extract taken from the MSWI plant contained 0.13 g L^{-1} of Cu(II). In order to additionally investigate the PAUF treatment of fly ash extract with a higher Cu(II) concentration, a second extract with 0.70 g L^{-1} of Cu(II) was achieved by adding $Cu(NO_3)_2{\cdot}3H_2O$ to the extract taken from the MSWI plant. The evaluation was focused on the elements with the highest concentration in the fly ash extract (Pb, Zn, Ca, and Cl^-, listed in Table 1).

Table 1. Concentration of elements Zn, Pb, Cu, Ca, and Cl^- in municipal solid waste incineration (MSWI) fly ash extract used for retention experiments. * Achieved by addition of $CuNO_3$.

Zn	Pb	Cu	Ca	Cl^-	
2.4	0.78	0.13/0.70 *	6	81	g L^{-1}

2.4.3. Pilot Plant Filtration Data

The permeate flux and specific power consumption (pressure and cross-flow pump) of the pilot plant were determined at 40 °C and a transmembrane pressure (TMP) of 5 bar, depending on the polymer concentration and tangential velocity in the filtration circuit. The pilot plant was filled with fly ash extract and operated in recirculation mode (permeate was recirculated into the feed reservoir). Pretreated HB-PEI solution was stepwise added to the feed reservoir and fed into the filtration circuit.

2.4.4. Selective Cu(II) Retention

The filtration circuit was filled with a 4 g L^{-1} solution of pretreated HB-PEI in water. To start the retention experiment, fly ash extract was continuously fed into the filtration circuit, with the pH value kept at 3.0. Samples were taken from the filtration circuit and analyzed by ICP-OES. Retention experiments were ended after 53 L (extract with 0.13 g L^{-1} Cu) or 21 L (extract with 0.7 g L^{-1} Cu).

2.4.5. Rinsing of Cu(II) Loaded Preconcentrates with Water

Rinsing of the Cu(II) loaded polymer concentrate was performed using concentrates from the Cu(II) retention experiments (Section 2.4.4). The filtration circuit was filled with polymer containing fly ash extract (4 g L^{-1} HB-PEI, 2.0 g L^{-1} Zn, 0.7 g L^{-1} Pb, 4.7 g L^{-1} Ca) and the pilot plant was operated at pH 3.0 with water as feed for 1 h. Samples were taken from the filtration circuit and analyzed by ICP-OES.

Thickening of the polymer solution was achieved by feeding fly ash extract with 4 g L^{-1} HB-PEI to the filtration circuit until a polymer concentration of 25 g L^{-1} HB-PEI was reached.

Cu(II) retention during rinsing was optimized using 25 g L^{-1} HB-PEI loaded with 200 mg of Cu(II)/g polymer. Rinsing was carried out with water for 1 h at pH 4.0, 3.5, and 3.0. Samples were taken from the filtration circuit and analyzed by ICP-OES.

2.4.6. Recovery of Cu(II) and Regeneration of Polyethyleneimine

Recovery of Cu(II) from the filtration circuit was performed using a concentrate received after rinsing (4 g L^{-1} HB-PEI, 0.7 g L^{-1} Cu(II)). The pH was decreased to 1.0 and the pilot plant was operated with water. Samples were taken from the filtration circuit and analyzed by ICP-OES.

2.5. Monitoring Cu(II) Enrichment and Release by UV-Vis Spectroscopy

Polyethyleneimine and Cu(II) form stable tetraamminecopper(II) complexes, which absorb light within the visible spectral range and appear in blue. Therefore, we investigated a possible photometric control of the Cu(II) enrichment and release.

UV-vis spectra of feed samples resulting from the Cu(II) loading experiments with HB-PEI (see Section 2.2.5) were recorded with a Lambda 35 UV-vis spectrometer (Perkin Elmer, Waltham, MA, USA) using polymethylmethacrylate sample cuvettes (VWR, Darmstadt, Germany). Solutions of Cu(II) nitrate trihydrate in ammonium hydroxide and in ultrapure water were prepared and investigated, functioning as comparative samples representing Cu(II) ammine complexes without polymer.

3. Results and Discussion

3.1. Metal Retention of Different PEIs in MSWI Fly Ash Extracts

Three different polyethyleneimines were investigated regarding their retention behavior towards all metal species contained in the fly ash extracts from MVA Ingolstadt and KEBAG Zuchwil. Composition of the fly ash extracts was previously analyzed; the resulting main metal components and chloride concentrations are shown in Table S1.

Figure 2 shows the retention of Cd(II), Cu(II), Ni(II), Pb(II), Zn(II), Fe(III), and Sb(V) using HB-PEI (Figure 2a,b), PE-PEI (Figure 2c,d), and MOD-PEI (Figure 2e,f). HB-PEI contains primary, secondary, and tertiary amino groups. Each PEI was investigated using the fly ash extracts from both MVA Ingolstadt (Figure 2b,d,f) and KEBAG Zuchwil (Figure 2a,c,e).

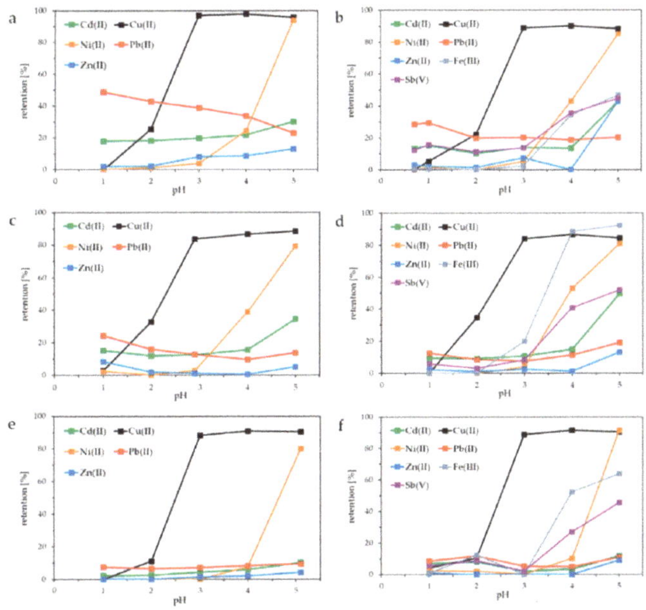

Figure 2. Influence of pH on the retention of Cd(II), Cu(II), Ni(II), Pb(II), Zn(II), Fe(III), and Sb(V) using fly ash extracts from KEBAG Zuchwil and MVA Ingolstadt: (**a**,**b**) hyperbranched polyethyleneimine (HB-PEI) = 10.0 g L^{-1} (Zuchwil), 7.9 g L^{-1} (Ingolstadt); (**c**,**d**) partially ethoxylated polyethyleneimine (PE-PEI) = 9.4 g L^{-1} (Zuchwil), 7.7 g L^{-1} (Ingolstadt); (**e**,**f**) modified polyethyleneimine (MOD-PEI) = 9.7 g L^{-1} (Zuchwil), 8.0 g L^{-1} (Ingolstadt).

As illustrated in Figure 2a–f, Cu(II) is the only metal ion retained at approximately 100% from pH 3.0 upwards. At this pH, Cd(II) and Pb(II) are retained at about 10% using PE-PEI (Figure 2c,d), and 15–30% using HB-PEI (Figure 2a,b). The fly ash extract from MVA Ingolstadt additionally contains Fe(III) and Sb(V), which are increasingly retained by the membrane starting from pH 3.0 on (Figure 2b,d,f).

As shown in Table 2, both fly ash extracts contained many more metal ions than depicted in Figure 2. The retention of Li, Na, K, Rb, Mg, Ca, Sr, Si, Al, and Mn was also investigated in the same PAUF experiments described in Figure 2. All were barely retained by the membrane, as can be seen in Table 2. Regarding the monovalent alkali metal ions that was to be expected, however, the nonbinding especially of divalent alkaline earth metal ions is a crucial result regarding the target Cu(II) selectivity in the treatment of highly saline fly ash extracts.

Table 2. Retention of Li(I), Na(I), K(I), Rb(I), Mg(II), Ca(II), Sr(II), Al(III), Si(IV), and Mn(II) using fly ash extracts from KEBAG Zuchwil and MVA Ingolstadt and HB-PEI, PE-PEI, and MOD-PEI polymers.

Elements	Retention					
	Fly Ash Extract from KEBAG Zuchwil			Fly Ash Extract from MVA Ingolstadt		
	HB-PEI	PE-PEI	MOD-PEI	HB-PEI	PE-PEI	MOD-PEI
Li, Na, K, Rb, Mg, Ca, Sr, Al, Si, Mn	<6%	<4%	<9%	<8%	<3%	<6%

The high Cu(II) retention observed is based on the interaction of Cu(II) with the respective PEI. Cu(II) forms tetraamminecopper(II) complexes with amine groups of the three PEIs. Ni(II) and Zn(II) are retained by forming ammine complexes as well but only at pH > 3.0.

To clarify the observed retention of additional metal ions (Figure 2a–f), ultrafiltration experiments were carried out using fly ash extract without the addition of PEI. The results obtained are shown in Figure S2.

To some extent, Pb, Fe, and Sb were retained in the PAUF experiments described above, but they also showed similar retention in the experiments without the addition of PEI (Figure S2). During these experiments, visible turbidity of the feed solutions was observed; the feed samples were centrifuged and the supernatants were analyzed. A continuously decreasing Fe concentration above pH 3.0 was observed: Between pH 3.0 and 5.0 Fe(II) and Fe(III) may both exist, but Fe(II)-hydroxide requires a pH > 7.0 for precipitation [3]. Therefore, Fe(III) was precipitated as solid iron(III)oxide hydrate and retained by the ultrafiltration membrane in the PAUF experiments (Figure 2b,d,f).

The Sb concentration of the centrifuged feed samples also decreased above pH 3.0. This could occur due to the precipitation of $Na[Sb(OH)_6]$, formed with sodium ions contained in the fly ash extract and also resulting from adjusting the pH with sodium hydroxide. Furthermore, Sb(V) and Sb(III) may coexist [31], enabling coprecipitation of $Sb(OH)_3$ with Fe(III) oxide hydrate. Similar retention progression of antimony and iron, shown in Figure 2b,f, indicates this.

Pb(II) was retained to different degrees using HB-PEI, PE-PEI, and MOD-PEI (Figure 2) and also without the addition of PEI (Figure S2). Dissolved Pb^{2+} forms lead chloro complexes such as $[PbCl]^+$, $[PbCl_3]^-$, $[PbCl_4]^{2-}$ and/or hardly soluble $PbCl_2$ in the presence of 60 g L^{-1} chloride [32], as contained in the fly ash extracts. Weibel et al., thoroughly investigated Pb-chloro complex formation in fly ash extract and identified $[PbCl_3]^-$ and $[PbCl_4]^{2-}$ as the mainly present species [3,6]. Presumably, a certain percentage of the Pb(II) binds to protonated amino groups of the PEIs via these negatively charged chloro complexes. Especially tertiary amino groups of polyethyleneimine may act as anion exchangers and therefore bind negatively charged chloro complexes. The lead retention observed in the experiments without PEI, however, may arise from lead precipitated as $PbCl_2$.

As shown in Figure 2a–d, Cd(II) was also retained to a small extent. This may be due to the formation of stable negatively charged Cd(II) chloro complexes binding to polyethyleneimine analogous to Pb(II) [33]. Weibel et al., found $[CdCl_4]^{2-}$ and $[CdCl_3]^-$ to be the dominant Cd-chloro complexes formed in fly ash extracts [6]. Cadmium, however, did not show any retention without PEI (Figure S2), because $CdCl_2$ is water-soluble, in contrast to $PbCl_2$.

Conclusively, at pH 3.0, Cu(II) was the only metal ion being bound to the investigated PEIs via metal complex formation with amine groups.

After enrichment, Cu(II) has to be released from the polyethyleneimine in order to achieve selective Cu(II) separation. At the same time, a complete metal release means regeneration of the polymer, which can then be reused in further enrichment cycles.

We investigated the Cu(II) release by decreasing the pH of the feed solution at the end of each PAUF experiment. For all six feed solutions shown in Figure 2, Cu(II) retention at pH 1.0 was 0% using fly ash extract from KEBAG Zuchwil (Figure 2a,c,e) and 4–5% using fly ash extract from MVA

Ingolstadt (Figure 2b,d,f). Therefore, Cu(II) was successfully released from all investigated PEIs at pH ≤ 1.

Summarizing the sorption behavior of MOD-PEI, Cu(II) was selectively separated at pH 3.0 from 16 different metal ions, including additional heavy metals, alkaline and alkaline earth metals. Moreover, the very high chloride concentration of 60 g L^{-1} in the highly saline fly ash extracts did not influence the pH-dependent Cu(II) chelating behavior. This also applies for HB-PEI [19] and PE-PEI [20]. The selective Cu(II) separation from real fly ash extract is even more exceptional, as the Cu(II) concentration in fly ash extracts is up to 380 times lower compared to interfering ions, such as alkaline (earth) metals or zinc. The highest selectivity regarding Cu(II) separation was achieved with MOD-PEI (Figure 2e,f). Compared to HB-PEI and PE-PEI MOD-PEI provides a lower number of amino groups. This leads to an early displacement of other competing metal ions by Cu(II) because Cu(II) forms the most stable transition-metal complex compounds among all metal ions present in the investigated fly ash extracts [34].

3.2. Copper Loading Capacity of HB-PEI, PE-PEI, and MOD-PEI Using Fly Ash Extract

Besides the qualitative characterization of the metal binding behavior of the PEIs, Cu loading capacity was also determined. During the stepwise addition of Cu(II) nitrate trihydrate to the feed solution of the PEIs in fly ash extract from KEBAG Zuchwil, each of the three feed solutions became deeper blue. The Cu loading was visible due to the formation of blue tetraamminecopper(II) complexes formed by dissolved Cu(II) and the amino groups of the polymers. As illustrated in Figure 3, HB-PEI shows the highest loading capacity for Cu(II) closely followed by PE-PEI and with a much higher distance MOD-PEI.

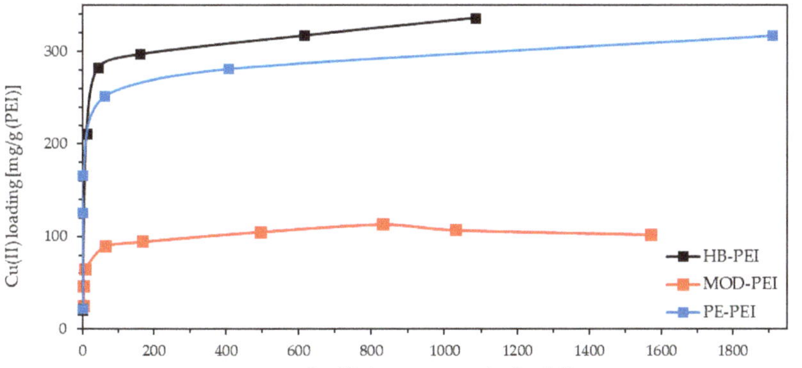

Figure 3. Cu(II) loading capacity for HB-PEI (4.6 g L^{-1}), PE-PEI (4.7 g L^{-1}), and MOD-PEI (3.7 g L^{-1}) using fly ash extract (KEBAG Zuchwil).

The determined Cu(II) loading capacities of HB-PEI and PE-PEI are in the same range as stated in [20]. However, the experiments described in this study were not carried out with real wastewater, but with a synthetic solution containing only copper sulfate. As a remarkable result, the very high concentration of interfering ions in real fly ash extract did not have any influence on the Cu(II) loading capacity of polyethyleneimines. The Cu loading capacity of PE-PEI is slightly lower compared to HB-PEI, because the amino groups are partially ethoxylated and therefore not available for Cu(II) binding. MOD-PEI is even more modified, which leads to an additional reduction of the Cu loading capacity: Even though MOD-PEI showed the highest selectivity for Cu(II) (see Section 3.1), its low Cu loading capacity restricts the applicability for Cu enrichment.

3.3. Viscosity of PEI Solutions

The viscosity of pretreated aqueous MOD-PEI, HB-PEI, and PE-PEI solutions was investigated. The membrane-permeant size fraction of the polymers was removed before measurement according to Section 2.2.3. Polymer concentrations up to 75 g L^{-1} were considered at a temperature of 30 °C.

The obtained data are given in Table 3. The viscosity of the aqueous solutions rose with increasing polymer concentration. Solutions of HB-PEI (Mw ≈ 25 kDa) showed higher viscosity than solutions of PE-PEI (Mw ≈ 70 kDa), but both remained fluid even at 75 g L^{-1} polymer. MOD-PEI (Mw ≈ 2000 kDa) showed far higher viscosity, with gel being formed at >20 g L^{-1} polymer.

Table 3. Dynamic viscosity of aqueous PEI solutions (30 °C). Membrane-permeant size fraction was removed before measurement. Gel was formed with >20 g L^{-1} MOD-PEI.

Concentration	-	1	5	10	25	50	75	g L^{-1}
MOD-PEI	0.78	1.7	3.6	5.5	-	-	-	mPa·s
PE-PEI	0.78	0.82	0.86	0.89	1.1	1.4	1.8	mPa·s
HB-PEI	0.78	0.84	0.89	0.95	1.2	1.9	3.1	mPa·s

Solutions of HB-PEI and PE-PEI provide a viscosity that is low enough for efficient treatment in tangential flow ultrafiltration, although if very high polymer concentrations in PAUF are desired, the use of PE-PEI may be advantageous [21]. Application of very-high-molecular-weight polymers in PAUF, such as MOD-PEI, is disadvantageous. Even at moderate polymer concentrations, the aqueous solutions cannot be treated properly in tangential flow ultrafiltration.

3.4. Operating Data of PAUF Pilot Plant Using HB-PEI

Operating data of the PAUF pilot plant were investigated by treating a fly ash extract from MVA Ingolstadt containing different concentrations of HB-PEI by tangential flow ultrafiltration. The pilot plant was operated at a transmembrane pressure (TMP) of 5 bar, and the temperature in the filtration circuit was 40 °C. The water-soluble polymer was trapped inside the filtration circuit and the polymer-free permeate was led back into the feed reservoir of the pilot plant.

The influence of polymer concentration and tangential velocity on the permeate flux and specific power uptake of the pilot plant are summarized in Figure 4. Power uptake included the power consumption of the pressure and the tangential flow pump. Performance of the tangential flow filtration clearly increased with increasing tangential velocity and decreased with increasing polymer concentration in the filtration circuit.

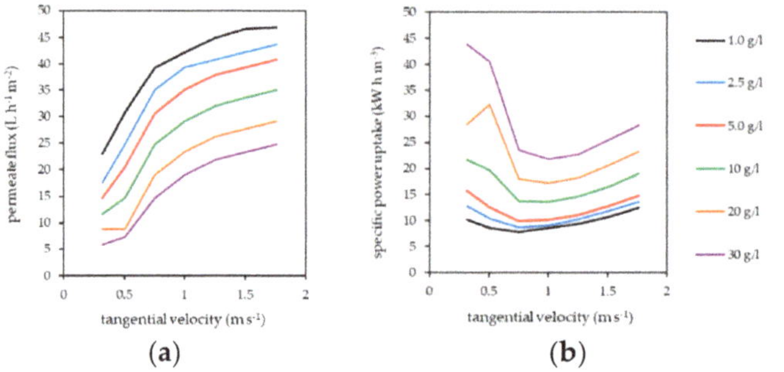

Figure 4. Influence of tangential velocity and polymer concentration on (**a**) permeate flux and (**b**) specific power consumption (pressure and cross-flow pump) of the pilot plant. HB-PEI diluted in fly ash extract. Transmembrane pressure (TMP) = 5 bar, T = 40 °C.

PAUF is influenced by gel layer formation [21–23], which occurs in the boundary layer on the membrane surface due to the retention of the water-soluble polymer. This results in additional flow resistance, decreasing the achievable permeate flux and increasing the specific power uptake of the process [35]. Gel layer formation is reduced by increased tangential velocity in tangential flow filtration, as turbulent flow conditions on the membrane surface reduce the thickness of the boundary layer [36]. Consequently, the permeate flux of the pilot plant increased with increasing tangential velocity.

The transition from laminar to turbulent flow regime in the channels of the tubular ceramic membranes appeared at tangential velocities between 0.3 and 0.8 m s^{-1} and was observable in a strong decrease of specific power uptake. Depending on the HB-PEI concentration, a minimum specific power uptake was observed at a tangential velocity of about 0.8 m s^{-1} (1–10 g L^{-1} HB-PEI) and 1.0 m s^{-1} (20 and 30 g L^{-1} HB-PEI). With further increase of tangential velocity, the specific power uptake rose moderately. This results from the logarithmic rise of the permeate flux, but a disproportional rise of pressure loss in the filtration circuit with increasing tangential velocity occurs.

The permeate flux achieved by the pilot plant is comparable to other work [15,21–23] but rather low. Reasons for this are the very high salt content of the treated fly ash extract, the high water-soluble polymer concentration used in the filtration circuit, and a progressive fouling of the ceramic tubular membranes that were used for the treatment of PEI containing fly ash extract. Irreversible fouling (i.e., by irreversible pore blocking), which cannot be reversed by counter-flushing of the membranes, was significant, but remained in a steady state. So, the pilot plant data given in this work represent realistic conditions using ceramic tubular membranes.

3.5. Multistage Process for Technical Scale Separation of Cu(II) from Fly Ash Extract

A PAUF-based process for the selective separation and purification of Cu(II) was investigated, as shown in Figure 5. In the retention stage, fly ash extract and HB-PEI solution were fed to an ultrafiltration plant. Cu(II) was bound by the polymer and retained in the filtration circuit. A preconcentrate containing Cu(II) loaded HB-PEI in fly ash extract was (continuously) ejected from the filtration circuit, and the Cu(II) depleted permeate was discharged as wastewater. In order to improve the selectivity of Cu(II) toward interfering ions in the fly ash extract, the preconcentrate was purified in another ultrafiltration stage. This can be done either by thickening the preconcentrate, where the concentration of Cu(II) loaded polymer and therefore Cu(II) in the filtration circuit is enriched or by rinsing the filtration circuit with water in order to displace fly ash extract from the polymer solution. Depending on the technical realization, the steps may be done in a different order or at the same time. For Cu(II) recovery and regeneration of HB-PEI, the pH is decreased and Cu(II) is released from the polymer. Cu(II) is rinsed from the filtration circuit, producing a polymer-free Cu(II) concentrate. The regenerated HB-PEI solution is recirculated into the retention stage.

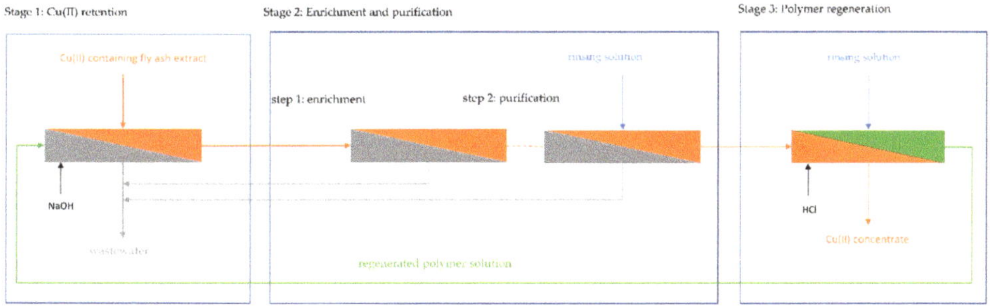

Figure 5. Multistep polymer-assisted ultrafiltration (PAUF) process for selective separation of Cu(II) from MSWI fly ash extract. A polymer-free Cu(II) concentrate is obtained in the process.

The technical implementation of the different steps (retention, enrichment/purification, regeneration) is investigated in the following sections. The cumulative volume of solutions fed into the pilot plant is given in ϴ (multiples of the filtration circuit volume). The concentration of elements in the filtration circuit is referenced to their concentration in the feed solution (in continuous operation) respective to their concentration in the filtration circuit at the beginning of the rinsing experiments (in batch operation).

3.5.1. Retention of Cu(II) in Continuous Operation

The retention of Cu(II) was investigated with two fly ash extracts containing 0.13 g L^{-1} and 0.7 g L^{-1} Cu(II). The pilot plant was filled with 4 g L^{-1} pretreated HB-PEI in water. After startup, fly ash extract was fed to the plant and treated at pH 3.0. Figure 6 shows the concentrations of Cu(II), Pb(II), Zn(II), and Ca(II) in the filtration circuit during both experiments.

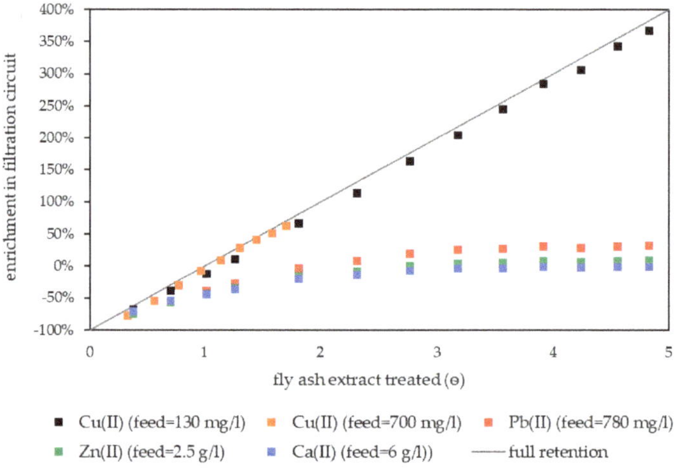

Figure 6. Enrichment of Cu(II), Pb(II), Zn(II), and Ca(II) in filtration circuit during continuous Cu(II) retention treating fly ash extracts from MVA Ingolstadt. The process was started with water in the filtration circuit, resulting in an arithmetically negative enrichment. Reference: concentration in the fly ash extract. HB-PEI = 4 g L^{-1}, pH 3.0, T = 40 °C.

At the beginning, the concentration of elements in the filtration circuit was lower than in the fly ash extract, resulting in an arithmetically negative enrichment factor. After the feed was switched to fly ash extract, the water in the filtration circuit was continuously displaced and the concentrations of Cu(II), Zn(II), Pb(II), and Ca(II) increased. After ϴ = 3, Ca(II) concentration in the filtration circuit was equal to the Ca(II) concentration in the feed solution and therefore the enrichment was zero. This implies that the displacement of water was completed and the filtration circuit from then on was only filled with fly ash extract and HB-PEI.

Ca(II) enrichment remained zero in further operation of the pilot plant. In accordance with the laboratory experiments, Ca(II) was not retained in PAUF. It represents further elements that were not bound by HB-PEI at pH 3.0 which also showed zero enrichment (including chloride).

For Zn(II) and Pb(II), a slight enrichment in the filtration circuit was observed, reaching a steady state at ϴ = 4. This slight enrichment was limited to 10% for Zn(II) and 30% for Pb(II). The latter already showed a slight binding by HB-PEI in laboratory experiments. This originates from Pb-chloro complexes interacting electrostatically with HB-PEI (see Section 3.1). At pH 3.0 Zn(II) can also form Zn-chloro complexes (weaker compared to Pb-chloro complexes) [6]. Both metal ions do not bind to

HB-PEI via metal complex formation at pH 3.0 and therefore showed no linear increase in the filtration circuit in continuous operation.

In contrast, the concentration of Cu(II) in the filtration circuit correlated linearly with the volume of fly ash extract treated, irrespective of whether a feed with 0.13 g L^{-1} or 0.7 g L^{-1} Cu(II) was used. With 4 g L^{-1} of HB-PEI, the effective Cu(II) concentration in the filtration circuit was limited to 0.8 g L^{-1} Cu(II) when loading of 200 mg Cu(II)/g polymer was accepted. Cu(II) exceeding this concentration was not sufficiently bound by HB-PEI and was lost in the following stages.

For Cu(II) retention in continuous operation, addition of sodium hydroxide solution to the filtration circuit is required, as otherwise the pH value drops and Cu(II) is no longer retained. The intense mixing due to the tangential flow and high buffer capacity of the polymer solution allowed perfect pH control in the filtration circuit. The process was able to selectively retain Cu(II) in the filtration circuit, even if strong fluctuation of the Cu(II) concentration occurred in the feed.

Though Cu(II) was enriched considerably, the achievable selectivity of Cu(II) toward Zn(II), Pb(II), and Ca(II) in a preconcentrate ejected from the filtration circuit was quite low (Table 4). In steady state, the Cu(II) loaded polymer was in the fly ash extract the filtration circuit was filled with. To improve Cu(II) selectivity, the fly ash extract had to be rinsed out from the preconcentrate in a second step.

Table 4. Concentration of Pb(II), Zn(II), and Ca(II) in feed solution and filtration circuit in steady state. Composition of concentrate ejected from retention step if loading of 200 mg Cu/g polymer is accepted.

	Cu(II)	Pb(II)	Zn(II)	Ca(II)	
Feed solution (fly ash extract)	0.13/0.70	0.8	2.5	6.0	g L^{-1}
Steady state (filtration circuit)	-	1.0	2.7	6.0	g L^{-1}
200 mg Cu/g polymer @ 4 g L^{-1} HB-PEI	0.8	-	-	-	g L^{-1}
Ejected from filtration circuit	0.8	1.0	2.7	6.0	g L^{-1}

3.5.2. Cu(II) Enrichment and Purification

Rinsing of the preconcentrate with water in the batch process was investigated. The purpose of this procedure is to displace fly ash extract from the preconcentrate gained in the retention step. Due to its high concentration in the preconcentrate, Ca(II) concentration serves as reference for further unbound species in the filtration circuit. As can be seen in Figure 7, the concentration of Ca(II) followed an exponential decrease during rinsing, which can be described by Equation (2). The concentration of an unbound species in the filtration circuit decreased from the initial concentration $c_{i,0}$ according to

$$c_{i,\Theta} = c_{i,0} \times e^{-\Theta} \tag{2}$$

with Θ giving the volume of rinsing water used (in multiples of filtration circuit volume). An evaluation of this equation is given in Table 5. Though slightly bound by the polymer in laboratory experiments and in the retention step, Zn(II) and Pb(II) were successfully rinsed from the filtration circuit equally to unbound species.

Table 5. Reduction of unbound species in the filtration circuit by rinsing with water (calculated by Equation (2)).

Rinsing Water Used (Θ)	0	1	2	3	4	5
Reduction of unbound species ($c_{i,\Theta}/c_{i,0}$)	100%	37%	14%	5%	2%	1%

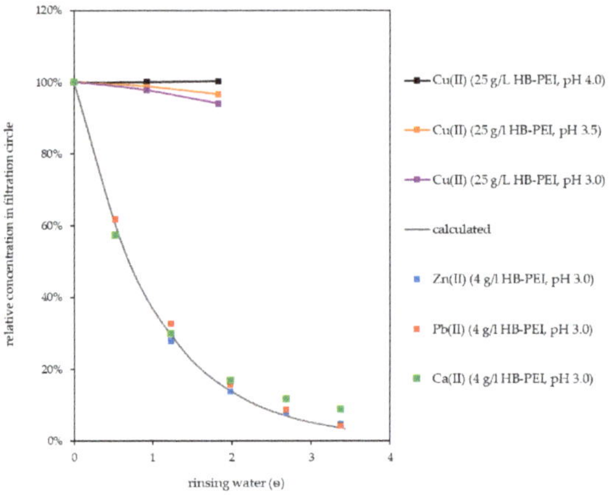

Figure 7. Rinsing of filtration circuit with water.

An alternative way to increase Cu(II) selectivity is to thicken the preconcentrate. This is achieved by feeding Cu(II) loaded preconcentrate exceeding the volume of the filtration circuit to the pilot plant. The polymer is retained by the membrane and Cu(II) is enriched linearly with increasing polymer concentration in the filtration circuit, while the concentration of unbound species passing the membrane remains equal. Combining the thickening step with purification of the concentrate obtained by rinsing with water, a substantial enhancement of Cu(II) selectivity toward unbound species could be obtained (Table 6).

Table 6. Enhancement of Cu(II) selectivity toward unbound species by increased polymer concentration (thickening) and rinsing of filtration circuit (with water); 4 g L^{-1} HB-PEI in the initial concentrate was considered.

		Rinsing with Water (ϴ)					
		0	**1**	**2**	**3**	**4**	**5**
c(HB-PEI) after thickening	4 g L^{-1}	1	3	7	20	55	148
	10 g L^{-1}	3	7	18	50	136	371
	25 g L^{-1}	6	17	46	126	341	928
	50 g L^{-1}	13	34	92	251	682	1855

When the preconcentrate was rinsed at an initial Cu(II) loading of 250 mg Cu(II)/g polymer, a significant loss of Cu(II) occurred. Therefore, rinsing was optimized by reducing the initial loading to 200 Cu(II)/g polymer. As can be seen in Figure 7, the Cu(II) concentration did not drop when rinsing was done at pH 4.0, and only slightly decreased when the pH value was lowered to 3.5 (96%) and 3.0 (94%) during rinsing (ϴ= 1.8).

The very high concentration of interfering ions in the preconcentrate requires intense rinsing of the filtration circuit. It was very efficient at the beginning but dilution of the fly ash extract in the filtration circuit resulted in a steady decrease of rinsing efficiency with increasing volume of rinsing water used. High purification of the Cu(II) preconcentrate was accompanied by a high consumption of rinsing water and, due to the application of tangential flow ultrafiltration, energy. Using PAUF for selective enrichment and separation requires processes for further metal recovery that tolerate remaining interfering ions in the PAUF concentrate in order to minimize the required rinsing.

3.5.3. Regeneration of the Polymer

Finally, regeneration of the polymer and gaining a polymer-free Cu(II) concentrate in a batch process were investigated. Cu(II) was released from HB-PEI by the addition of hydrochloric acid to the filtration circuit until pH 1.0 was reached. Unbound Cu(II) was then rinsed from the filtration circuit by water. The discharge of Cu(II) (Δ) from the filtration circuit, depending on the volume of rinsing solution used (Θ), is described with an adaption of Equation (2):

$$\Delta = 1 - \frac{c_{i,\Theta}}{c_{i,0}} = 1 - e^{-\Theta} \quad (3)$$

Experimental and calculated data for the Cu(II) discharge are given in Figure 8. Cu(II) was efficiently discharged from the pilot plant by rinsing with water, generating a polymer-free Cu(II) concentrate. The regenerated HB-PEI remained in the filtration circuit and was used in further separation cycles.

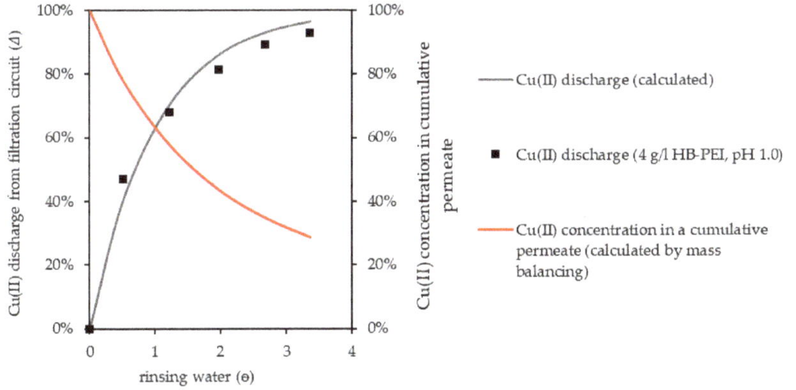

Figure 8. HB-PEI regeneration by pH decrease and Cu(II) discharge from the filtration circuit by rinsing with water. Cu(II) concentration in cumulative permeate is referenced to initial Cu(II) concentration in the filtration circuit.

Due to the decreased Cu(II) concentration in the filtration circuit, Cu(II) concentration in the accumulated polymer-free permeate steadily decreased (Figure 8, calculated by mass balancing) with increasing volume of rinsing water. As the regenerated polymer solution was supposed to be used for Cu(II) retention again, a sufficient discharge of Cu(II) from the filtration circuit was necessary, as remaining unbound Cu(II) in the polymer solution was recirculated to the retention step, reducing the performance of this step. Depending on the polymer concentration in the filtration circuit, Cu(II) concentrations in a polymer-free concentrate, as given in Table 7, can be achieved in a batch regeneration process. An initial polymer loading of 250 mg Cu(II)/g polymer, no Cu(II) retention during rinsing, and 90% Cu(II) discharge are assumed.

Table 7. Final Cu(II) concentration in cumulative permeate collected during polymer regeneration and rinsing of the filtration circuit with water. No Cu(II) retention and 90% discharge ($\Theta = 2.3$) are assumed. Calculation according to Equation (3) and by mass balancing.

c(HB-PEI)	**4**	**10**	**25**	**50**	**100**	**g L^{-1}**
Initial Cu(II) in filtration circuit	0.8	2	5	10	20	g L^{-1}
Final Cu(II) in cumulative permeate	0.3	0.8	2.0	3.9	7.8	g L^{-1}

3.6. Photometric Control of the Cu(II) Enrichment and Release

For pilot scale operation, control of the Cu(II) enrichment and release in the filtration circuit is highly beneficial. An inline measurement of the Cu(II) enrichment enables an automated pH decrease by acid addition as soon as the Cu(II) loading of the polymer is completed. Then again, as soon as the Cu(II) release is completed, the pH is automatically increased and a new enrichment cycle can be started.

Therefore, the specific change in light absorption of PEI depending on its complex formation can be utilized. Due to the formation of Cu(II) complexes with HB-PEI (tetraamminecopper(II)) [37] and H_2O, photometric control of the enrichment and release during PAUF experiments [23] with fly ash extract was investigated.

Figure 9a shows the increasing Cu(II) loading of HB-PEI at pH 4.0 (feed solutions in cuvettes 1–10), followed by a clearly visible color change representing the Cu(II) release at pH 1.0 (feed solution in cuvette 11). Decreasing the pH value, the feed solutions turned colorless to light blue again, indicating the formation of hexaaquacopper(II) and thus the regeneration of HB-PEI.

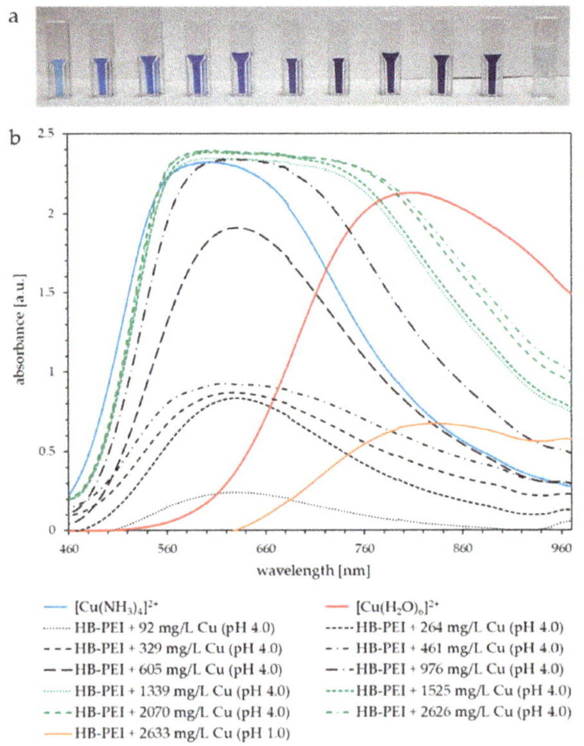

Figure 9. (**a**) PAUF feed samples of Cu(II) enrichment and final Cu(II) release (far right) from HB-PEI using fly ash extract (KEBAG Zuchwil), HB-PEI = 4.6 g L^{-1}. (**b**) Corresponding UV-vis spectra of increasing Cu(II) loading (gray, black, and green graphs) and release (orange graph). Comparable solutions of $[Cu(NH_3)_4]^{2+}$ (blue graph) and $[Cu(H_2O)_6]^{2+}$ (red graph) were additionally analyzed.

Figure 9b shows the UV-vis spectra of the PAUF feed samples with increasing Cu(II) concentration from gray to black to green graphs. The prepared solution of $[Cu(NH_3)_4]^{2+}$ (Figure 9b, blue graph) has an absorption maximum at 605 nm. The copper(II) ammine complex formed with HB-PEI absorbs at very similar wavelengths. The intensity of the absorption bands increases until the maximum loading capacity is reached at around 976 mg L^{-1} of Cu(II). Additional Cu(II) forms Cu(II) aqua complexes. Therefore, UV-vis spectra are shifted more to the right toward the absorption maximum of $[Cu(H_2O)_6]^{2+}$ (810 nm, red graph) and a mixture of ammine and aqua complexes occurs in the feed solution. At pH 1.0, Cu(II) is released from HB-PEI and the Cu(II) aqua complex exists exclusively, showing an absorption maximum at 830 nm (Figure 9b, orange graph).

It is easy to distinguish between the absorption maxima of the ammine and aqua complexes formed within the Cu(II) enrichment and those obtained during the release from HB-PEI. This enables calibrated photometric control of the selective Cu(II) separation from real fly ash extract by PAUF.

4. Conclusions

In this work, the application of polymer-assisted ultrafiltration (PAUF) for selective retention and separation of Cu(II) from MSWI fly ash extracts was investigated at laboratory and pilot scale. The following conclusions can be drawn from these investigations:

- The results from laboratory experiments concerning the heavy metal binding of water-soluble polymers can be successfully transferred to a pilot scale process. The overall instrumental design of the PAUF process is quite simple, allowing estimations of the different process stages using basic equations. This enables an easy scale-up of PAUF processes based on laboratory and pilot experiments.
- Hyperbranched and ethoxylated PEI with moderate molecular weights are highly suitable for the selective retention of Cu(II) from fly ash extract. PAUF allows perfect pH control and highly selective Cu(II) retention in a continuous pilot scale process. In combination with photometric control, fluctuation of the Cu(II) concentration in the feed solution can be handled, allowing for a superior process control.
- In PAUF, Cu(II) retention occurs directly in the fly ash extract. The Cu(II) concentration is limited by the polymer concentration which in turn is efficiently treatable by ultrafiltration. Due to the very high concentration of interfering ions in MSWI fly ash extracts, a subsequent ultrafiltration step is needed, in which the fly ash extract is rinsed from the Cu(II) loaded polymer solution.
- Rinsing the polymer solution after the retention stage is an important but laborious step to increase Cu(II) selectivity. Additionally, thickening the polymer concentrate can be done for an effective increase of Cu(II) selectivity and overall Cu(II) concentration in the polymer solution. However, for metal recovery from PAUF concentrates, processes should be used that tolerate remaining fly ash extract in the concentrate, as rinsing should be minimized as far as possible.
- After regeneration by pH decrease as described in this work, polyethyleneimine can be reused for Cu(II) retention. A polymer-free Cu(II) concentrate is obtained, which can be used for metal production.

Supplementary Materials: The following are available online at http://www.mdpi.com/2227-9717/8/12/1662/s1: Figure S1. Schematic illustration of laboratory ultrafiltration system. Figure S2. pH dependent metal retention using fly ash extract from MVA Ingolstadt without addition of polymer. Table S1. Average metal concentrations (>1 mg L^{-1}) of fly ash extracts from MVA Ingolstadt and KEBAG Zuchwil. Chloride concentration of both extracts is approximately 60 g L^{-1}.

Author Contributions: Conceptualization, C.H. and S.F.; methodology, C.H. and S.F.; experimental investigation, C.H. and S.F.; data curation, C.H. and S.F.; writing—original draft preparation, C.H. and S.F.; writing—review and editing, C.H., S.F. and M.S.; visualization, C.H. and S.F.; project administration, M.S.; funding acquisition, M.S. All authors have read and agreed to the published version of the manuscript.

Funding: This research was funded by the German Federal Ministry of Education and Research (Bundesministerium für Bildung und Forschung, BMBF), grant number 033R140.

Acknowledgments: The authors would like to thank Anton Perfoll and Michael Funk (MVA Ingolstadt) and the staff members of KEBAG Zuchwil for providing fly ash extract samples for our research. The authors would also like to thank BASF SE for providing the polymers used in this study.

Conflicts of Interest: The authors declare no conflict of interest. The funders had no role in the design of the study, in the collection, analyses or interpretation of data, in the writing of the manuscript or in the decision to publish the results.

References

1. Huber, F.; Simon, F.-G.; Hyks, J.; Braga, R.; Biganzoli, L.; Costa, G.; Funari, V.; Grosso, M. Metal recovery from incineration bottom ash: State-of-the-art and recent developments. *J. Hazard. Mater.* **2020**, *393*. [CrossRef]
2. Jiao, F.; Zhang, L.; Dong, Z.; Namioka, T.; Yamada, N.; Ninomiya, Y. Study on the species of heavy metals in MSW incineration fly ash and their leaching behavior. *Fuel Process. Technol.* **2016**, *152*, 108–115. [CrossRef]
3. Weibel, G.; Eggenberger, U.; Schlumberger, S.; Mäder, U.K. Chemical associations and mobilization of heavy metals in fly ash from municipal solid waste incineration. *Waste Manag.* **2017**, *62*, 147–159. [CrossRef]
4. Haberl, J.; Koralewska, R.; Schlumberger, S.; Schuster, M. Quantification of main and trace metal components in the fly ash of waste-to-energy plants located in Germany and Switzerland: An overview and comparison of concentration fluctuations within and between several plants with particular focus on valuable metals. *Waste Manag.* **2018**, *75*, 361–371. [CrossRef]
5. Fabricius, A.-L.; Renner, M.; Voss, M.; Funk, M.; Perfoll, A.; Gehring, F.; Graf, R.; Fromm, S.; Duester, L. Municipal waste incineration fly ashes: From a multi-element approach to market potential evaluation. *Environ. Sci. Eur.* **2020**, *32*, 1–14. [CrossRef]
6. Weibel, G.; Eggenberger, U.; Kulik, D.A.; Hummel, W.; Schlumberger, S.; Klink, W.; Fisch, M.; Mäder, U.K. Extraction of heavy metals from MSWI fly ash using hydrochloric acid and sodium chloride solution. *Waste Manag.* **2018**, *76*, 457–471. [CrossRef] [PubMed]
7. Ferraro, A.; Farina, I.; Race, M.; Colangelo, F.; Cioffi, R.; Fabbricino, M. Pre-treatments of MSWI fly-ashes: A comprehensive review to determine optimal conditions for their reuse and/or environmentally sustainable disposal. *Rev. Environ. Sci. Bio/Technol.* **2019**, *18*, 453–471. [CrossRef]
8. Luo, H.; Cheng, Y.; He, D.; Yang, E.-H. Review of leaching behavior of municipal solid waste incineration (MSWI) ash. *Sci. Total Environ.* **2019**, *668*, 90–103. [CrossRef] [PubMed]
9. Schlumberger, S.; Schuster, M.; Ringmann, S.; Koralewska, R. Recovery of high purity zinc from filter ash produced during the thermal treatment of waste and inerting of residual materials. *Waste Manag. Res.* **2007**, *25*, 547–555. [CrossRef]
10. Antuñano, N.; Cambra, J.; Arias, P. Hydrometallurgical processes for Waelz oxide valorisation—An overview. *Process. Saf. Environ. Prot.* **2019**, *129*, 308–320. [CrossRef]
11. Sverdrup, H.U.; Ragnarsdottir, K.V.; Koca, D. On modelling the global copper mining rates, market supply, copper price and the end of copper reserves. *Resour. Conserv. Recycl.* **2014**, *87*, 158–174. [CrossRef]
12. Elshkaki, A.; Graedel, T.; Ciacci, L.; Reck, B.K. Copper demand, supply, and associated energy use to 2050. *Glob. Environ. Chang.* **2016**, *39*, 305–315. [CrossRef]
13. Haberl, J.; Schuster, M. Solubility of elements in waste incineration fly ash and bottom ash under various leaching conditions studied by a sequential extraction procedure. *Waste Manag.* **2019**, *87*, 268–278. [CrossRef] [PubMed]
14. Crini, G.; Morin-Crini, N.; Fatin-Rouge, N.; Déon, S.; Fievet, P. Metal removal from aqueous media by polymer-assisted ultrafiltration with chitosan. *Arab. J. Chem.* **2017**, *10*, S3826–S3839. [CrossRef]
15. Huang, Y.; Feng, X. Polymer-enhanced ultrafiltration: Fundamentals, applications and recent developments. *J. Membr. Sci.* **2019**, *586*, 53–83. [CrossRef]
16. Geckeler, K.E.; Shkinev, V.M.; Spivakov, B.Y. Liquid-Phase Polymer-Based Retention (Lpr)-A New Method for Selective Ion Separation. *Sep. Purif. Methods* **1988**, *17*, 105–140. [CrossRef]
17. Chou, Y.-H.; Choo, K.-H.; Chen, S.-S.; Yu, J.-H.; Peng, C.-Y.; Li, C.-W. Copper recovery via polyelectrolyte enhanced ultrafiltration followed by dithionite based chemical reduction: Effects of solution pH and polyelectrolyte type. *Sep. Purif. Technol.* **2018**, *198*, 113–120. [CrossRef]
18. Kochkodan, O.D.; Kochkodan, V.M.; Sharma, V.K. Removal of Cu(II) in water by polymer enhanced ultrafiltration: Influence of polymer nature and pH. *J. Environ. Sci. Heal. Part. A* **2018**, *53*, 33–38. [CrossRef]
19. Molinari, R.; Gallo, S.; Argurio, P. Metal ions removal from wastewater or washing water from contaminated soil by ultrafiltration–complexation. *Water Res.* **2004**, *38*, 593–600. [CrossRef]

20. Llanos, J.; Pérez, A.; Canizares, P. Copper recovery by polymer enhanced ultrafiltration (PEUF) and electrochemical regeneration. *J. Membr. Sci.* **2008**, *323*, 28–36. [CrossRef]
21. Llanos, J.; Perez, A.; Canizares, P. Water-soluble polymer ultrafiltration process at pilot scale: Study of hydrodynamics and factors limiting flux. *J. Membr. Sci.* **2009**, *341*, 37–45. [CrossRef]
22. Canizares, P. A semi-continuous laboratory-scale polymer enhanced ultrafiltration process for the recovery of cadmium and lead from aqueous effluents. *J. Membr. Sci.* **2004**, *240*, 197–209. [CrossRef]
23. Canizares, P.; De Lucas, A.; Pérez, Á.; Camarillo, R. Effect of polymer nature and hydrodynamic conditions on a process of polymer enhanced ultrafiltration. *J. Membr. Sci.* **2005**, *253*, 149–163. [CrossRef]
24. Baticle, P.; Kiefer, C.; Lakhchaf, N.; Leclerc, O.; Persin, M.; Sarrazin, J. Treatment of nickel containing industrial effluents with a hybrid process comprising of polymer complexation–ultrafiltration–electrolysis. *Sep. Purif. Technol.* **2000**, *18*, 195–207. [CrossRef]
25. Kadioglu, S.I.; Yilmaz, L.; Aydogan, N.; Ozbelge, H.O. Removal of Heavy Metals from Multicomponent Metal Mixtures by Polymer Enhanced Ultrafiltration: Effects of pH, Ionic Strength and Conformational Changes in Polymer Structure. *Sep. Sci. Technol.* **2010**, *45*, 1363–1373. [CrossRef]
26. Gallo, D.; Acosta, E.J.; Scamehorn, J.F.; Sabatini, D.A. Pilot-scale study of Polyelectrolyte-Enhanced UF for arsenic removal. *J. Am. Water Work. Assoc.* **2006**, *98*, 106–116. [CrossRef]
27. Geckeler, K.E.; Volchek, K. Removal of Hazardous Substances from Water Using Ultrafiltration in Conjunction with Soluble Polymers. *Environ. Sci. Technol.* **1996**, *30*, 725–734. [CrossRef]
28. Schulte-Bockholt, M.; Schuster, M. Removal enrichment and recovery of Ni(II), Zn(II) and phosphate from phosphation rinsing waters with liquid-phase polymer-based retention technique. *Sep. Purif. Technol.* **2008**, *63*, 172–178. [CrossRef]
29. Tavares, C.R.G.; Vieira, M.; Petrus, J.; Bortoletto, E.; Ceravollo, F. Ultrafiltration/complexation process for metal removal from pulp and paper industry wastewater. *Desalination* **2002**, *144*, 261–265. [CrossRef]
30. Petrov, S.; Nenov, V. Removal and recovery of copper from wastewater by a complexation-ultrafiltration process. *Desalination* **2004**, *162*, 201–209. [CrossRef]
31. Krupka, K.M.; Serne, R.J. Geochemical Factors Affecting the Behavior of Antimony, Cobalt, Europium, Technetium, and Uranium in Vadose Zone Sediments, 2002, doi:10.2172/15004491. Available online: http://www.pnl.gov/main/publications/external/technical_reports/PNNL-14126.pdf (accessed on 28 November 2020).
32. Hagemann, S. Thermodynamische Eigenschaften des Bleis in Lösungen der Ozeanischen Salze. Ph.D. Thesis, Technische Universität Carolo-Wilhelmina, Braunschweig, Germany, 1999.
33. Comans, R.N.J.; Van Dijk, C.P.J. Role of complexation processes in cadmium mobilization during estuarine mixing. *Nat. Cell Biol.* **1988**, *336*, 151–154. [CrossRef]
34. Irving, H.; Williams, R.J.P. Order of Stability of Metal Complexes. *Nat. Cell Biol.* **1948**, *162*, 746–747. [CrossRef]
35. Vela, M.C.V.; Álvarez-Blanco, S.; Lora, J.; Gozálvez-Zafrilla, J.M.; Rodríguez, E.B. Modelling of flux decline in crossflow ultrafiltration of macromolecules: Comparison between predicted and experimental results. *Desalination* **2007**, *204*, 328–334. [CrossRef]
36. Rai, P.; Rai, C.; Majumdar, G.; Dasgupta, S.; De, S. Resistance in series model for ultrafiltration of mosambi (Citrus sinensis (L.) Osbeck) juice in a stirred continuous mode. *J. Membr. Sci.* **2006**, *283*, 116–122. [CrossRef]
37. Tauler, R.; Casassas, E. Spectroscopic resolution of macromolecular complexes using factor analysis: Cu(II) -polyethyleneimine system. *Chemom. Intell. Lab. Syst.* **1992**, *14*, 305–317. [CrossRef]

Publisher's Note: MDPI stays neutral with regard to jurisdictional claims in published maps and institutional affiliations.

MDPI
St. Alban-Anlage 66
4052 Basel
Switzerland
Tel. +41 61 683 77 34
Fax +41 61 302 89 18
www.mdpi.com

Processes Editorial Office
E-mail: processes@mdpi.com
www.mdpi.com/journal/processes

www.ingramcontent.com/pod-product-compliance
Ingram Content Group UK Ltd.
Pitfield, Milton Keynes, MK11 3LW, UK
UKHW060123010826
14089UKWH00006B/124

9 783036 537559